ASPECTS OF LEADERSHIP
ETHICS, LAW, AND SPIRITUALITY

Edited by
Carroll Connelley and Paolo Tripodi

Marine Corps University Press
Quantico, Virginia

DISCLAIMER

The views expressed in this book are solely those of the authors. They do not necessarily reflect the opinions of the organizations for which they work, Marine Corps University, the U.S. Marine Corps, the Department of the Navy, the U.S. Army, the U.S. Department of Defense, the Judge Advocate General's Corps, the U.S. government, the French Ministry of Defense, the French Air Force, or the International Committee of the Red Cross.

Published by

Marine Corps University Press

3078 Upshur Avenue

Quantico, VA

22134

www.tecom.usmc.mil/mcupress

1st Printing, 2012

PCN 10600008900

ISBN 978-0-16-091368-6

── Contents ──

LAW

SPIRITUALITY

— FOREWORD —

Aspects of Leadership is a new book edited by Lieutenant Colonel Carroll Connelley and Dr. Paolo Tripodi, the Ethics Branch head, and comprised of essays by a group of 20 leadership and ethics scholars and practitioners. The Lejeune Leadership Institute Ethics Branch, Marine Corps University, provides current and relevant scholarly research and instruction on ethics and moral leadership and the law of war.

This collection of essays provides timely and insightful views on current leadership behavior and ethical concerns that men and women of the armed forces face in the demanding, complex, and prolonged decade of armed conflict.

This book offers significant and relevant perspectives on the historical underpinnings as well as from recent deployments and operations that affect ethics and law of war on individual Marines, units, and the populations engaged in conflict. The various chapters are the result of thoughtful perspectives and exhaustive research by intelligent and committed scholars and practitioners who apply their knowledge and wisdom with the passion to make clear what right ethical behavior by service members should look like.

Dialogues addressing leadership from various aspects are captured and shared in this collection and should benefit not only the formal schools at Marine Corps University, but also fellow service professional military educational institutions. This book will also be relevant for those dedicated officer and enlisted personnel in the operating forces looking to broaden their understanding of leadership as they confront the difficult environments our Marines currently and will continue to face in the foreseeable future. It seems fairly settled within the Marine Corps that the study of leadership is something to be revered. Every commandant from the very earliest years of our founding emphasized education in leadership, understanding that leaders were made, not born.

Dr. James van Zummeren
Director, Lejeune Leadership Institute
Quantico, VA

Acknowledgments

You can hold this book in your hands today because of the hard work of many people, and even more than that, because of their infinite patience in dealing with me. Those I am about to mention have provided great professional contribution to this project, and at the same time have endured my stress. I am not sure whether I should thank them, or offer them apologies. Likely, I should do both!

Carroll Connelley provided the horsepower for the book. It has been a great experience working with him. Very likely on several occasions he might have regretted agreeing to work with me on this project. I am sure that, had he not been a Marine gifted with a unique work ethic, he might have quit and this project would not have been completed. I am confident that his beautiful family shared some of the Tripodi-induced stress, and they deserve much recognition; it makes me happy to be popular with his children and to have them know me as the "count."

It is very likely that the contributors to this book had to deal with a process that might have appeared pedantic at times. Peer reviews, initial revisions, editorial revisions, manuscript editor's revisions . . . I am sure that anytime they thought "ok, we are done" and saw my name in their inbox, they had to find special willpower to open the message. They have been outstanding at every moment of this process. Carroll and I will be indebted to them for a long time to come.

Nearly all chapters in the book have been read and copy edited several (no mistake, several!) times by Andrea Hamlen and Stase Rodebaugh. Megan Hennessey skillfully edited four chapters, and Dr. Patrice Scanlon served as a second reader. Andrea and Stase are two important "pillars" of Marine Corps University, where they are responsible for the Leadership Communication Skills Center. They have been extremely professional and provided great guidance.

My final victims in this process were the MCU Press managing editor, Shawn Vreeland; manuscript editor, Andrea Connell; designer, Robert Kocher; and the MCU Press senior editor, Ken Williams. They have provided tremendous expertise to finalize the book. Carroll and I are particularly grateful to Ken Williams and Steve Evans, who believed in this project from the very beginning and responded in a positive and encouraging way to our proposal.

It has been my wife, Jenny, who has had to cope with the greatest amount of my stress. I am sure that she is even happier than I am to see the book completed. When I felt overwhelmed by the process, I was comforted by my son, Antonio. I thought he was the one who best understood me—he listened to me, and after a short while looked at me and asked, "Can I watch a little Elmo now?" He and Jenny are the source of my strength and happiness.

Paolo Tripodi,
Richmond, VA

— INTRODUCTION —

CARROLL CONNELLEY

But if you are ferocious in battle remember to be magnanimous in victory . . . I know of men who have taken life needlessly in other conflicts, I can assure you they live with the mark of Cain upon them.

-Lieutenant Colonel Tim Collins' Address to the Royal Irish Guards, 22 March 2003[1]

Lieutenant Colonel Collins' entire address to his unit just prior to sending his forces into battle on the eve of Operation Iraqi Freedom has been cited[2] as one of the most inspirational modern-day-warrior speeches. In full, the guidance to his soldiers struck the perfect balance between encouraging the utter destruction of the opposing forces and inspiring his troops to accomplish the task in a manner that would allow them to return home with honor.

---

1 Joseph J. Thomas, "We Go to Liberate: LtCol Tim Collins' Address to the Royal Irish Guards" in *Leadership Explored: Lessons in Leadership from Great Works of Literature*, ed. Joseph J. Thomas (Virginia Beach, VA: Academx Publishing Services, 2006), 332–33.

2 Ibid. Thomas adds, "Collins' speech was reprinted in the *London Times* along with a photo of its author, cigar clenched in teeth. His words drew comparisons to Lincoln's *Gettysburg Address* and Churchill's towering war speeches." See also John Keegan, *The Iraq War* (New York: Knopf, 2004), 167. Lieutenant-Colonel Tim Collins would become famous by making an inspiring eve-of-battle speech to his troops, which President George W. Bush had displayed on a wall in the Oval Office at the White House.

Without being strident, Lieutenant Colonel Collins appealed to his soldiers' ethical, legal, and spiritual dimensions in his speech.

While leadership has many aspects, this book focuses on the ethical, legal, and spiritual because of their foundational nature. As one of the course readings for the Marine Corps Command and Staff College states, "Leadership is the driving force of all war actions, and ethics is the foundation of leadership. Therefore, it is important for military leaders to become lifelong students of these two interwoven concepts."[3] In order to support this precept, we compiled this collection to provide insight on how ethics, law, and spirituality can enhance a leader's ability to create a command climate to accomplish the mission and support his or her Marines.

A study of combat leadership conducted by the U.S. Army found that the "sine qua non of almost every successful commander was the unquestioned integrity concerning his duties, coupled with a solid ethical foundation in matters of dealing with combat or warfare."[4] Ethics, according to the military, appears to be the bedrock on which good leadership is built. But ethics is a broad field that has been considered throughout history and is studied in many different forms, beginning with the virtue ethics of the Greeks, moving through stoicism, onto consequentialism, then Kant, and more recently, applied ethics, to name but a few. For those of us in the military, while different approaches to ethics continue to be debated by scholars, at its core, ethics is the question of making the morally right decision at the most difficult time. However, to only reflect on the ethical dimension and ignore the legal and spiritual components that inform and help military

3 U.S. Marine Corps, Command and Staff College Distance Education Program, *MAGTF Expeditionary Operations, Course Book and Readings 8906, Volume I, AY12*, 1–2.
4 The U.S. Military Academy History Department, "Leadership in Combat: An Historical Appraisal" in *West Point's Perspectives on Officership*, ed. Timothy Spurlock (Dallas, TX: Alliance Press, 2001), 257.

members cope with difficult decisions would leave this work incomplete.

The study of leadership for warriors begins with the classics of the ancient world and runs through the accounts of our most recent conflicts. Vice Admiral James Stockdale was a student of that history and a modern heroic leader who, after deep study and careful reflection, was able to successfully employ his education under the most difficult circumstances. Admiral Stockdale's extended time as a prisoner of war not only tried his ethical stamina, but also challenged the legal responsibilities he owed to his country and men, as well as tested his spiritual fortitude. Having been through such trials, Admiral Stockdale is an enormous repository of reflection on how the ethical, legal, and spiritual aspects of leadership help form a leader.

In his 13 July 1979 speech[5] to the cadets at the United States Military Academy concerning duty, keeping one's word, and making right choices, Admiral Stockdale referred to the 17th-century philosopher John Locke's book, *An Essay Concerning Human Understanding*.[6] Admiral Stockdale pointed to Locke's conception of three different answers to the question of "why a man must keep his word" and proposed that the answers were "as applicable today as they were then."[7] Admiral Stockdale went on to describe the answers:

> First, said Locke, a [spiritual] man will say, "Because God who has the power of eternal life and death requires it of me that I keep my word." Secondly, said Locke, if one takes the Hobbesian view of life, he will say, "Because society requires it and the

<hr>

5 James B. Stockdale, "A Vietnam Experience, Duty," in *West Point's Perspectives on Officership*, 263.

6 John Locke, *An Essay Concerning Human Understanding* (Charleston, SC: Nabu Press, 2010).

7 Stockdale, "A Vietnam Experience, Duty," in *West Point's Perspectives on Officership*, 263.

state will punish you if you don't." (Hobbes was a very practical kind of hardnosed guy.) And thirdly, John Locke observed that had one of the old Greek philosophers been asked why a man should keep his word, the latter would say, "Because not to keep your word is dishonest, below the dignity of man, the opposite to virtue (arête)."[8]

Consequently, when asked to describe how to make morally right decisions, Admiral Stockdale referred to the spiritual, legal, and ethical foundations of making decisions and leading. All three aspects are taken into account when dealing with the most vexing moral dilemmas. Further, Admiral Stockdale insisted to the prospective officers that it would be their "duty to be moralists."[9] Defining a "moralist not as one who sententiously exhorts men to good, but *one who elucidates what the good is*,"[10] Admiral Stockdale concluded his speech by appealing to a final duty he believed military officers have:

> You must be able to act as philosophers in your careers in order to explain and understand the lack of moral economy in this universe. Many people have a great deal of difficulty with the fact that virtue is not always rewarded nor is evil always punished. To handle tragedy may indeed be the mark of an educated man, for one of the principal goals of education is to prepare us for failure. When it happens you have to stand up and cope with it, not lash out at scapegoats or go into your shell.[11]

While this speech was given over thirty years ago, it still rings true today. In the midst of an ongoing conflict with an enemy that is difficult to identify, does not adhere to the law of war, and appears to operate with a different set of moral standards, a leader must be able

---

8 Ibid, 263.
9 Ibid, 265.
10 Ibid.
11 Ibid, 266.

to intelligently explain to his subordinates why following the law and making the morally right decision at those most difficult times is essential.

Where law is concerned, General James N. Mattis explains the essential connection to leadership and good decision making. In the foreword to the *Marine Corps Reference Publication on War Crimes*, General Mattis notes that "compliance with the Law of War is not only required under the UCMJ, but is also absolutely essential to mission accomplishment."[12] Despite the continuing consternation of the overlegalization of combat, one of the most respected and venerated warriors of our day has unmistakably posited that the law is in fact a force multiplier for leaders and not a constriction, as has so often been described. It is a matter of leaders understanding the law—not deferring to judge advocates—and making the link between law and mission accomplishment that is paramount. Simply put, the law is another support for assisting Marines in making those morally right decisions in the most difficult circumstances.

Spirituality, too, is an oft-misunderstood aspect of leadership. It is considered here because all indications identify the element of spirituality as being able to assist a command and its Marines to make measured decisions and promote a positive command climate that effectively deals with postcombat effects. In his seminal work, Marine Corps Vietnam veteran and Rhodes scholar Karl Marlantes describes the connection among spirituality, adherence to the law, and successful reintegration into society upon return from combat.

> Warriors deal with death. They take life away from others. This is normally the role of God. Asking young warriors to take on that role without adequate psychological and spiritual prepa-

---

12 Department of the Navy, MCRP 4-11.8B, *War Crimes* (Washington DC: Department of the Navy, 2005), i., http://www.marines.mil/news/publications/Pages/MCRP%204-11.8B.aspx#.T3S47o5191A.

ration can lead to damaging consequences. It can also lead to killing and infliction of pain in excess of what is required to accomplish the mission. If warriors are returned home having had better psychological and spiritual preparation, they will integrate into civilian life faster and they and their families will suffer less.[13]

The current combat environment is replete with demanding situations that perplex even the Corps' most senior leaders. It is our belief that when the aspects of leadership discussed in these pages are used properly, they promote a positive command climate and ultimately nurture Marines both in combat and upon their return. Ultimately, this collection is presented in the earnest hope that it will spark discussions that will assist future leaders in reflecting on the ethical, legal, and spiritual challenges under the most difficult circumstances. Part I of this collection addresses the difficulty of ethical decision making in the current counterinsurgency environment. To begin the examination, Lieutenant Colonel Brian Christmas, USMC, and Dr. Paula Holmes Eber address the complex ethical challenges found in operating among foreign populations while trying to understand and work with diverse legal, religious, and moral systems. Using case examples from operations in Marjeh, Afghanistan, during 2010, the authors demonstrate how a clear understanding of Islamic practices and local leadership can assist in dealing with conflicts that affect the unit, Coalition partners, and the local populace.

Geoffroy Murat, in his chapter, suggests a tool that leaders can rely on in order to be successful in operations: looking to "care ethics" or stakeholder theory—which states that personal development comes from being interdependent with others and encourages placing others at the center of one's actions—to help guide decision making.

---

13 Karl Marlantes, *What It Is Like To Go To War* (New York: Atlantic Monthly Press, 2011), 1.

Major Clinton Culp, USMC (Ret.) tackles the question of which pedagogical method is best suited to influence the ethical conduct of service members. The chapter primarily examines the current pedagogical methods that affect the majority of new officers who receive pre-commissioning ethics education, briefly touches on enlisted ethics education, and examines the institution's role in facilitating moral conduct.

Captain Emmanuel Goffi, French Air Force, addresses the difficult conflict that leaders must resolve in terms of their competing responsibilities toward their subordinates, their nations, and the international community. Used as an example is the French involvement in the conflict in Afghanistan, in which French troops have ostensibly been deployed with the primary purpose of defending French national interests; however, Paris's declared justification for involvement has been premised primarily on humanitarian concerns.

Dr. Peter Bradley explores what happens when loyalty to a fellow service member conflicts with a professional military obligation. This chapter examines the military duty-personal loyalty dilemma from a psychological perspective, drawing on research from the areas of decision making and military cohesion to illustrate how an individual's internal forces interact with situational influences to affect the choices military personnel make in these cases.

Dr. Paolo Tripodi draws on the historical account of the Holocaust to offer reflection for leaders on the issue of dissent. This chapter explores the motivations that led to disobedience and organizes them into two categories: subjective (the result of the individual's inability to perform a certain task) and objective (result of an individual's opposition to perform a certain task).

The chapter by Dr. Clyde Croswell Jr. and Lieutenant Colonel Daniel L. Yaroslaski, USMC, concludes the ethical consideration of leadership by positing that the need for global leadership and ethical mindfulness has never been more apparent. Their stated goals are ambitious as they attempt to arm the discerning reader with a new vocabulary; create an expanded understanding of the ethical nature of leadership complexity, sense making, and exercise of authority; and ground tactical, operational, and strategic thinking all within the biological, natural principles of living systems

Part II of this collection addresses various issues that have coalesced around the law of armed conflict and the responsibility leaders have for this issue that, at times, has been relegated to judge advocates. Major Winston Williams, USA, begins the conversation discussing training the rules of engagement (ROE) for the counterinsurgency fight, focusing on the delicate balance between exercising the inherent right of self-defense and winning the support of the local populace.

Professor Laurie Blank, Esq., then examines and refocuses the debate about rules of engagement to analyze the critical intersection of law, strategy, and leadership that ROE represent in armed conflict. This chapter addresses the challenge of how civilian and military leaders can translate the important role that ROE contribute to mission accomplishment.

Professor Jamie A. Williamson, Esq., then takes a broader perspective on the law of armed conflict as he discusses the importance of humanity in war and the critical role a commander plays in providing this perspective to his subordinates.

Lieutenant Colonel Chris Jenks tackles the complicated topic of balancing the degree and manner by which risk is borne between Afghan civilians and U.S. military members as a result of the ongoing counterinsurgency operations. The author, by utilizing the evolving tactical directive on the use of force, explores the agency of risk between Coalition military and the civilian population in Afghanistan.

Lieutenant Colonel Kenneth Hobbs examines command responsibility when soldiers from multiple nations participate in coalition operations under a unified command structure. This chapter suggests that if a multinational commander is held criminally liable for the actions of subordinates from partner nations in the same manner as those from the commander's own national forces, then a fundamental change should take place in the way coalitions are established.

Part III of this collection considers the warrior's spiritual dimension and how leadership can effectively support this. Commander David Gibson and Lieutenant Commander Judy Malana begin the conversation by examining the spiritual dimension of invisible injuries sustained on the battlefield. While spirituality may be individually practiced, and because it can affect the entire unit, they suggest a conceptual framework for intervention and healing on both the individual and unit level.

Colonel Franklin Eric Wester examines the results of a survey of U.S. land forces in the combat zone of Iraq collected in the summer of 2009. Named the Army's Excellence in Character, Ethics, and Leadership (EXCEL) Survey, it measured spirituality as one of the individual variables among soldiers. Colonel Wester analyzes the data to identify statistically significant correlations between higher scores of spirituality and measures of ethics and resilience.

Lieutenant Colonel Jeffrey Wilson discusses in detail the meaning and definition of the term "spirituality" within the military context. He argues that since spirituality is the philosophical underpinning of any logically coherent construction of ethics and law for both individuals and organizations, both legal and ethical education must begin with an education in spiritual matters. Lieutenant Colonel Wilson draws upon philosophy, behavioral science, and military history to craft a narrative of a uniquely American sense of spirituality that serves as a nexus for ethics and law in the American armed forces.

Dr. Pauletta Otis tackles the stimulating topic of the impact religion has on military leadership by considering questions such as if the personal morality of incoming personnel differ in any measurable way with regard to religious identity; if a service member's faith system or belief in personal responsibility to a "higher authority" makes him or her a more ethical or moral professional; and if spiritual or religious commanders are more ethical in their decision making. In this essay, Dr. Otis focuses on the relationship between personal spirituality and ethical behavior, religious identity and ethical behavior, and how either individual faith or religious institutions relate to the mission and requirements of the military.

Finally, Major General Arnold Fields concludes the discussion of spirituality by proposing that spiritual leadership is the amalgamation of natural, learned, and spiritual qualities necessary to influence ethical behaviors of self and others in the interest of morally acceptable objectives and outcomes. In this personal essay, Major General Fields reflects on his own experiences in combat and posits that fighting and killing in the name of freedom remain a moral dichotomy for the nation to ponder and the individual American warrior to resolve in his or her own spiritual realm.

PART I
— ETHICS —

Leadership, Ethics, and Culture in COIN Operations

Case Examples from Marjeh, Afghanistan

Brian Christmas and Paula Holmes-Eber

Many of the more complex challenges in operating among foreign populations stem from trying to understand and work with people who follow different religious, political, and legal systems. Around the world, notions of leadership and authority, methods for judging and resolving disputes, and concepts of morally and religiously acceptable behavior vary radically from country to country or even from region to region within a country. Misunderstandings and conflicts between the local population and outside military forces can quickly arise due to different religious or cultural interpretations of events and actions. When handled poorly, such misunderstandings may even escalate to serious hostilities on the part of the local population, resulting in violent conflict and a widening gap between the population and military or security forces in the area.

Tensions and hostilities from these misunderstandings are not inevitable, however. As this chapter demonstrates, military decision makers can often effectively resolve conflicts with local populations by partnering with religious, tribal, or other locally recognized leadership to leverage the community's traditional dispute resolution mechanisms. We argue that by working with local leadership to apply culturally accepted processes for resolving conflict, tense situations can be resolved peacefully and success-fully without the need for more kinetic approaches. Since these solutions have a "local face," the community is more likely to accept the outcome, local leaders are empowered and legitimized, and U.S. and Coalition forces build stronger relationships with the community in the long run.

Using three vignettes from 3rd Battalion, 6th Marine's (3/6) oper-ations in Marjeh, Afghanistan, in 2010, this chapter demonstrates how a clear understanding of local leadership patterns, Islamic beliefs, and culturally accepted methods for resolving disputes enabled the battalion to resolve local conflicts that affected the unit, Coalition partners, and the local populace. Our analysis combines in-depth case studies of actual operational events with cross-cultural research on conflict resolution. This synthesis of conflict resolution theory and operational experience allows us to expand our analysis beyond the specific case examples presented here to understand the larger cultural lessons learned from each vignette.

The annals of recent Marine Corps engagements in Iraq and Af-ghanistan have many examples of unsuccessful cross-cultural interactions. This may teach us what to avoid in that specific culture, but failures are not particularly helpful in identifying the specific factors that lead to success. Because our goal is to identify positive lessons that can apply to military operations, not only in Afghanistan, but in many cultural contexts, we have selected three vignettes of conflicts that were resolved successfully. The theoretical literature on cross-cultural conflict resolution is then applied to explain the underlying cross-cultural principles that led to the successful outcomes.

Religion, Law, and Conflict Resolution in Cross-Cultural Perspective

In working in foreign cultural contexts, misunderstandings can often arise when a service member assumes that other cultures resolve conflict and evaluate behavior according to Western standards. In the United States and Europe, conflicts are officially resolved by using formal court systems based on Roman and British common law practices; applying secular rather than religious standards to behavior; and relying on state-funded political and security systems to enforce peace.

Cross-cultural research, however, demonstrates that people in both Western and non-Western nations employ a large range of dispute resolution mechanisms to resolve conflict—many of which are not based on formal state institutions. Furthermore, in many regions—particularly those that have had an external leadership imposed upon them (through colonial rule or conquest, for example)—state-sanctified leadership may not have the power or legitimacy to enforce laws or motivate the local population to comply with these laws. Rather than viewing state leadership, law, and the courts as the only system for resolving conflicts, scholars note that people around the world rely upon a variety of mechanisms for resolving disputes. This *legal pluralism*[1] provides people with numerous avenues for resolving differences. In the United States, for example, many conflicts never reach the courts but are resolved through mediation, arbitration, or simply by pressure and intervention from family or community members.

Military leaders, therefore, need to be aware that in many countries there are parallel and even competing systems of justice and dispute resolution. This is particularly common in Muslim countries, in which secular European judicial systems operate side-

1 Olivia Harris, *Inside and Outside the Law: Anthropological Studies of Authority and Ambiguity* (London: Routledge, 1996).

by-side with religious Islamic law. Typically in these countries, secular law governs business, international relations, and criminal law (all issues in which the state desires control), while religious law is relegated to family and domestic issues.

Ideally, secular and Islamic legal systems do not overlap, but in most Muslim countries there are grey areas where cases could be tried in either court system. The decision as to where to bring a complaint often reflects local values, as well as a pragmatic assessment of which legal system is likely to be most favorable to the complainant. Frequently, local Muslim legal and religious scholars and leaders (such as a *mullah*, *imam*, *qadi*, etc.) hold a considerable amount of power, authority, and respect from the local population due to their influential roles in resolving conflict. This explains why, in most Muslim countries, there are two parallel (and sometimes competing) power systems: one based on state authority and the other on religious authority. Operationally, when working in such areas, military leaders will be most successful if they can work with leaders of both systems.

Religious and secular legal systems are not the only forms of conflict resolution, however. Frequently, indigenous forms of leadership (such as tribal leaders, business leaders, and so on) exert much influence over the community; as a result, the population may turn to these leaders for intervention in and resolution of local conflicts. Military leaders may overlook these oftentimes powerful mechanisms for resolving disputes, since they do not have the familiar trappings of a Western-style political and judicial system. Alternative nonstate political-legal systems can be based on tribal precedent; other legal/moral philosophies (e.g., the Pashtun code of honor and revenge, *Pashtunwali*, in Afghanistan); or other historical traditions established long before the arrival of European or foreign legal systems.[2] The resulting indigenous dispute

2 See Thomas Barfield, *Afghan Customary Law and Its Relationship to Formal Judicial Institutions* (Washington, DC: United States Institute for Peace, 2003); and Louise Anderson, "The Law and the Desert: Alternative Methods of Delivering Justice," *Journal of Law and Society* 30 (2003): 120–36.

resolution mechanisms can range from a formal council meeting of elders (e.g., the Iraqi tribal *shuras* and the Afghan *jirgas*) who listen to complaints and mete out judgment, to more individual forms of conflict resolution, such as the use of mediators hired to assist in guiding negotiation between the offended parties. In some cultures, a respected community leader, religious leader, or elder may adjudicate complaints, sitting on his front porch each morning and "holding court." In others, members of the community may band together to form a vigilante group, hunting down and killing individuals they consider to be criminals or violators of an accepted cultural rule.

By understanding and recognizing the variety of avenues for resolving disputes in an area of operation, military leaders can select conflict resolution methods that are most likely to lead to successful outcomes. The following analysis applies cultural research on Afghanistan to analyze the vignettes from Marine Corps Battalion 3/6's operations in that area. The specific vignettes are then evaluated for the larger cross-cultural lessons, illustrating how military leaders can resolve crises through the use of local leadership and resolution methods.

LEADERSHIP AND CONFLICT RESOLUTION IN SOUTHERN AFGHANISTAN

Studies of Afghanistan reveal a two-tiered legal dispute resolution system, reflecting the difficult and often ineffective relationship between the state and regional areas. First, there is a state-regulated legal system composed of state-appointed courts and judges. Over the past 30 years, depending on the government in place at the time, this legal system has been more or less founded on a state-mandated interpretation of Islam, theoretically based on the Hanafi *fiqh* (jurisprudence).[3] Although loosely tied to Sunni

---

3 See Barfield, *Afghan Customary Law;* and Nazif Shahrani, "Local Politics in Afghanistan: Dynamics of State-Society Relations in Perspective" (keynote address presented at the symposium on "Beyond the State-Local Politics in Afghanistan, Bonn, Germany, Center for Development Research, 26–28 February 2009).

sharia (Islamic law), as Barfield[4] notes, this connection is more an ideal than a reality due to the inadequate religious and legal education of the majority of judges and lawyers in the country.

In theory, the Afghan state legal system is the only recognized method for resolving disputes in the country. However, due to inadequate funding and lack of skilled personnel, the Afghan court system is effectively limited to the major urban centers of the country and a select set of higher-level courts in the outlying provinces. Furthermore, given the uneasy relationship between the state and outlying tribal and ethnic groups, even where state courts do exist, the local population is reluctant to use the state legal system, viewing it as a structure the state exploits to exert control in the region.[5]

Thus a second—parallel and informal—tier of conflict resolution exists in outlying provincial areas. This is particularly evident in the Pashtun-dominated region of southern Afghanistan. Today, as in the past, local Pashtun tribal groups tend to continue to follow their own system of justice based on a mixture of religious and cultural leadership and sanctions. The result is a syncretic mix of a local folk interpretation of Islam combined with community application of the tenets of Pashtunwali.

Recent research on Afghan village areas reveals a balanced system of conflict resolution and community leadership composed of the following elements: (a) tribal councils or jirgas; (b) the village religious/legal leader or mullah; and (c) the village representative or *malik*.[6] Each of these three parallel systems works in conjunction, either simultaneously or sequentially, to resolve community disputes and issues. According to a survey conducted by Brick,[7]

---

4 Thomas Barfield, *Afghan Customary Law.*
5 Shahrani, "Local Politics in Afghanistan."
6 Jennifer Brick, "The Political Economy of Customary Village Organizations in Rural Afghanistan" (paper presented to the Annual Meeting of the Central Eurasian Studies Society, Washington, DC, September 2008).
7 Brick, "The Political Economy."

villagers were equally as likely to go to each of these different leaders to resolve an issue. This finding is critical, since it suggests that no one leadership system is dominant. Rather, a system of checks and balances is in place so that religious (mullah), community (jirga), and political (malik) leadership must work together effectively in order to maintain harmony in the village.

This preference, particularly among rural Pashtuns, for an egalitarian approach to decision making and conflict resolution is evident throughout the literature on Afghanistan.[8] The egalitarian village leadership system is in clear contrast (and perhaps in response) to the top-down, state-centralized efforts to control local areas. As the three following vignettes illustrate, by understanding and working with all three parallel leadership systems in Marjeh, Afghanistan, under the leadership of Lieutent Colonel Brian S. Christmas, 3rd Battalion, 6th Marines was able to successfully resolve local conflicts and build effective long-term positive relationships in the region. In order to emphasize the perspective of the leader in the cases, all three vignettes are narrated in the first person by Lieutenant Colonel Christmas. All case summaries and the lessons learned continue to be presented by both authors.

3RD BATTALION, 6TH MARINES: OPERATIONS IN MARJEH, AFGHANISTAN, 2009–2010

In December of 2009, 3rd Battalion, 6th Marines was informed that they would be deploying to Afghanistan in January of 2010

---

8 See David Edwards, "Learning from the Swat Pathans: Political Leadership in Afghanistan, 1978–97," *American Ethnologist* 25 (1998): 721–28. Also see Brick, "The Political Economy"; Nazif Shahrani, "War, Factionalism, and the State in Afghanistan," *American Anthropologist* 104 (2002): 715–22; Shahrani "Local Politics in Afghanistan"; Thomas Barfield, "Weapons of the Not So Weak in Afghanistan: Pashturn Agrarian Structure and Tribal Organization for Times of War and Peace" (paper presented at the Agrarian Studies Colloquium Series on "Hinterlands, Frontiers, Cities and States: Transitions and Identities," Yale University, 23 February 2007); and Barfield, *Afghan Customary Law.*

under the 2d MEB (Marine Expeditionary Brigade) and Regimental Combat Team 7 (RCT-7).[9, 10] They were assigned the mission of clearing the northern area of Marjeh in Helmand Province, Afghanistan. They partnered with an Afghan National Army (ANA) *kandak* (battalion) and, following the initial clearing, also partnered with various Afghan National Civil Order Police (ANCOP) and Afghan Uniformed Police (AUP) units. The deployment lasted over seven months and consisted of both conventional and counterinsurgency operations.

---

9 The narration of three cases is provided in the first person by LtCol Christmas to illustrate his perspective. In contrast, all case summaries and analyses are provided by both authors.

10 In 2004, I (LtCol Christmas) served as the Battalion Landing Team (BLT) operations officer during operations in the Oruzgon Province of Afghanistan. To the outsider looking in, our actions would seem to be purely kinetic and focused on the enemy. But a closer look reveals a BLT commander who had a clear understanding of the culture and the people that significantly influenced our successful day-to-day operations. From Jan Mohammed, the province governor and aggressive leader of the Afghan Militia Forces and his soldiers, to the local leaders in the countless villages that we fought in or traveled through, LtCol Asad Khan's proactive, sincere, and knowledgeable approach to these individuals ensured our success. Never relinquishing his authority but understanding when to give and take strengthened our relationships and promulgated the confidence of those that he was trying to influence. My leadership education continued as the executive officer of a battalion deployed to Fallujah, Iraq, in 2005 following the initial clearing of the city. LtCol Bill Jurney was assigned the daunting task of piecing a large portion of the city and the outlying areas back together. This included the reintegration of the civilian populace, the re-establishment of the local government and authorities, and the rebuilding of the infrastructure that was damaged or destroyed during the initial fight. All of this was completed while engaging an insurgent enemy—a task that has become common on today's battlefield and requires engaged and informed leadership. LtCol Jurney, like LtCol Khan, provided me the example of developing and maintaining relationships with all of the "players" involved in ensuring success. In this case, LtCol Jurney had a clear handle on relations with the leadership of the two Iraqi battalions that were operating in our AO, the political and religious leaders, and the local leaders, all of whom influenced the ability to repopulate, rebuild, and enhance what was previously a prosperous and viable community. My experience with these two sincere, informed, and proactive leaders provided me with strong role models to emulate. Coupled with the education provided at Command and Staff College, I was prepared to handle the challenges faced during operations in Marjeh, Afghanistan.

Northern Marjeh included a mixed terrain consisting of plush farms with canals formed in grids as well as desert with deep sand. There were both small and large villages and comparable markets and bazaars. The majority of the narrow roads throughout the area of operations (AO) required one-way traffic and very careful driving along the canals. Many areas were only accessible by foot traffic or helicopter. During the beginning of the deployment, the Marines and sailors endured freezing temperatures, volatile rain storms, and hail the size of golf balls; by the end of the deployment, the Marines and sailors were enduring 120 degree temperatures.

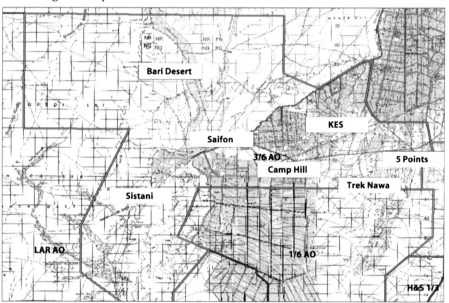

MAP 1. **Map of 3/6 Area of Operations in Marjeh, Afghanistan.**

In late January, elements of the battalion (partnered with the ANA down to the squad level) inserted via ground and air into the Bari Desert, where they began to conduct shaping operations. In the second week of February, in conjunction with adjacent units, 3/6 conducted a ground and heliborne assault on northern Marjeh in order to seize named objectives and ultimately clear their designated area of operation.

As the Coalition forces continued to clear the enemy from Marjeh, the "government-in-a-box" was put in place. The "box" consisted of Haji Zahir, the appointed subgovernor, and a few administrative officials. Haji Zahir was a brave, proactive, and engaged individual who wanted to make a difference, but he was challenged by a shortage of resources and experience. The governor, along with many U.S. and ANA officials, did not seem to understand the complexity of the area that was now designated "Marjeh." It included all of the areas to the north, up to the canal boundary with the British and Nadi-Ali, through portions of Trek-Nawa, to the west and northwest into the Bari Dessert inclusive of the Sistani Peninsula, and south out of Marjeh proper toward Camp Dwyer. The focus tended to be on the central area where the big bazaars were and where the "box" was placed. As a result, it took engaged and proactive leadership to court the governor and to ultimately help him to realize the importance of addressing this entire area. Without the Afghan government's direct engagement, the mission, again at all levels of warfare, would not have been successful.

The importance of building and maintaining relationships with the political, religious, military, and community leaders in an area of operations cannot be overstated. Unfortunately, often these relationships are typically handled in an extreme manner. In some cases, military leaders approach the situation with apprehension and treat local leaders with "kid gloves." At the other end of the spectrum are cases in which the situation is mishandled by a leader who does not understand its complexity or who is too arrogant to see the importance of local leadership. There is a fine line that a leader must walk so that he will not give too little or too much latitude and/or attention to the detriment of his mission as the following three vignettes illustrate.

Vignette 1: Using a Mosque as an Insurgent Weapons Storage Facility

During our initial push into northern Marjeh, following successful shaping operations in the Bari Dessert through January and into February of 2010, Captain Bill Hefty, along with the Marines and sailors of Company I and their ANA partners, cleared through an area that was heavily laden with improvised explosive devices (IEDs). In one instance, the Coalition force came to a mosque that contained a large depository of IEDs and manufacturing materials along with small arms and ammunition. This was discovered following a search conducted by the company's ANA partners.

The mosque and surrounding area were secured and the incident reported to the battalion. Understanding the importance and complexity of the issue, especially being early into our time in Marjeh, we informed the regiment, who then informed higher command. Ultimately, this resulted in delaying further actions in the area until the Afghan director of the Hajj, along with a council of mullahs and other officials, could see the weapons stash for themselves.

My ANA counterpart, Lieutenant Colonel Farouq, and I arrived on the scene shortly after the reports, followed by a CNN crew sent down from the Marine Expeditionary Brigade. A few hours later the director of the Hajj and officials from the Helmand Province government, including the subgovernor, Haji Zahir, flew in. They were accompanied by Afghan television and newspaper reporters.

Realizing the importance of this event and the effect of first impressions on likely future relations, I ensured that my role and Lieutenant Colonel Farouq's roles were equal as we addressed the visiting religious and provincial government leaders. Throughout their time inspecting the mosque, these leaders were given full access to the area and were treated with the utmost respect. At times this was annoying, as the company was ready to push, as was I, and the amount of time it was taking to prove that a mosque was being used in this manner seemed unnecessary. However, reflecting upon my previous experiences in Afghanistan in 2004, I recognized the importance of working with the director of the Hajj, so I initiated a conversation with him, acknowledging his value without demeaning my own position and authority. That night, the director of the Hajj denounced insurgents using local mosques as sanctuaries and IED manufacturing and storage facilities. His speech was covered by CNN and the Afghan networks.

Lessons Learned

This highly sensitive incident was not resolved by conventional military tactics but by allowing local and regional leadership to take the lead in condemning the illegal cache. By taking the time to let the local leaders conduct their inspection, we demonstrated our understanding of their culture and the importance of Islam's role for them. The result was that a highly respected religious leader, the director of the Hajj, spoke out publicly on television against the insurgents. Because Afghan religious officials (and not U.S. Marines) publicly stated that the actions were contrary to Islam, the local population was more willing to accept the evidence.

Equally important, this incident established a partnership between local leadership and Battalion 3/6, providing legitimacy to the actions of the battalion in the eyes of the populace. The efforts taken to work with the Afghan religious leaders, the ANA, and the press were valuable at all levels of war (strategic, operational, and tactical). The newly established relationship with the director of the Hajj would prove beneficial throughout continued operations over the next five months, and our ANA counterparts experienced our dedication to our partnership. Ultimately, the time delay during the tactical execution resulted in an operational and strategic victory and would prove highly beneficial in the long term, as the next two vignettes will emphasize.

Vignette 2: False Accusations of Marines Desecrating a Quran

With the initial clear of "Marjeh proper" nearly completed, and Company I's recent push into the Sistani Peninsula, we continued to capitalize on our success, the positive press (Afghan and U.S.), and the development of our relationships with local political and religious leaders. One of our first steps was to hold a shura with the tribal leaders to discuss our plans to position our forces throughout the Sistani Peninsula. Sistani is home to many tribes including the Norzai, Ichsokzi, and Alikozi, and this particular shura proved educational and improved our ability to handle the sensitive incidents that we would encounter in future operations.

The shura was to be the first of many we held in the region, and I gained a true appreciation for the importance of the culture, knowledge of the local leadership, and the value of organizations like the Human Terrain Teams (HTT).[11] Prior to the shura, while walking around and assessing the ground, I had my jump Marines and sailor pull out our "shura-in-a-bag"[12] to supplement the company's resources. During this time, John, the senior member of the HTT approached me with a small piece of paper in his hand. He asked if he could speak with me for a moment. My education began the minute John started to talk. He handed me the slip of paper and described the complexities of Sistani, the multiple tribes represented throughout, where they were situated, and where they stood on the totem pole. He provided me with the names of their leaders and then described what they were likely to discuss and what was most important to each. This information

---

11 Human terrain teams are primarily an Army program designed to provide battalion and regimental commanders with a greater understanding of the culture and needs of the local population. The controversial program deploys a team of civilians (who are intended to have cultural, linguistic, or operational knowledge of the region) with battalions and regiments. Their role is to talk to the population and advise commanders on the sociocultural situation in the region.
12 "Shura-in-a-bag" consisted of a large pack that housed a tea set, with extra glasses; a teapot; Afghan serving dishes; cashews; and other assorted nuts, dried fruit, and candy. It also included large bottles of water, sugar, "good" tea, spoons, napkins, and a poncho line to sit on. In my vehicle, I also kept an Afghan rug.

on local leadership was instrumental in our ability to execute the shura successfully.

In order to provide sufficient forces to continue operations in "Marjeh proper," we established an agreed-upon presence on the outskirts of Sistani. We agreed to conduct security patrols, provide an active civil affairs program, and rely on the local leaders and people to manage daily activities occurring within the peninsula. As a result of our continued relationships with the local population and their leaders, we were able to successfully achieve our limited goals in this part of our area of operations.

Threatened by our successful operations, our enemy attempted to sway the positive local opinions of our Coalition force against us by creating disorder. After a few months into our operation, a very young and animated mullah told the locals that following a recent search of a home, a Coalition patrol had desecrated a Quran. The "mullah" gathered a small mob and headed across the Sistani Desert to the district center into Marine Corps Battalion 1/6's area of operations to protest. Although the protest was not in my AO (Battalion 1/6 had to deal with this first mob), I departed for Sistani in order to investigate.

When I arrived, I immediately stepped to the side, allowing the new ANA commander, Lieutenant Colonel Mohammed,[13] to take the lead as he questioned local people in the area. We found that the home where the desecration allegedly occurred was abandoned. All indications were that the accusations were false. We questioned several members of the mob who stated that ANA soldiers, with the help of Marines, had beaten them. They stated that there were many individuals with broken limbs, severe bruising, and other injuries. However, nobody with such injuries could be produced. The emotions tied to the alleged desecration of a Quran

13 LtCol Farouq's ANA Battalion rotated out following the main surge into Marjeh. His battalion was relieved by LtCol Mohammad's battalion that would remain with 3/6 until its relief-in-place.

and word of mouth, especially from a religious leader, were very powerful and evoked a strong emotional reaction among the people.

During the questioning, what struck us the most was that the mullah was not someone we had ever met or known. This was a place where we had established relations with local religious and community leaders, where we had agreed with the community about how we would operate in the area, and where we had already conducted various civil affairs-related efforts, including providing school supplies to the three local *madrassas* (religious schools). Frankly, it did not make sense.

After we were informed of a second mob forming the following day, I called upon the subgovernor, Haji Zahir. I then picked him up at the district center, along with other religious officials and government personnel, and in a similar fashion to the mosque incident, provided them full access to the house and the local area. We walked around and spoke to some of the local population, and it became evident that the claims of a damaged Quran were unfounded.

The incident came to a head when Haji Zahir asked the "mullah" to take a walk with him and to lead their group in prayer. As they returned from washing at the stream and praying, the subgovernor and his entourage were smiling and laughing. Having the relationship that I did with the governor, I inquired what was so amusing. He exclaimed, "That man is no mullah, he does not even know how to lead us in prayer." Ultimately, we gathered as many locals as we could and Haji Zahir gave the floor to the mullah, who, with great emotion and detail, explained the supposed atrocity and condemned the efforts of our Coalition forces. The locals were visibly angry afterwards and wanted vindication.

Then, Haji Zahir and a Marjeh religious leader spoke to the people, describing what they had observed during their visit. The leaders then dispersed all involved, providing Qurans to the local community and dispelling the accusations made. What was amazing about the whole situation is that the first mob was made up of individuals who did not even come from the area of the alleged desecration. Clearly, this was the enemy's attempt to instigate problems as a result of our continued successful operations and relationship with the locals, the government, and the military.

Lessons Learned

By taking a role as a Coalition partner supporting the governor and the religious leaders, I was able to show a presence without usurping the appropriate local authority for resolving disputes. As a result, the religious and political leadership in the area was able to use clear, culturally accepted evidence (that the mullah was a "fake mullah" who did not know how to pray) to challenge the false accusations. Furthermore, these leaders were able to provide the culturally appropriate solution to the problem: dispensing additional Qurans to the population. Equally important, however, by working as partners with the local leaders, our role as legitimate protectors of the peace was reinforced in the eyes of the community.

The relationships mentioned above include the trust and mutual respect required between a U.S. commander and his Coalition counterpart and with that partner's subordinates. One's Coalition partners need to be willing to listen, make informed decisions, and at times, be willing to follow one in the heat of battle. We had this relationship with our first ANA partners as was evidenced by our initial clearing operations in Marjah. We maintained a mutual trust and respect, capitalizing on each others' strengths and providing support mechanisms to manage or overcome our weaknesses.

As parts of our battle space began to transition from the clear to the hold phase, our ANA Battalion also transitioned, requiring our attention to ensure these types of relationships were fostered with our new partners. The need for this relationship transition was made evident upon our having to deal with the Marines' unintended upsetting of a Quran during the search of a home.

Vignette 3: A Mishandled Quran

In the southeast corner of our area of operations, where "Marjeh proper" and Trek Nawa border each other, we established a platoon-sized position to address the infiltration lines of the insurgents. It was a volatile area that guaranteed contact with the insurgents multiple times throughout the day. The platoon assigned to this area was partnered with an ANA counterpart, which was the pattern throughout our operations.

After some time, we began to receive reports of ANA soldiers possibly collaborating with the enemy, providing the numbers and likely route of outgoing patrols. To address this issue, I spoke with my new ANA counterpart, Lieutenant Colonel Mohammed. At first, Lieutenant Colonel Mohammed's reaction was defensive, but once I explained that the information was gained through unquestionable intelligence collection sources, he became willing to find a solution.

Having learned from my previous experiences that working through local leadership led to greater legitimacy in our efforts, my recommendation was that Lieutenant Colonel Mohammad should handle the situation. I provided him the identification of the suspects, emphasizing that since we were Coalition partners, our information and intelligence collection should also be partnered. I suggested that he explain to his subordinates that the knowledge of the actions by the perpetrators was gained through his own intelligence means.

This was acceptable to him and we made our first trip down to the small base. While he addressed the issue of collaboration with the enemy, I focused on inspecting the facility with the platoon commander and talking to the Marines, sailors, interpreters, and the ANA who were not directly involved with their commander. It was a good visit that accomplished a great deal, especially toward building our relationship at multiple levels. This would not be our last challenging visit to this post, as we soon discovered.

Within a few weeks, we were forced to return. Marines were again accused of upsetting and damaging a Quran during a search of a home. This time, however, the accusations were true. Yet what made this event different from the previous incident was that it was not the locals who were protesting the event, rather, it was our ANA partners on the Trek Nawa border. Tensions with their partnered Marines and sailors were literally on the verge of outright aggression.

To set the stage, the Marines and sailors were aware of the disloyalty of some of the ANA as a result of our previous visit. Both Marine and ANA forces were led by strong platoon commanders; however, they were both frustrated by dealing with constant contact with the enemy, oppressive heat, and restricted terrain. So on this particular day, the ANA had decided that they were not going to participate in the patrols and required searches.

That morning, the Marines and sailors went out on the search alone. They were disgruntled, but maintained as much composure as could be mustered throughout the execution of the patrols. While the Marines and sailors continued their search, an older Afghan man approached the ANA soldiers who had remained behind. He was carrying a Quran that appeared to have fallen apart. He was not overly emotional, but wanted to know why the ANA would mistreat the Quran during their search of his house.

The ANA's unexpected reaction was outrage, fits of anger and emotion, and the requirement to restrain and disarm some of their own soldiers. Everyone returned to the patrol base, where emotions remained high. These emotions ultimately incited the majority of the ANA soldiers to begin questioning their partnership with the Marines and to threaten to avenge the alleged desecration. The instigators within the group easily capitalized on this situation as tensions mounted exponentially.

Upon hearing reports of the incident, I immediately called Haji Zahir and explained that a group of Marines allegedly desecrated the Quran. I stated that I was on my way to investigate the situation with Lieutenant Colonel Mohammed and that we would let him know the results. I wanted him to hear about the incident from me; I wanted him to know that we were addressing it immediately; and I wanted him to know that we understood its importance and his important role in the situation's ultimate reconciliation. The drive was a short distance from the forward operating base (FOB); therefore, we arrived on the scene very quickly.

Indeed, emotions were high. The Marine leadership and the ANA leadership had done a good job separating the two forces. Marines remained on the posts of the patrol base. Before arriving at the scene, the only information I had about the situation was that there was an alleged desecration. As I walked in, I cordially greeted an old man sitting calmly by the side entry point. As the conversation continued, he was whisked away by two ANA soldiers. I found this a bit odd, but knew that I had other things to worry about as the noise emitting from the inside of the compound was akin to wailing Iraqis at a funeral. Both Lieutenant Colonel Mohammed and I understood the value of not overreacting, and we both calmly approached our separated forces to investigate the circumstances. As he spoke to his leadership, I spoke to the Marine platoon commander and platoon sergeant. They explained that Marines were conducting searches, while their ANA partners refused to participate. In one case, a Marine was looking at a book, when another walked in the room, knocking the book out of his hands out of frustration. The book separated and the pages were dispersed across the floor. Not realizing that the book was a Quran and not showing their usual respect for the owner and his property, the Marines moved on to the next house.

As the lieutenant colonel went on, I realized who the old man was upon my arrival and that there was more going on than just the desecration of the Quran. Several issues needed to be addressed and the first started with my Marines and sailors. I spoke to them in full sight of the ANA. I spoke of the Bible and our general understanding and then I spoke of the Quran and their understanding. I explained that to Muslims, the type in the Quran is the words spoken by God to Mohammed and that there is only one version of the Quran that is sacred, spoken by God, and written by the illiterate prophet who neither wrote nor read until God gifted him. Most already knew this, but needed a reminder. They clearly understood the passion of their partners, as this passion is similar to that shown by our Marines and sailors for their own religion.

I approached Lieutenant Colonel Mohammed, who clearly had the hardest task and was unable to restrain the emotions of his platoon. I asked if he would mind if I spoke with his men. Clearly frustrated and skeptical, he supported my request. Rather than using my interpreter, I asked one of the platoon's two interpreters, a clear-headed man who had gained the respect of the ANA, to translate for me as I addressed the ANA soldiers. He obliged. As I began to address the soldiers, I was immediately attacked with harsh words and witnessed the uncontrolled emotions of others in the background. Remaining calm, I responded, acknowledging their claims of our mistake. I then explained our intent of addressing the situation by working with the local community and religious and governmental leaders to ensure appropriate steps would be taken to rectify the mistake.

There were three ANA soldiers who were especially vocal, one with an ability to articulate his thoughts coherently and with persuasion. This became a challenge, as he and his two assistants were clearly instigating and taking advantage of a highly sensitive situation. In speaking with Lieutenant Colonel Mohammed, I asked for his recommendation. He stated that we needed to remove the three individuals as well as the other interpreter. I happily concurred and my counterpart took actions to ensure removal of the instigators without causing additional conflict.

My task was not complete, however, as I followed up with Haji Zahir and explained all that happened. Ultimately, Lieutenant Colonel Mohammad and I drove down to the district center and spoke with the subgovernor. Haji Zahir called the director of the Hajj, explained the circumstances, and asked for guidance. We were told that we needed to replace the Quran as well as provide a cow to be slaughtered and given to the people of the town. Based on this guidance, Haji Zahir, Lieutenant Colonel Mohammed, and I drove to the patrol base and had the ANA, partnered with Marines, lead the foot patrol back to the compound where the incident occurred.

We were attacked by insurgents throughout the movement. Despite the constant firing and ultimate danger, the subgovernor understood the importance of his role, maintained his composure, and willingly continued on with the patrol. Upon our arrival, I asked Haji Zahir if it was best to give the cow to the community as a gift or to provide the community with the money to purchase the cow. Out of sight of others, I provided Haji Zahir the money for the cow and we proceeded to greet the owner of the compound. The old man was elated at the subgovernor's willing visit, especially as shots ricocheted around us. He was even more ecstatic when presented with two Qurans to replace his damaged one. Haji Zahir explained to him and to others standing within earshot that he had spoken to the director of the Hajj on our behalf and that he recommended replacing the Quran and providing a cow for a feast to be held in the village. He then handed the old man the money and expressed the importance of the work that the Coalition forces were conducting. He stated that the Americans do not understand everything about their culture, but have good hearts, have a willingness to learn, and want to help Afghanistan succeed.

The cow supplement seemed to shock the recipient, who thanked the subgovernor with enthusiasm and thanked both me and Lieutenant Colonel Mohammed with equal respect and appreciation. A good portion of the ANA soldiers witnessed this exchange, which had its added benefits. The ANA platoon commander was given the responsibility of following up later with the village elders to ensure that the money actually went toward the feast, further integrating them into the community's solution of the problem.

Lessons Learned

By turning to the district religious and political leadership systems to help resolve the conflict, an extremely volatile situation was averted. The local leadership came up with a culturally appropriate solution requiring payment of a cow and a feast to the community. This solution, which is based on *Pashtunwali* notions of reinstating honor (*nang*), is a culturally accepted way of resolving conflict in Pashtun areas of Afghanistan. The Marines paid a social debt for dishonoring the elderly man, the religious community, and the ANA, thus absolving any hostilities between the parties.

Equally important, both religious and political district leadership visited the home of the elderly man in person. Their presence not only added authority and validity to the Marines' actions but also reinforced the support of the community leaders for the presence of the Marines and the ANA in the area. The parallel presence of both religious and political leaders during all three of the vignettes described here reflects the Afghan village preference for balanced leadership in which no single individual is given undue authority. By including all important leaders in these events, rather than working with only one, this balance was maintained and strengthened throughout the Marines' deployment to the area.

Ultimately, a potentially disastrous event that could have soured the partnership between the ANA and the Marines actually had the reverse effect. The relationship between the Coalition partners was strengthened both at the platoon and the battalion levels; the local community gained confidence in the ANA and the local Afghan government; the ANA gained confidence in the local Afghan government; and the relationship between the Coalition partners and government, religious, local, and military leaders continued to grow.

THE REST OF THE STORY

Operations in northern Marjeh continued to progress and over the seven months within the AO, all involved witnessed strengthened relationships and mission accomplishment. The subgovernor held office hours on Saturdays in the central portion of our northern AO and conducted shuras with local leaders in all corners of the AO. Government of the Islamic Republic of Afghanistan (GIRoA) officials had begun to show the necessary understanding of Marjeh and the need to expand governance to all parts of the area. Demonstrating his commitment to his role in the community, the director of the Hajj accepted an invitation to lead a shura of the local leaders and to meet with the local mullahs immediately afterwards. This balanced approach to community leadership proved instrumental in our continued success, which was evidenced by a decrease in IED incidents and an increase in civilian-assisted IED finds and cell phone hotline tips.

In addition, after an initial shura in the Camp Hill area, a local mullah declined assistance for his mosque, stating he would let me know when he was ready. His lack of support and trust was disappointing. However, several months into the operation, the mullah pulled me aside to tell me that he was ready. He had changed his mind and wanted to be part of the solution.

We realized the extent of the affection and respect that we had earned in the area by the end of our tour. During clearing operations early in our mission, we had bestowed our battalion coin upon the director of the Hajj. Several months later, the director revealed the contents of his pocket and there was the battalion coin—a symbol of our friendship—carried as a daily reminder of the bond that existed between the Marines and the local community leaders.

Conclusions

Through the example of these three vignettes, the military leader gains an appreciation for the importance of making a concerted effort to work with and through Coalition partners, host nation government, religious leaders, and local officials in order to build trust and a strong relationship. The vignettes reveal the operational value of employing culturally acceptable solutions to the potentially volatile issues that inevitably arise throughout a unit's deployment. By including local leadership and using indigenous approaches to conflict resolution, these solutions are more likely to be accepted by the local population. Furthermore, host nation leadership is given ownership of the process, since solutions are generated by a unified effort of host nation entities with minimal involvement from the foreign Coalition partner. As a result, local leadership is strengthened; a solid foundation of trust and cooperation is built between Coalition partners and the host nation power brokers; and the local population gains confidence in the ability of Coalition and local leadership to provide stability in the area.

ETHICS AND IRREGULAR WARFARE
THE ROLE OF THE STAKEHOLDER THEORY AND CARE ETHICS

GEOFFROY MURAT

For a long time, military ethics has been associated with protection ethics: service members' main mission was to protect a territory and national security. However, protection ethics are more difficult to implement in modern conflicts in which Western armies are deployed to areas far from the national territory they are supposed to protect. In addition, the connection between national security and several interventions abroad, such as the deployment of French soldiers in Afghanistan or in Chad, is far from obvious. Therefore, it is important to consider whether applying care ethics over protection ethics would help troops to establish better and more effective connections with a broad range of local actors including inhabitants of the villages they come across; nongovernmental organizations (NGOs) that are involved on the battlefield; or local, national, and international mass media.

The evolution of war has made a significant impact on and transformed the scope of troops' assignments. For these service members, the goal of these assignments is not only to be the most powerful, but also, and more importantly, to be able to understand a complex environment where politics and military affairs are interconnected. Military personnel should also understand and take into account the role all stakeholders play in the theater of operations. In irregular wars, destroying the enemy is only one objective, but not the most important; in order to succeed, it is essential to convince the enemy to enter into a peaceful political process of conflict resolution.

In such an environment, one has to question the very training of troops. The culture of military institutions based on the ethics of virtue does not accurately prepare them for the conflicts of our time. In order to define a military culture that would be more appropriate for the new strategic environment, it is necessary to explore and analyze the features that are specific to a soldier or Marine's job. These include spirit of comradeship and solidarity, which can go as far as sacrificing one's life to protect fellow troops.

The underlying values of these specific behaviors are related to a moral theory called "the ethics of care." This theory states that personal development comes from being interdependent with others and encourages placing others at the center of one's actions. The objective of care ethics is to build confidence among individuals. Thus, care ethics stresses what we owe each other rather than how we should behave as individuals.[1] Such a culture suggests that individuals should try to fulfil the needs of those they are responsible for. This is what Marines and soldiers do when they try to help each other on the battlefield. It would therefore be possible to build military ethics based on the feeling of care that soldiers share among themselves and to extend it to all stakeholders involved in a conflict. Because a soldier's mission today is to understand the expectations of all actors in a conflict, care ethics has great potential to be a powerful guide in these types of missions.

1 For more details on this idea, see Virginia Held, *The Ethics of Care* (Oxford: Oxford University Press, 2004). For more detail on the ethics of care, see also Annette Baier, *Postures of the Mind* (Minneapolis: University of Minnesota Press, 1985); Carol Gilligan, *In a Different Voice* (Cambridge: Harvard University Press, 1990); and Joan Tronto, "Beyond Gender Difference: To a Theory of Care," *Signs* 12 (1987): 644–63.

A potent illustration of how service members may implement care ethics is provided by Colonel Benoît Royal when he referred to an incident that happened to a French unit in Somalia in 1993. As the unit engaged General Aideed's violent militia, despite a very tense situation, French soldiers opted to clear buildings room-by-room and did not use artillery, which could have killed non-combatants. They advanced in enemy-held areas street-by-street, building-by-building and did not hesitate to put their lives at risk while they made sure that they would not harm noncombatants. When General Aideed later met international troops, he refused any negotiation with the international community but said he respected the French troops because they showed that they knew how soldiers should fight.[2] Often, fighting with honor means taking risks in order to save lives and accepting the risks that are inherent to irregular warfare. It is wrong to believe that insurgents do not have their own ethical values—though those values may differ from our own.

In the confrontation with Aideed's militia, French troops showed care for the local population. They accepted risks in order to save noncombatant lives, although it would have been much safer for them to just use artillery on the buildings. In a conflict situation where there is little certainty regarding who and where the enemy is and with the constant concern for local populations, troops have to make decisions that are likely to be influenced by the stress generated by the environment. To understand how soldiers and Marines react, it is important to consider the role played by emotions. Feelings—such as sympathy, empathy, or sensitivity—could even help service members fulfil their missions. Indeed, these feelings allow troops deployed in a country they are not familiar with to care for local players in the conflict.

---

2 Benoît Royal, *L'éthique du soldat français* (Paris: Economica, 2008), 35.

Emotions play a key role in establishing vital connections in the fight against terrorism. They need to be taken into account to understand troops' ethical behavior and how they think on the battlefield. Thus, care ethics is particularly relevant when analyzing military ethics. Care ethics can contribute to military ethics as the concept addresses irrationality, emotions, and feelings, all of which are overlooked in many other theories regarding morality that focus mainly on rationality. For instance, utilitarianism uses mathematical calculus to determine the amount of welfare from which each actor can benefit. Likewise, deontological approaches deny that emotions have any influence in the analysis of a situation.[3] Even Aristotle in his *Nichomachean Ethics* associates morality with rational behavior. According to Aristotle, the only way to find the good is to forget one's feelings and follow only one's own individual rationality.[4] Yet if military ethics requires service members to rally all actors around them, it is important for them to understand their environment and the situation of other actors in the conflict. This is a condition to convince these actors to consider a political settlement of the conflict, as opposed to using military force.

Among the many authors who have examined counterinsurgency operations, Pierre Chareyron offers a perspective that is particularly relevant for this chapter. Chareyron identifies three factors that are of great importance in counterinsurgency (COIN) operations: (1) "The more you protect yourself, the more you are in danger"; (2) "The bigger the military activities, the longer the war"; and (3) "Tactical irregularity, ethical regularity."[5]

Believing that troops can take an overly protective approach in COIN operations and be successful could be misleading. Insur-

---

3 For more details, see in particular John Rawls, *A Theory of Justice* (Cambridge, MA: Belknap Press, 1971).
4 Aristotle, *Ethique à Nicomaque* (Paris: Vrin, 2007).
5 Pierre Chareyron, "La contre-insurrection à l'épreuve du conflit afghan." *Politique étrangère* 1 (2010).

gents often do not want to destroy the enemy army; they are aware that they are not strong enough to do so. However, by adopting a highly protective approach and engaging in other activities characteristic of conventional warfare, regular armies make their moves much easier to read. This tactic is dangerous because a highly protective approach makes it difficult for regular armies to connect with local populations.

As time goes by, foreign interventions conducted by regular armies might be seen as invasions. People who were neutral at the beginning of the occupation are more likely to get involved in the conflict and to turn against the foreign army, which will make it more difficult to initiate a reliable political process.

If the war is not fought according to just war rules, unethical behavior from the regular army can easily lead to failure. Unethical behavior compromises the legitimacy of the intervening force. This is problematic because COIN operations need legitimacy if they are to convince the parties involved in the conflict to negotiate with each other. Events such as Abu Ghraib lead local civilians to believe that the foreign army does not respect them or care about their traditions and history.

One of the most significant weaknesses of modern insurgencies is that they do not hold the same legitimacy as did national liberation movements in the past. The cases of Vietnam and Algeria in the 1950s and 1960s are good examples. In Iraq, al-Qaeda was not supported by many local tribes, whereas Ho Chi Minh in Vietnam and the "Front de Libération Nationale" (FLN) in Algeria succeeded in winning the support of the majority of the population. If insurgents were to gain a similar legitimacy, armies conducting COIN operations probably would never reach a settlement of the conflict, no matter how many battles they won. Clearly the war for legitimacy is closely linked to the ethical behavior of the troops operating in a COIN campaign.

Chareyron's three assumptions shape the basic context of COIN operations. In many situations, their implementation is far from being complete as regular armies find it difficult to accept risks to gain the trust of the local populace. The idea that emerging technologies will help Western democracies prevent casualties in asymmetric conflicts is a dangerous illusion if one considers Chareyron's reflections. While these new technologies may save lives in the short term, an overreliance on technology on the irregular battlefield can be counterproductive in the long term because it might prevent troops from understanding the way the insurgents and their local supporters perceive the conflict and their environment.

When Marines and soldiers operate in a COIN environment, they might have to play a humanitarian role. They may also need to act as a police force and establish connections with local tribes. Troops are central to a network of local actors with whom they have to build strong relations; the success of a COIN campaign will depend on the strength of the connections they are able to establish.

Thus a key issue is to explore how troops can make sound decisions on the battlefield and ensure that such decisions successfully challenge the negative perceptions of the local actors. Care ethics and Edward Freeman's stakeholder theory provide interesting reflections.

The stakeholder theory, developed in the 1980s in the field of business ethics, has significantly changed the way we look at corporate life. In the 1980s, the business world was based on a dominant paradigm that a business organization's main goal was to make profits to be distributed to its shareholders. Freeman questioned this paradigm and introduced the idea that shareholders are of course to be considered, but they do not constitute the most important objective of a business. According to Freeman, shareholders only represent a small component of all stakeholders. A company that intends to achieve sustainable profits should care about all stakeholders: clients, suppliers, its employees, and the overall environment. This approach is indeed a condition for sustainability and, ultimately, success.[6]

Freeman's theory can find application beyond business ethics; indeed, military ethics can benefit from adopting some of Freeman's reflections. The asymmetrical wars that are the military's main focus of effort today involve numerous stakeholders acting as networks with different interests and rationales. Troops may achieve their goals by dealing efficiently with their stakeholders. Within this context, "stakeholders" are all persons or organizations affected by the troops' presence and operations.

In the past, many theorists have neglected the role of sound ethical decisions in relation to both military and business environments. For instance, Milton Friedman, an opponent of the shareholder theory, believed that the business world is amoral and that it should not bother with ethics.[7] Clausewitz, echoing this sentiment, contended that morals and warfare would be an absurd mix since conflict is an outburst of violence that can resort to any means to achieve its goals.[8] These theoretical approaches raised ethical concerns, as some practitioners tried to understand how morals may be of some importance in the business world or in warfare.

Freeman's stakeholder theory deals with two main factors: the purpose of a corporation and the reason the corporation exists. In the objective waging of wars, two similar questions should be answered: first, what is the object of war? And second, what

---

6 Edward Freeman, J.S. Harrison, A.C. Wicks, B.L. Parmar, and S. de Colle, *Stakeholder Theory: The State of the Art* (Cambridge: Cambridge University Press, 2010).

7 Milton Friedman, "The Social Responsibility of Business Is to Increase Its Profits," *The New York Times Magazine*, 13 September 1970.

8 Carl von Clausewitz, *On War* (Ware, UK: Wordsworth Editions Ltd, 1997).

is its purpose? The aim of waging war is to convince or coerce the enemy to negotiate and, to this end, it is required that all stakeholders involved in the conflict be taken into consideration. Local populations, public opinions in countries at war, the media, NGOs, and many other parties are all stakeholders whose contributions must be dealt with to reach an agreement aiming at sustainable peace.

Stakeholders are not only numerous, but differ significantly from each other. Knowing how to deal with such diversity is of primary importance. Stakeholder theorists have tried to determine analysis patterns that would enable one to understand how to properly deal with requests coming from each stakeholder. Such an approach provides guidelines as to how to meet the requirements of every stakeholder. In addition, such a categorization makes it possible to prioritize tasks and actions.

There have been many attempts to determine stakeholders' interests and to prioritize ways of satisfying those interests. Mitchell, Agle, and Wood developed one of the most well-known models used to determine why one stakeholder is more visible than another in a decision-making process. This model takes three factors into account: power, legitimacy, and degree of urgency of a stakeholder's needs. Power is defined as the ability to influence the decision-making process. Legitimacy is regarded as the status, function, or entitlement that enables a stakeholder to appear as legitimate in playing a role in the decision-making process. The degree of urgency of a stakeholder's needs refers to the fact that a stakeholder will apply pressure on a decision-making process all the more when its requirements are in urgent need of being met.[9] These determinants are also relevant to military operations

---

9 Ronald K. Mitchell, Bradley R. Agle, and Donna J. Wood, "Toward a Theory of Stakeholder Identification and Salience," *The Academy of Management Review* 22 (1997): 853–86.

in which various stakeholders will play on these three character-istics to draw the decision maker's attention. For instance, A is ordered to bomb a building in Iraq, and if one attempts to deter-mine the pattern of his stakeholders according to Mitchell's meth-odology, the following dynamics should be considered:

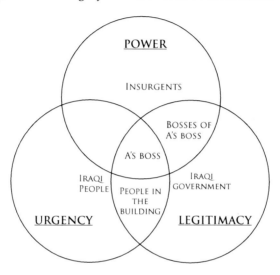

FIGURE 1. **Example of Mitchell, Agle, and Wood's Stakeholder Methodology.**

In figure 1, A's boss would indisputably be an important stake-holder from the perspective of A—he has an urgent need, i.e., making A execute his order; he is legitimate, as A is supposed to obey his orders; and he is powerful, as he is A's boss. That is why A's boss is placed at the center. If one looks at the people living in the building, they have an urgent need for the building to remain intact; however, they have little power over A. They do have a legitimate expectation that they will not be targeted, as they do not participate directly in the conflict.[10] That is why, in figure 1, they are located in the circles labeled "urgency" and "legitimacy"

10 The law of armed conflict gives them the legitimacy not to be targeted.

but not in "power." The insurgents have neither legitimacy nor needs from the perspective of A; however, they have power and leverage since they can take action to prevent or delay A from achieving his objectives.

Locating actors in figure 1 provides a tool for analyzing the situation and could help with decision making. Yet this type of categorization has a few shortcomings. To begin with, some stakeholders may easily be inadvertently neglected. For instance, in the case presented above, the influence from the media and Western public opinion can play a role in A's decision-making process. It is equally difficult to determine the legitimacy, leverage, or even degree of urgency of the needs of all stakeholders. In practice, information is always incomplete, and it would be unreasonable to believe that all stakeholders around A could be mapped out and precisely assessed in terms of legitimacy, power, and degree of urgency of their needs. Furthermore, this assessment is done at one point in time and in a specific environment. Assuming that it should hold in the medium to long term is unreasonable. It is difficult to understand what dynamics are responsible for the way stakeholders will evolve. In Freeman's view, categorizing the entire spectrum of stakeholders is impossible. Freeman suggested that it is unlikely, if not impossible, that a single pattern may apply to managing one's stakeholders in any setting and under any circumstances.[11]

Mitchell, Agle, and Wood's classifications help to clarify this important discussion, but they are insufficient as far as enabling a decision maker to choose a certain course of action. In some cases, Mitchell, Agle, and Wood's approach may help actors understand a given situation, but in others in which the distribution of power is not clearly defined, the theory may be less helpful. Much depends on the specific situation a stakeholder is dealing with.

---

11 Freeman, *Stakeholder Theory,* 72–75.

In addition, Freeman argues that it does not fall within the duties of philosophers or theorists to tell practitioners how to act. Many officers have succeeded in leading their troops without consulting either with philosophers or ethicists. This is why a more relevant approach would be to bring practice and ethics closer together and to use descriptive and normative approaches in training and education programs.

For Freeman and his proponents, the best way to manage a stakeholder is to adopt a pragmatic approach. Freeman suggests that the most effective way to do this is to develop a personal relationship with stakeholders, which Freeman and McVea call a "names and faces" approach.[12] Behind each stakeholder is a person, whether the stakeholder is an individual, an association, an organization, or an institution. In order to grasp the stakeholder's needs, it is important to learn and understand its character. Many individuals are mistaken if they strive to meet the needs of stakeholders immediately without spending time bonding with them in person. Failing to spend time with the stakeholders and to fully identify their needs increases the likelihood of misunderstanding and misinterpretation, resulting in their needs being incorrectly met. The needs of stakeholders may be very specific, clearly defined, and profoundly diverse, thus making it impossible to exhaustively itemize all the needs that stakeholders may express. When considering perception—and in particular the fact that a relationship of trust is often built on how people perceive each other—any decision or any use of technology should require a thorough analysis of its perception by local populations, regardless of the benefits it brings to the military campaign. What is truly important in the end is not whether troops did the right thing, but if the *stakeholders* of the conflict think that the troops indeed did the right thing.

12 John F. McVea and R. Edward Freeman, "A Names and Faces Approach to Stakeholder Management," *Journal of Management Inquiry* 14 (2005): 55–69.

Care Ethics and Stakeholder Theory are Complementary Approaches

Care ethics and Freeman's vision of "pragmatism" can be brought together. Both visions promote the idea that close links must be developed with the environment before individuals are in a position to make a sound decision. Both stress the importance of acquiring an understanding of the needs of others before acting. More specifically, care ethics rests on a paradigmatic change. It sets out that the individual is not really autonomous, but one is dependent on the relations that one is able to build with others. This is why care ethics implies that in a COIN operation, a soldier or Marine should pay constant attention to the expectations and needs of others. However, it is impossible to address all stakeholders' needs, especially since some of their interests can be contradictory. In these cases, troops have to properly prioritize the way they will address stakeholders' demands. This requires beginning the decision-making process with an analysis of the stakeholders' needs. Troops should focus on how to most effectively help local actors.

There are many examples of service members who show concern to the local stakeholders through their actions. Often such an approach is a condition for the success of any mission. Yet the way armies' "values" are identified in many national militaries and the way ethics is taught in military academies[13] tend to focus on individuals or groups of service members rather than on a holistic understanding of the entire range of stakeholders in a conflict.

Therefore, a proper application of care ethics should go beyond military units and should apply to all stakeholders. This is where the stakeholder theory and care ethics interestingly complement each other. Care ethics provides a framework that allows troops

13 Paul Robinson, Nigel de Lee, and Don Carrick, *Ethics Education in the Military* (Farnham, UK: Ashgate, 2008), 7.

to try to develop a feeling of regard, while stakeholder theory enables them to understand that they must consider all parties involved in their decision-making process if they want to obtain their goal, that is, to convince the enemy to join a political conflict-resolution process.

Similarly, using troops in their home countries to help the local population in cases of natural disasters promotes behaviors based on a sense of care for stakeholders outside the military. This activity may allow troops to consider stakeholders other than those within military institutions, a skill they will need to exercise when participating in operations abroad. What is frequently described as a secondary task should be embedded in a service member's mission statement, as he or she will subsequently have to perform humanitarian and law enforcement tasks in foreign countries.

CARE ETHICS ENABLES US TO QUESTION VIOLENCE DURING WARTIME

Discussing violence in relation to care might seem contradictory. Yet care ethics theorists have looked into the phenomenon of violence. For instance, Daniel Levine provides an enlightening analysis of the connection between care ethics and counterinsurgency methods.[14]

It would be a mistake to think of care ethics as being far or completely removed from the use of force and troops' ability to coerce certain stakeholders. Indeed, it is possible to take care of people while forcing them to behave in a certain way. The case of the communist uprising in Malaysia after World War II provides a good example. British troops' original strategy was to split the insurgents from the villagers by founding new villages where ci-

14 Daniel H. Levine, "Care and Counterinsurgency," *Journal of Military Ethics* 9 (2010): 139–59.

vilians had to relocate. This action did not yield any results per se, as Levine shows in his article. Only when this act was perceived as one of British benevolence did the communist uprising begin to lose ground. The British successfully established a relationship based on trust by implementing humanitarian programs to care for and feed the local population while stressing that the newly founded villages would possibly strengthen their inhabitants' safety. In addition, they promised independence to Malaysia once the risk of a communist regime taking power was addressed. As Gérard Chaliand explains, the British approach was probably only possible due to the very specific conditions of the communist uprising, which involved mainly Chinese segments of the population, whereas the majority of the population was Malaysian or indigenous.[15]

The origin of care ethics lies in an analysis of the parent-child relationship. Carol Gillingam studied the feeling of care that unites a mother and her child; she claims that in her relationship with her child, a mother is always tempted to use violence at one point or another.[16] Children are indeed vulnerable creatures who like (and do not hesitate) to test their parents' boundaries. The difference in physical strength between a mother and her child is such that she could easily kill her child by hitting him or her too hard. Furthermore, she often finds herself alone with her child, having complete control over the situation, and her child is fully vulnerable. Yet the sense of trust that mothers build with their children is strong enough that they may punish their children through means other than the use of force. Punishment is then accepted by children and regarded as legitimate. For instance, a mother may ground her child until authorizing him or her to be released from punishment. Using coercion and constraint is not unrelated to

15 Gérard Chaliand, *Les guerres irrégulières: XXème-XXIème siècles* (Paris: Gallimard, 2008).
16 Gilligan, *In a Different Voice*.

expressing a feeling of care. According to Levine, troops involved in counterinsurgencies are presented with the following type of situation:

> The counterinsurgent's moral assignment is not to limit the use of violence appropriately but to understand why *these* civilians support *these* insurgents and how they can be made to trust *this* government, which is supported by counterinsurgent troops.[17]

The main purpose of a counterinsurgency operation is to unite civilian populations behind the legitimate government. Most importantly, a major objective is to create the conditions for insurgents and the government to enter into a reliable conflict-resolution process. To this end, it is essential that civilian populations are taken care of and are paid attention to, as Levine emphasizes.

Levine also stresses the importance of moderation when using force. Force should be used only if it will be perceived as legitimate or it will cause a breach of confidence between civilians and counterinsurgents. Levine agrees with Freeman's theory that stakeholders have divergent interests and suggests that creativity is the key to solving conflicts while satisfying these different interest groups. A similar logic applies to insurgents and a regular army confronting them. For instance, in the case of Afghanistan, there is little chance that either side will prevail over the other in a decisive way. The idea is no longer to determine who must surrender to whom but to see whether there are courses of action that would allow both sides to move on together and resolve their differences. In Levine's words,

> If an analogy is required, then the relation between counterinsurgents and insurgents is probably closest to the one between members of a family on bad terms with each other. In such circumstances, a certain sense of trust has to develop for the well-

---

17 Levine, "Care and Counterinsurgency," 140.

being of people with whom these individuals on bad terms with each other are related (i.e., the other members of the family), even if they are faced with deep-rooted animosity, anger, or moral disagreement. Concurrently, most relations between members of a family on bad terms with one another do not exclude the possibility of reconciliation even if no one has been looking forward to this so far.[18]

Insurgents and counterinsurgents are somehow compelled to understand each other. Since both counterinsurgents and insurgents are unable to destroy each other, the path to negotiation must always be left open. Care ethics and stakeholder theory, if brought together with the clear objective of convincing the enemy to negotiate rather than fight, provides an analytical tool that can be used to understand the role of troops in counterinsurgency campaigns.

Troops can and should show concern and care for the local stakeholders in a conflict. The vignettes offered by Lieutenant Colonel Christmas in this book are an illustration of this concept. Care ethics and stakeholder theory provide service members with a framework that will enable them to better deal with ethical dilemmas that are typical of asymmetric warfare. Yet military ethics training and education programs make no reference to care ethics. Instead, ethics curricula often cover mainly lessons about the law of war and rules of engagement. Many in the military associate these deontological rules with ethics and neglect a care approach and its importance in asymmetric warfare.

In his vignette about a mosque that was a suspected weapons storage facility, Lieutenant Colonel Christmas was faced with a difficult dilemma: investigate the facility and risk upsetting the local population or risk losing service members by leaving a potential threat unchecked. Lieutenant Colonel Christmas chose to take the time to have religious leaders involved in the inspection

18 Ibid., 157–58.

of the mosque. Had he only considered military necessity, he would have inspected the mosque himself right away. He might have consulted his rules of engagement, only to discover that these rules would not have been specific enough to guide him to a decision. Indeed, this scenario reveals the great limitations of a rules-based ethics approach, and this approach fails to predict the situations troops might face. Lieutenant Colonel Christmas made a critical, effective decision because he practiced two concepts: first, he placed himself in the local stakeholders' situation and analyzed the possible consequences of having only Marines conduct an inspection of the mosque; and second, he recognized that such a course of action would most likely be perceived as a lack of respect for Islam. He decided that the claims of the local population were important. By taking the time to allow local leaders to conduct their inspection, he demonstrated his understanding of their culture and the important role of Islam for them.[19]

Lieutenant Colonel Christmas also considered all the stakeholders in his decision-making process. The decision to wait for the religious leaders to get involved in the inspection of the mosque allowed him to strengthen the relationship with the Afghan National Army (ANA) and with the local press. In the long term, the religious leaders could also become precious allies in other situations—there was a huge benefit in showing them that ISAF treated them with respect. He stated, "The newly established relationship with the director of the Hajj would prove beneficial throughout continued operations over the next five months. And our ANA counterparts experienced our dedication to our partnership. The result was that a highly respected leader, the director of the Hajj, spoke out publicly on television against the insurgents. Ultimately, the time delay during the tactical execution resulted in

19 See the chapter in this book by Brian Christmas and Paula Holmes-Eber, "Leadership, Ethics, and Culture in COIN Operations: Case Examples from Marjeh, Afghanistan."

an operational and strategic victory and would prove highly beneficial in the long term." Indeed, the relationship the lieutenant colonel created with the religious leaders allowed him to navigate two other challenging situations he describes in his chapter.

Two other examples illustrate the links between Lieutenant Colonel Christmas's actions in Marjeh and stakeholder theory. First, he did not consider all the members of the ANA forces the same way. He took the time to build a close relationship with Lieutenant Colonel Muhammed. He acted on a "names and faces approach," that is to say, he put a name and a face to a larger institution or organization. Thanks to the special connection with Lieutenant Colonel Muhammed, Lieutenant Colonel Christmas had an ally during some difficult moments in his mission, in particular, when there were doubts about the betrayal from several members of the ANA. Had he not taken the time to build this trust relationship with his ANA counterparts, Lieutenant Colonel Christmas would have had difficulty resolving the tense situation. Second, by keeping a broad view of the situation and identifying all the stakeholders involved, he was able to find an innovative solution to conflicts that seemed to have none.

The theories presented in this article, as well as Lieutenant Colonel Christmas's experience, provide a framework that may help troops deal more effectively with difficult dilemmas. Experience is the key to handling a large variety of situations, yet training and education programs should provide Marines and soldiers with the skills they need to operate successfully in asymmetric warfare environments. Learning the rules of engagement is important in order to understand the context and goals of the mission; however, it is not enough information with which to make a sound decision. In a complex environment, stakeholder theory is a tool that can help service members analyze the situation and assess the best, most effective course of action. Additionally, a care approach, through which positive relationships are created among as many stakeholders as possible, including military allies, religious leaders, and the media, is necessary to alleviate the many moments of tension that troops will face. When military leaders are deployed on a mission such as the one Lieutenant Colonel Christmas experienced in Marjeh, they will not think only about the rules of engagement, but above all, about their overall responsibility. This responsibility requires an ability to analyze and build long-term relationships with all the stakeholders around them. The stakeholders analysis and care approach may be natural for many military leaders, but Marines and soldiers should learn to apply these theories early in their careers before deploying into asymmetric warfare situations.

A PEDAGOGY OF PRACTICAL MILITARY ETHICS

CLINTON A. CULP

As iron sharpens iron, so a man sharpens the countenance of his friend.

-Proverbs 27:16

Given today's fast-paced and chaotic battlefield where the enemy is not easily identifiable and the leaders are young, the need for moral military members has never been greater. This statement does not imply that there was not a need for an ethical military in the past, but merely states the obvious fact that in today's world, the U.S. military requires young leaders to exercise more autonomy than they've had in the past. For this reason, a young military member's conduct may have moral ramifications that affect not only those around him; his conduct may also have ethical implications that could affect an entire nation.[1] As a result, the U.S. Marine Corps, as well as the other branches of the armed services,

---

1 Within the context of this paper, the term "morals" is defined as practices and customs of a group or person, while "ethics" is the principles and rules that are explicitly stated and held by that group or person. Richard M. Fox and Joseph P. DeMarco, *Moral Reasoning: A Philosophic Approach To Applied Ethics*, 2nd ed. (Fort Worth, TX: Harcourt College Publishers, 2000). This text will use both "ethics" and "morals" interchangeably and trust that the reader is able to distinguish between the two based on the context. The text will refer to "character education" as a holistic educational approach where one's moral conduct advances in sophistication and is brought in line with the praiseworthy ethical standard of the organization, in this case, the U.S. Marine Corps.

has instituted precommissioning ethics education for about half of its officers.[2] By and large, this ethics education instills U.S. military values in its members.

This chapter investigates current ethics programs and evaluates which pedagogical method is best suited to influence the ethical conduct of the serviceperson. As such, the chapter will primarily examine the current pedagogical methods that affect the majority of new officers who receive precommissioning ethics education, briefly touching on enlisted ethics education, and will examine the institution's role in facilitating moral conduct. Specifically, it will focus on some common barriers to the pedagogy of ethics (e.g., curriculum, use of ethical theories and case studies, conduct of the classroom, lack of horizontal integration) and will offer methods for minimizing barriers. Lastly, the chapter will examine what the junior officers in the operating forces can do to influence ethical conduct within their units.

CURRENT PEDAGOGICAL METHODS

Each branch of the U.S. military services has its own pedagogical method, and within each branch different methods are used for entry-level training and follow-on for professional development. To complicate matters, there are differences in pedagogical methods between enlisted and officer training and education. Because there is a wide variation in the moral sophistication of individuals, presumably, the appropriate method is used at the proper time to facilitate moral development. The variation of moral sophistication is seen at all levels within the military. For instance, two 18-year-old enlisted persons can be at different levels of moral reasoning, and the same can be said of 22-year-old newly

2 DoD, "Table B-30. FY 2009 Active Component Officer Accessions By Source Of Commission, Service And Gender," Department of Defense, http://prhome.defense.gov/MPP/ACCESSION%20POLICY/PopRep2009/appendixb/b_30.html.

commissioned officers and seasoned veterans. However, it can be generally stated that the older (i.e., more life experience) one is and the more education one has (i.e., critical thinking and reasoning skills), the more sophisticated his moral reasoning will be.[3]

Enlisted Character Education

At the enlisted level the ethics education system is primarily a rules-based system of punishments and rewards; its intent is to instill habitual and immediate obedience to orders. While this method is a good initial start, one has to question whether 9 to 13 weeks can have an effect after 18 years of life experience. U.S. Army Lieutenant Colonel Kenneth Williams' study of moral reasoning indicated that, at least for a nine-week initial entry training program, moral reasoning is in fact not affected and in some cases actually decreases.[4] After initial entry training and occupational specialty schools, there is no formal ethics training until the service member attends a noncommissioned officer's course. In this respect, the military seems to take a pragmatic approach (i.e., caught, not taught) to teaching ethics to its enlisted members.

Officer Character Education

The pedagogical methods used for teaching ethics to officers is also varied, and more so than in enlisted ethics training. In part, this variation is due to the multiple programs through which officer candidates can receive a commission. Each service academy has its own highly developed ethics education programs. There are hundreds of Reserve Officers Training Corps (ROTC) programs,

3 Thomas Lickona, *Educating For Character: How Our Schools Can Teach Respect And Responsibility* (New York: Bantam Books, 1991); Lawrence Kohlberg, *The Philosophy Of Moral Development: Moral Stages And The Idea Of Justice: Essays On Moral Development*, vol. 1 (San Francisco: Harper & Row, 1981); James Rest, et al., "A Neo-Kohlbergian Approach to Morality Research," *Journal of Moral Education* 29 (2000).
4 LtCol Kenneth Williams used the Defining Issues Test 2 (DIT2) as his measure of moral reasoning. Kenneth R. Williams, "An Assessment Of Moral And Character Education in Initial Entry Training (IET)," *Journal of Military Ethics* 9 (2010).

all of which base their ethics education on their respective service academy's model. However, the ROTC programs also rely on the liberal education that cadets or midshipmen receive from their prospective university. The final major commissioning sources are Officer Candidate and Training Schools (OCS/OTC) and the Direct Appointment (DA) process where the prospective officer's precommissioning ethics education is fully dependent on the particular degree program within his respective university.

Much has been said about the pedagogical methods used at the academies with regard to ethics education and how to improve an already successful system.[5] The academies arguably have an advantage in time, curriculum, and pedagogical methods over ROTC programs due to the lack of "undue" influences (e.g. fewer contact hours with the midshipman, fraternal organizations, and a general partying lifestyle that is prevalent on most college campuses) that the cadets and midshipmen are exposed to while at the academies.[6] The academies are also able to take advantage of a multimodal pedagogical method, similar to the approach used in initial entry training for enlisted personnel. Typically, the methods used at these academies vary from the authoritative boot camp model to motivational speakers and role models to formal classes and case studies during the students' junior and senior years. Formal classes in military ethics at the academies are taught mainly by civilian professors, most of whom have degrees in philosophy, some of whom have a military background. These classes are intended to provide a broad overview of several ethical theo-

---

5 J. Joseph Miller, "Squaring The Circle: Teaching Philosophical Ethics in the Military," *Journal of Military Ethics* 3 (2004); Paul Robinson, Nigel De Lee, and Don Carrick, eds., *Ethics Education in the Military* (Farnham, UK: Ashgate, 2008); J. Carl Ficarrotta, *Kantian Thinking About Military Ethics* (Farnham UK: Ashgate, 2010).

6 An argument can be made that providing ROTC cadets and midshipmen with a liberal arts education exposes them to a broader spectrum of ethics that lends itself to experiences that are more practical.

ries in an environment that allows for an open dialogue on ethical issues. However, what occurs most often is a move from a dialectic to a didactic method in which the student seeks the approved solution to the problem at hand.

If this move occurs at the service academies, where educated philosophy professors teach formal classes, one has to wonder how ethics education is being conducted within the more prevalent ROTC programs, which 30 percent of new officers attend. To date, little research has been conducted on the effectiveness of these programs with regard to ethics training.

ROTC Ethics Education

Senior military leaders assume that a senior company grade or a field grade officer who passes the requirements to be assigned to an ROTC program is capable of teaching ethics. Certainly, the officers have the capacity to lead by example, but character education requires much more. While certain individuals are more than capable of teaching ethics, few are able to do so without formal education. Just as not every rifle expert makes a good marksmanship coach, and not every engineer makes as a good engineering educator, not every good leader will be effective in teaching ethics. The military is accustomed to training, and many officers have been trainers prior to being assigned to an ROTC unit. However, rarely have officers been educators; this role requires a skill set that most officers do not acquire in the field or in training. Indeed, training does not equal education. J. Joseph Miller summarizes this dichotomy by contrasting "the technician's method" (i.e., asking "how" questions) with the "philosopher's method" (i.e., asking "why" questions)[7] when teaching military ethics. The military sends excellent trainers who are well versed in the technician's method to ROTC units. The technician's method pays off,

7 Joseph Miller, "Squaring The Circle: Teaching Philosophical Ethics In The Military," *Journal of Military Ethics* 3 (2004).

to some degree, as the instructor trains young soon-to-be officers in the rules and how to use formulas for making ethical decisions within those rules. This method serves the novice well, as the method is easy to train and allows a young officer to arrive at an ethical decision with relative ease. However, most ethical issues are complex and cannot be neatly "plugged" into a technical formula. A few pitfalls, which often occur at ROTC programs, are classes in formal ethical theories, the tendency to apply an ethical theory to a particular "real world" problem, overreliance and improper use of case studies, overuse of the didactic method, and lack of horizontally integrating ethics education. I will now discuss each of these pitfalls in greater detail.

Formal Classes

The military sends generally well-meaning and morally upright individuals to be instructors at ROTC programs; however, few, if any, are educated in teaching ethics or ethical theory. A typical syllabus for a leadership and ethics ROTC course includes readings on act and rule utilitarianism ethics, Kantian ethics, virtue ethics, *jus ad bellum* and *jus in bello* and other ethical theories. Each of the theories requires more than a passing glance, and, often ROTC instructors are ill equipped to teach an in-depth study in ethical theory. As such, the theories presented are often misunderstood and misused.

It is important to note that ROTC courses are taken in addition to courses that are required for the midshipmen's or cadet's degree. This means that midshipmen and cadets are required to participate in physical training courses and leadership billets in addition to their course work, which generally results in an ROTC student carrying a heavy course load, typically 15 to 18 hours per semester. If the student is pursuing a technical major, this course load can easily exceed 18 hours per semester, and often is as high as 21 hours per semester.

Given the amount of reading, reflection, writing, and discussion required to properly understand ethical theories, it is easy to see that it would be difficult, if not impossible, to cover all the ethical theories in the text during one semester. ROTC programs require leadership and ethics courses; however, they are not typically required for a diploma. Therefore, given the emphasis on graduating and receiving one's commission from ROTC leadership, cadets and midshipmen tend to prioritize their assignments in required courses, which are needed for students to receive their college diploma, over courses that are treated as supplementary. This system may suggest to students that high GPAs and graduation (i.e. meet ROTC accession goals) are more important than having men and women of exemplary character in the operating forces.

Applying the Theory

A natural result of the formal classes is a misuse of the various ethical theories. This often manifests itself in questions such as "what would Mill say about "gundecking" maintenance records?" These types of questions often result in a misapplication of theory, which typically seeks ways to morally justify why one should (or presumably should not) falsify or cheat on the maintenance record. This line of questioning usually digresses into a "battle" between Mill and Kant (or another ethical theorist), with each one defending a side, when, in reality, both would agree that gundecking is wrong.

Case Studies

Case studies are "easy"; in fact, case studies tend to be the default method used to teach ethics courses in the ROTC curriculum. While case studies are an effective method for teaching about ethical decisions, typically one of two things happens when using case studies. First, the student often uses the "applying the theory" approach to the case, allowing the particular theorist to do the heavy lifting, resulting in a misplaced or shallow application of the particular theory. This approach often fails to tie "core values" with the particulars of the specific case. Second, when case studies are used in group discussions, the group usually arrives at an answer through consensus building. This particular method looks and sounds like moral reasoning is occurring; however, the answer often might depend on the dominant individual within a group and not on sound moral reasoning.

The Didactic Classroom

In most military situations, one often defers to authority, and the classroom is no exception. It takes special effort for both the student and the educator to overcome this phenomenon. As soon as the educator exerts his authority or indicates that he has the right answer, students quickly seek the approved answer over the reasoned answer. Worse yet, the educator "lectures" or "preaches" on the proper application of ethics, as if he were the gatekeeper of such knowledge.

Lack of Horizontal Integration

Teaching ethics can, and often does, occur outside the classroom, yet instructors rarely take advantage of these opportunities. When an ROTC student does something right or wrong, his/her conduct is rarely linked to the organization's values and principles. A pat on the back for a job well done, without mention that their conduct exemplified the organization's values, often results in a missed opportunity to reinforce right conduct. Likewise, punishment is rarely linked to the values that are the foundation of the rule that was broken, again missing an opportunity to reinforce right conduct.

The list in this section is by no means an all-inclusive list of potential pitfalls that occur during character education in ROTC programs, but the list helps identify key areas in which the U.S. military can improve character education.

SUGGESTIONS FOR IMPROVEMENT

Formal Classes

The military's intent is to increase the practical moral reasoning skills and general character of its personnel. Perhaps a different curriculum for entry-level training, both officer and enlisted, would better serve this purpose. A curriculum that focuses on critical thinking skills and increasing moral sensitivity, moral empathy, and open mindedness, while linking the organizations' values and principles to military rules and regulations, would better serve that purpose than a broad-brush class in metaethics and ethical theory. These skills are critical to moral reasoning and under the best of circumstances are difficult to master. To some, these skills may be viewed as "soft" and thought of as counter-productive to the military mission. However, without these "soft" skills, moral reasoning cannot occur. Leaders should welcome healthy debate over ethical issues, especially prior to the point of action.

Offering a process for evaluating a situation and determining its ethical implications can be problematic, as many individuals may treat the process as a formula, one of the very things I criticize about the current method for teaching ethics courses at ROTC programs. Yet a process-based method rather than a rules-based formula is often useful for both the novice educator and student.[8] The intent of this process is to answer the "why" questions as opposed to the "how" questions. Most military personnel, even those in entry-level officer training and education, are familiar

8 Reid Hastie and Robyn M. Dawes, *Rational Coice In An Uncertain World: The Psychology Of Judgment And Decision Making* (Thousand Oaks, CA: SAGE, 2010).

with U.S. Air Force Colonel John Boyd's decision-making cycle, the Observe, Orient, Decide, Act (OODA) Loop. While Colonel Boyd's OODA Loop was originally intended for fighter pilots, it has been adapted for other decision-making domains including business, medical professions, and ground combat. Using such a familiar decision-making tool will aid the individual through the difficulties of moral decision making.

The OODA Loop is easily adaptable to practical moral reasoning skills. During the *observation* phase, one must be able to identify situations that are morally relevant, i.e., one must have moral sensitivity. This requires military personnel to identify the relevant social values, moral values, and principles involved; the organization's values and principles must also be considered. What is the role of the individuals and the organization as a whole? Is there an existing code of ethics or rule governing the situation? *Orienting* oneself to the morally relevant situation requires critical thinking skills, moral empathy, and an open mind. During the next phase of the OODA Loop, *decide*, one has to determine whether or not social values take precedence over moral values. The individual should assess whether the values and principles being expressed in the situation are in conflict or congruent with those of the organization. If a conflict exists, does it constitute an exception? The decision requires moral reasoning skills. If one has evaluated the situation honestly and critically, he has completed the moral reasoning process; however, in the end, the decision must be implemented—that is, one must *act*.

The Ethical OODA Loop

The following example of how to implement the OODA Loop in an ethical context is adapted from the conflicting loyalties scenario in *Ethics for the Marine Lieutenant*.[9] This scenario was intention-

9 William T. Stooksbury, ed., *Ethics For The Marine Lieutenant* (Annapolis, MD: Center for the Study of Professional Military Ethics, 2002).

ally chosen to illustrate the application of the ethical OODA Loop to an all-too-common noncombat moral dilemma. This process is easily adaptable to any morally relevant situation, including combat situations.

After a successful field operation at Camp Lejeune, the Battalion commander authorizes a 72-hour liberty. One of the platoon commanders in Alpha Company did particularly well, receiving an "attaboy" from the company and battalion commanders. The lieutenants of Alpha Company decide to have an "O" call over in Greenville. Being a new dad and wanting to go home, you try to beg off, but succumb to the pleas of your fellow officers, who remind you of the importance of being a "team player," and the need for "unit cohesion."

After a few drinks, you switch to soda in preparation for your drive home; however, the company executive officer (XO) is not ready to leave. You offer him a ride home, but he says he is "feeling lucky and might hook up with an ECU gal and won't have to drive back to camp swampy tonight." Knowing the XO is a big boy, and senior to you, you decide to head home.

At 0200 the next morning, you receive a phone call from the XO and he certainly was not "lucky." He was arrested for DWI and wants you to come and pick him up from the Greeneville police. Hesitantly, you head back to Greeneville to pick up the XO. After picking him up, you head back to Camp Lejeune, and, on the way back the distraught XO runs through his options: say nothing, i.e., cover it up; wait and see, plea-bargain down to reckless driving, and even then no one needs to know; or "come clean" and tell the company commander.

Before we delve into this issue, it is necessary to orient ourselves to the "core values" of the Marine Corps: honor, courage, and commitment.

Honor is comprised of integrity, responsibility, honesty, and tradition. Being a Marine of integrity, one consistently demonstrates the highest standards of right conduct, both legal and ethical. Responsibility is accepting the consequences, both good and bad, for one's decisions and actions. Honesty is summed up in that Marines do not lie, cheat, or steal and they seek fairness in all actions. Tradition is also an important aspect in that Marines respect the customs, courtesies, and traditions that bind Marines together. It also means having a respect for the traditions and heritage of others.

Courage is the moral, mental, and physical strength to endure hardship, resist opposition, and face danger. Courage is comprised of self-discipline, patriotism, loyalty, and valor. Self-discipline is the ability to police one's conduct, maintaining moral, mental, and physical fitness. Patriotism is having devotion to defend the Constitution of the United States and to America herself. Loyalty is the strong feeling of allegiance to the United States of America, the U.S. Marine Corps, one's command and fellow Marines, and to one's family and self. Valor is having determination to face danger, both in and out of battle, with boldness.

Commitment is completing one's pledge or promise, and being committed to a goal requires one to self-identify with the goal and act in a manner that supports that goal. Commitment is comprised of competence, teamwork, selflessness, concern for people, and spiritual heritage. Competence is maintaining and improving one's skill level in order to achieve one's goals—the goal of being a Marine. Teamwork is the understanding that, while on rare occasions individual effort achieves the goal, it is more often the case that the individual's effort supports his or her team members in achieving the goal. Selflessness is placing the welfare and needs of the nation, the Corps, and others before one's personal welfare. Concern for people is an understanding that all people (even our enemies) are of value, and as such, Marines treat all people with respect and dignity. Spiritual heritage is an understanding that our nation was founded on religious and spiritual principles, that all people are endowed by their Creator with inalienable rights of life, liberty, and the pursuit of happiness. Maintaining spiritual health enables Marines and their families to endure the hardship that is the life of a Marine.

Now that we have briefly examined the core values of the Marine Corps, let us turn our attention back to the situation at hand. The ethical nature of the XO's actions can be evaluated using the OODA Loop. This process first requires a person to *observe* the scenario, highlighting that there are several morally relevant issues: first is the DWI itself, second is the possible cover-up that the XO is contemplating, third is coming clean, and fourth is the possible action the platoon commander may need to take if the XO does not come clean. Although interconnected, this analysis of the situation will deal with each issue separately. This requires us to determine what the XO values. One can assume that the XO values his oath of office and the core values of the Marine Corps. The XO also values camaraderie, drinking, and chasing women, which are not inherently bad nor do they necessarily constitute an ethical issue. However, when taken to excess and conflict with the core values of the Marine Corps, there are ethical problems. For instance, honor, and specifically integrity and responsibility, means demonstrating the highest standard of consistent adherence to right, legal, and ethical conduct. Responsibility means personally accepting the consequences for one's conduct and coaching subordinates to make moral and ethical decisions that follow the organization's code of conduct. (In this context one must exhibit responsibility in order to coach it.)

Clearly, there is a conflict between what the XO personally values (having a good time, drinking, and chasing women) and the moral values of integrity and responsibility. Here the XO has at least two distinct roles: that of a leader of Marines and that of a driver. What effect will his conduct have in both roles? If the XO places precedence on his personal values (having a good time, drinking, and chasing women), which he did by turning down a ride and driving while intoxicated, he has violated the moral values of integrity and responsibility. By violating these values, the XO has also let down the Marines he leads and violated the core values of the Marine Corps.

It's also necessary to consider his role as a driver. By driving drunk, the XO jeopardized the lives of the drivers who shared the road with him, thereby violating the common trust that one will not drive while drunk. By employing moral empathy, we can speculate as to how the XO would feel if he were the one being violated. Arguably, he would not wish to be on the receiving end of his conduct. In this situation his personal values (having a good time, drinking, and chasing women,) do not constitute an exception to his moral values, integrity, and responsibility. The *decision* and *action* in this case should have been made prior to the XO getting in his car and driving while intoxicated. Because it is easy to apply hindsight, it's necessary to briefly consider a future action: a possible cover-up. The XO provides several courses of action, two of which appear to be different but in fact are the same, namely not coming clean with the hope his problem will go unnoticed and waiting to tell his CO in hopes of pleading down to a lesser charge. Both options are a deliberate cover-up and the latter adds equivocation to the first. As such, these decisions will be treated as equivalent moral issues.

Moving to the second step of the OODA Loop, *orient* requires an evaluator to apply critical thinking skills, moral empathy, and an open mind to determine which courses of action could be considered ethical. First, it is clear that the XO values his role in the

Marine Corps; if he did not value such a role, there would be no reason for him to be distraught. The XO also values loyalty, as seen by his organizing the "O" call and in his implicit and explicit communication to the platoon commander that "no one needs to know." The XO's self-preservation is not inherently immoral, nor is his valuing loyalty; however, when examining his actions, the XO's priorities are called into question. The XO's self-preservation is in conflict with the core value of commitment, namely selflessness, where selflessness is taking care of one's subordinates and families before themselves. Loyalty is also in conflict. The XO's view of loyalty—that the platoon commander should be loyal to him—is mistaken. The XO should know that loyalty is due to higher principles such as the Constitution of the United States and the Marine Corps, and not to an individual. The XO is also avoiding the consideration that loyalty (when not in conflict to one's oath, the Constitution, and the Marine Corps) goes both ways. That is, the XO should show loyalty to the platoon commander as well.

There is a clear conflict with the XO's values of selfishness and selflessness, as well as that of misplaced loyalty (i.e., loyalty to the XO before the organization). In this situation, this misplaced loyalty also brings up a conflict between loyalty and honesty. If the platoon commander is to remain loyal to the XO, the platoon commander will necessarily violate the core value of honesty, either directly by having to lie about the XO's conduct if asked or by not reporting the XO's conduct if not asked, both of which are dishonest. Using moral empathy, would the XO want to be put in the platoon commander's position (i.e., forced to choose between competing loyalties or having to be dishonest about what he knows)? What would the XO do if he were the company commander? If the XO was consistent, this would not constitute an exception; he could not justify his "cover-up" as taking precedence over core values (i.e., his choosing lying and misplaced loyalty over the core values of proper loyalty and honesty). Nor could he justify an exception to loyalty to the organization (i.e., not putting the platoon commander in this ethical position).

After evaluating the XO's predicament, it is easy to see that the morally right *decision* is to "come clean" and tell the company commander as soon as possible what happened over the weekend. This decision places moral values over personal values and having a good time (i.e., the selfless act over the selfish act and honesty over misplaced loyalty). It also takes into account the moral position of others who are affected by the forthcoming moral action.

Unfortunately, the OODA Loop process is rarely used; at best, moral decision making is linked to leadership (which indeed it is) with a matter-of-fact statement such as "this is poor leadership by example." At worst, it is tritely summed up with the statement "if you drink and drive, you'll get busted" (i.e., it is against the rules, and if you break the rules, you will be punished). The more important question of "why" is it poor leadership is rarely asked. In other words, the links to Marine Corps core values and why it is morally important to do the right thing does not receive enough attention.

Applying the Theory

Ethical theories are not necessarily essential in practical ethics. A basic understanding of the classic ethical theories would certainly be helpful, but it is not necessary and could be potentially misleading if not understood fully. Those who are teaching formal classes in practical military ethics should have a solid background in ethical theories. However, they should be cautious when teaching ethical theories to entry-level personnel. If they are not formally educated in ethics, instructors should focus more on the practical application of ethics rather than on theoretical application.

Case Studies

Case studies should emphasize basic moral reasoning skills, such as the ability to think critically, develop moral sensitivity, moral empathy, and open mindedness to alternative possible courses of action. The focus of case studies should be to determine the values, principles, and potential rules used in making unethical decisions and then to identify those elements that should have been used to make the ethical decision. Positive cases should also be studied to identify sound reasoning that resulted in proper conduct. These positive case studies are often given in the form of positive role model stories of military legends. Nevertheless, a concerted effort should be made not only to tell the stories of military lore but also to link proper conduct to core values and principles.

A proper focus on core values should eliminate the theoretical "battle royal." A duty of the educator is to prevent groupthink from prevailing over sound moral reasoning. When a group easily agrees with one another and myopically seeks "how" to apply the rules to a particular case without asking the difficult "why" questions, the learning that occurs within the group is of little benefit. Therefore, the educator must guide the students, keeping them on task and focusing on the relevant issues, values, and principles.

The Dialectic Classroom

Classroom conduct is perhaps where the most gains can be made with regard to the development of moral reasoning, but it is also the most challenging aspect to change. The dialectic classroom can be problematic, remembering the tendency not to challenge authority in the military, especially at the entry level, when the very purpose of the class is to have the students challenge belief systems. Typically, the senior member of the ROTC unit teaches the ethics courses, and this can be a distracting factor if not properly managed. The teacher must make it clear that, barring extenuating circumstances, nothing within the dialogue will be used against the student. If students do not trust the educator, learning will not occur. Whether or not it is intentional—and I suspect most times it is not—the mere fact that the educator is a ranking officer can generate a barrier against open and honest dialogue. It is the educator's responsibility to remove this barrier.

If possible, the educator could have the class meet in civilian attire. While this may seem trivial, the fact is that being at ease in the learning setting does make a difference from a student's point of view and may facilitate dialogue.[10] If the class is held in a military setting (e.g., a typical military classroom with a podium and desks neatly covered and aligned), the educator may wish to consider moving the site of the class to a nonmilitary setting. If the number of students permits, circular or semicircular seating should be used. This allows for everyone to be engaged in the dialogue and also helps to reduce the "sage on the stage" effect.

Silence in the dialectic classroom is not "dead time." Questions, if framed properly, should create thought, and thought takes time. This again can be an impediment for the military educator because military officers often take movement and talking as making progress, i.e., all the students are conversing on the subject matter, therefore, learning must be occurring. Unless the movement and discussion are purposeful, it is wasted time and effort. Most officers, having been trained as technical experts, are expected to quickly master skills and provide answers in a timely manner (where "timely manner" typically means as fast as one can). Finding the patience to wait for an answer is a skill that most military educators will have to learn. The educator also needs to be open to different points of view; this does not mean

10 Jerry H. Gill, *Learning To Learn: Toward A Philosophy Of Education* (Atlantic Highlands, NJ: Humanities Press, 1993).

that the educator has to agree with those points of view, but he must respect the student. It is from this open-mindedness that the educator is able to provide meaningful feedback that facilitates dialogue and challenges the student to consider a different point of view. This feedback can be both verbal and nonverbal. In particular, educators should avoid closed-ended feedback and questions—that is, answering student questions with a simple yes or no, or asking yes or no question to students—as this does not facilitate dialogue. Educators should also be sensitive to the fact that nonverbal communication such as body language and eye contact speaks as loudly as verbal communication.

The educator should be aware that not all students will develop the same level of moral reasoning. Indeed, it is the educator's job to identify a student's level of personal moral reasoning skills, and then to ask questions that challenge the student at his level of moral reasoning. By asking a higher-level moral reasoning question to one who reasons at a lower level, the educator runs the risk of placing the student in distress. This moral distress is where the student perceives the dialogue as a threat and might resist further discussion. In order to be an effective educator and to properly assess a student's level of moral reasoning, however, the educator must be familiar with the levels of moral reasoning.[11]

Lawrence Kohlberg developed three levels (each with two stages) of moral reasoning in 1981 in his seminal work *Essays on Moral Development: Volume One: The Philosophy of Moral Development.* Kohlberg's six stages have been the signpost for the pedagogy of

11 See seminal works on schema and levels of moral reasoning, including: Kohlberg, *The Philosophy Of Moral Development:* Thomas Lickona, *Educating For Character: How Our Schools Can Teach Respect And Responsibility* (New York: Bantam Books, 1991); James Rest, "Moral Judgment Research And The Cognitive-Developmental Approach To Moral Education," *The Personnel and Guidance Journal* 58 (1980); Stephen J. Thoma and James R. Rest, "The Relationship Between Moral Decision Making And Patterns Of Consolidation And Transition," *Developmental Psychology* 35 (1999).

moral development. The following is a brief overview of Kohlberg's levels and stages of moral development:

Level one is the preconventional level and consists of two stages: stage one, punishment and obedience, and two, individual instrumental purpose and exchange. In stage one of the preconventional level, a person defines moral correctness as a literal obedience to rules and authority. Thus, a person who has a stage one understanding of morality does not consider the interests of others, nor does he recognize that he is separable from others. When an individual enters stage two, he defines what is morally right as serving one's own or others and understands the reciprocity of concrete exchange.

Level two is the conventional level and has two stages: stage three, mutual interpersonal expectations, relationships, and conformity; and stage four, social system and consciences maintenance. In stage three, being morally right requires playing a good role and maintaining loyalty and trust by being concerned about the feelings of others in the group. In stage four, morally correct actions are those that uphold social order and welfare of the society.

Level three is the postconventional and principled level and consists of stage five, prior rights and social contract or utility, and six, universal ethical principles. In stage five an individual upholds the basic values, rights, and legal contracts of society, even when they differ from actual rules and laws of society. Stage six is where one is guided by universal ethical principles that everyone should follow. Justice is Kohlberg's universal ethical principle: respect of the equal dignity of individual human beings requires treating people as an end in themselves and not as a means to an end.[12] Stage six is the pinnacle of moral reasoning; as such, it should be everyone's goal. However, this does not mean that if one's moral

12 Kohlberg's stages, especially stage six, is a very Kantian notion. Kohlberg, *The Philosophy Of Moral Development.*

reasoning is at a lower stage, he or she is an immoral person. It only means the individual's moral reasoning is less sophisticated.

In 2000, James Rest, Darcia Narvaez, Stephen Thoma, and Muriel Bebeau revised Kohlberg's stages. Such a revision allowed for a more flexible interpretation of an individual's level of moral reasoning and consolidated the levels into three schemas thereby bringing the theory in line with modern cognitive science theory. Rest's moral schema theory includes the personal interest schema (Kohlberg's stages two and three), maintaining norms schema (Kohlberg's stage four), and postconventional schema (Kohlberg's stages five and six). People within the personal interest schema justify a morally right decision by appealing to the personal stake they have in the consequences of their action. People within the maintaining norms schema (stage four) perceive a need for social norms to govern the group. These norms apply to the entire group and should be plain, consistent, and definite; there should be reciprocity of the norms within the group. The group must be organized into some hierarchical chain of command and one must obey authorities within the chain of command out of respect for the group. Within this schema, social norms define morality. The last schema is the postconventional schema, where people reason first from a moral criterion.[13] There is an appeal to a moral ideal (e.g., beneficence, honesty, responsibility, justice)—these ideals are shared within the community, and there is full reciprocity of the ideal within the community.

Understanding the levels or schemas of moral reasoning allows the educator to formulate the appropriate "why" question, be it

---

13 Rest's moral schema theory allows for a more flexible moral criterion in which to form the foundations of one's moral reasoning. Where the moral criterion for Kohlberg's stage theory is justice, the criterion for Rest's moral schema theory may be justice but it could also be care (beneficence), honesty, or responsibility. The moral criterion used is dependent on the individual and the situation; however, it is not to be confused with relativistic or situational ethics. Moral schema theory relies on a primary moral value such as justice, honesty, beneficence, or responsibility at the postconventional schema.

a question designed to bring the student from a personal interest schema to a maintaining norms schema or from the maintaining norms schema to the postconventional schema. Educators should avoid saying what the "approved" answer is, and they should also avoid providing their own personal answer to a particular problem. The goal is to ask the appropriate "why" question that draws a reasoned answer from the student. This requires patience on the part of the educator, especially an educator with a military background who is accustomed to a quick answer. Once the student has articulated his answer, the educator should ask follow-up questions that relate the student's answer to the values and principles of the organization. The educator's purpose is to ask questions that create dissonance.

In summary, the dialectic classroom and process is nothing new; it is the maieutic method, more commonly known as the Socratic Method. Knowing at what level or schema a student is reasoning helps frame the line of questions. The crux of this method is the follow-up question; the goal is to "stress" the current belief, without putting the student in "distress." Self-reflection, writing one's thoughts down, and follow-up discussion challenge or strengthen the validity of the newly reasoned conclusion. This process is thought to reinforce, create, or rewire neural pathways and schemas within the brain, thereby increasing reasoning skills.[14]

Horizontal Integration

The connection between personal conduct and the core values of the Marine Corps should be made evident. This approach allows for positive reinforcement of conduct as opposed to negative reinforcement. Positive reinforcement acknowledges someone for exhibiting right conduct rather than only instructing someone on

14 Michael S. Gazzaniga, *The Ethical Brain* (New York: Dana Press, 2005); Laurence Tancredi, *Hardwired Behavior: What Neuroscience Reveals About Morality* (New York: Cambridge University Press, 2005).

what the rules are and telling them not to break them. In other words, it is important that the instructors value a service member's honesty in a particular situation. Such an approach will reinforce the service member's belief about honesty. Linking conduct to organizational values in this manner reinforces the importance of both the value and the individual's right conduct.

Intuitive Ethics

Evidence suggests that the majority of daily judgments and behaviors (including moral ones) are intuitive; they appear in one's consciousness without their origin being known. Benjamin Libet's 1999 study developed theories of intuitive decision models of moral conduct. Libet found that increased brain activity occurred about 300 milliseconds before the subject became consciously aware of impending action, followed by 150 milliseconds of consciousness before the brain sent the signal to act, and another 50 milliseconds for that signal to get to the muscle to initiate movement.[15]

Jonathan Haidt, in his social intuitionist model of moral conduct, suggested that most moral reasoning is a post hoc rationalization of intuitive behavior after one receives feedback from his or her social group. Haidt offers four reasons why his model provides a strong explanation for moral conduct. First, there are dual processes that drive our conduct, both conscious and unconscious. Second, the agent acts more like a defense lawyer than a judge; we seek to morally defend our actions. Third, often we cannot explain why we do the things we do; we manufacture reasons post hoc when pressed for answers. Fourth, several studies indicate moral action may be linked with moral emotion to a greater extent than it is linked with moral reasoning.[16] Indeed, moral conduct is often

15 Benjamin Libet, "Do We Have Free Will?," *Journal of Counciousness Studies* 6 (1999).
16 Jonathan Haidt, "The Emotional Dog And Its Rational Tail: A Social Intuistionist Approach To Moral Judgment," *Psychologcal Review* 108 (2001).

a result of an intuitive emotional reaction as opposed to a rational reasoned action. The following are a few studies in cognitive neuroscience that seem to support a social intuitive link to moral conduct.

Anthony Greenwald and colleagues, using their Implicit Association Test (IAT), also seem to support an intuitive process for moral conduct.[17] Greenwald and Farnham compare reaction times between congruent and noncongruent concepts and attributes to determine if there is an implicit association between the two when compared to an explicit measure of the concept and attribute. The classic example uses the concepts of a flower and an insect with attributes of pleasant and unpleasant. A congruent task links the concept of a flower with a pleasant word and the concept of an insect with an unpleasant word, while an incongruent task links the concept of a flower with an unpleasant word and the concept of an insect with a pleasant word. Reaction times for identifying a pair as congruent or incongruent are measured, and typically incongruent pairs take longer to identify, thereby indicating an implicit association with the congruent pair. Concepts used with the IAT have been religion, gender, race, and ethnicity, all of which have moral implications. When comparing explicit measures of stereotyping within those concepts, there is little correlation between the two; in other words, one might not stereotype a concept such as ethnicity on an explicit measure, but when given the IAT, there are implicit associations with certain ethnicities. The results of Greenwald's IAT indicate that there is both an explicit

---

17 Anthony G. Greenwald and Shelly D. Farnham, "Using the Implicit Association Test to Measure Self-Esteem and Self-Concept," *Journal of Personality and Social Psychology* 79 (2000); Anthony G. Greenwald, Debbie E. McGhee, and Jordan L. K. Schwartz, "Measuring Individual Differences in Implicit Cognition: The Implicit Association Test," *Journal of Personality and Social Psychology* 74 (1998); Anthony G. Greenwald, Colin T. Smith, N. Sriram, Yoav Baranan, and Brian A. Nosek, "Implicit Race Attitudes Predicted Vote in the 2008 U.S. Presidential Election," *Analyses of Social Issues and Public Policy* 9 (2009).

and implicit mechanism at work when one makes judgments to include judgments with moral implications.

One might conclude that there is something more going on in the gap (Libet's 150 milliseconds) that allows "dishonest" subjects to override their propensity to cheat. Clearly, the growing evidence that there might be something going on "under the hood" needs to be taken seriously. Our intuitions play a greater role in our moral conduct than we previously thought.

As a result of the increasing evidence that much of our moral conduct is based on our intuition, an effort should be made to integrate intuitive ethics into military training and education. Military personnel need to be put into as many "real world" morally ambiguous situations as possible and be required to make intuitive judgments, decisions, and actions in order to resolve the situation. These situations or vignettes could easily be integrated into existing training scenarios. However, in order for ethics education to increase good intuitive ethics, there needs to be a link between the implicit and explicit. One way to make the implicit known, and thereby reinforce right conduct and correct wrong conduct, is by continuing the dialogue into the after action review for those vignettes that can be evaluated in terms of morality. There should be a determined effort to question the morally relevant conduct and to link this conduct to the core values of the Marine Corps.

Educating the Educators

In addition to sound leadership and role modeling, educators need to be skilled in practical moral reasoning skills (e.g., critical thinking, moral empathy, open-mindedness, and the ability to link conduct with the values of the organization). Certainly, the educators who will be teaching entry-level ROTC classes need to be educated on how to conduct character education. Very little formal education on how to conduct character development is provided to ROTC instructors. Perhaps there should be a formal education process for those who instruct at ROTC programs. The U.S. Marine Corps can remedy such a shortcoming at The Basic School (TBS); however, other services do not have this opportunity.

The military/DOD/USMC should also identify those responsible for character education in the operating force and educate them.

The captain, lieutenant, and staff noncommissioned officer are in the ideal positions to conduct character education in the operating forces. Yet as character education does not equal training, perhaps there should be a deliberate effort to equip these young service members to make the needed links between one's conduct and the core values of the organization. If character development is important, and arguably it is as important as any technical skill such as firing a rifle, there should be a concerted effort to educate educators at TBS, Expeditionary Warfare School (EWS), and at the Staff Noncommissioned Officers Course (SNCOC) and other postcommissioning schools.

Regardless of the method of educating the educator, the Marine Corps should invest time in discussing ethical issues prior to the point of action. War-gaming tactical situations are an expected and common practice in the troop-leading steps, as is the Marine Corps Planning Process (MCPP); incorporating ethical issues in this process will be beneficial to develop moral reasoning skills.

CONCLUSION

In order to achieve maximum effectiveness in character education, the Marine Corps should integrate both intuitive and maieutic methods in ethics education programs. More importantly, the Marine Corps should set the conditions for success within the classroom and integrate both reason and intuition outside the classroom. Unethical behavior should not necessarily be punished in training; rather, Marines who display unethical conduct should be remediated by spending time discussing the connection between core values and moral conduct. Those who educate Marines about making morally correct decisions must take advantage of proven pedagogical methods in order to enrich the conversation and facilitate greater moral development throughout the Corps.

Leadership in a World of Blurred Responsibilities

Emmanuel R. Goffi

The new international environment, new forms of conflict, new expectations from public opinion, the powerful role played by the media, and the conflicting agendas of great nations have put service members and, particularly, military leadership in situations in which exercising leadership is extremely difficult. The changing nature of the missions that military leaders are asked to execute and the increasing demands of their responsibilities have placed a heavy burden on them. Often these new responsibilities are confused and blurred, and as a result military leadership is more complex. The "fog" of responsibilities is part of the framework in which service members are asked to operate.

Leaders deal with at least three levels of responsibility: toward the profession, the nation, and the international community. Assuming that through comradeship and esprit de corps, the profession of arms is comparable to a family, these three levels can be compared to Emile Durkheim's "three loyalties" consisting of the loyalty toward "the family in which one is born, the nation or the political group, and humanity."[1]

Many thanks to Dr. Paolo Tripodi and Lieutenant Colonel Carroll Connelley (USMC) for the tremendous work they have done to carry this project to its end. I also acknowledge Ms. Stase Rodebaugh for her compelling and thoughtful comments which helped me make this paper clearer. I am grateful to Major James Gingras (USAF) and Heather Richards for the time they spent reading these lines and for their insightful feedback. All my thanks and love to my wife Christine and our son Thomas.

1 Emile Durkheim, *Moral Education* (Mineola, NY: Courier Dover Publication, 2002), 74.

Military leaders are asked to assume the responsibilities related not only to their rank and position, but also to their profession. Assuming responsibility is the path for leaders to build their legitimacy. In this paper, legitimacy is defined as the combination of three elements: tradition, charisma, and rationality-legality.[2] Indeed, in the military, these three essential elements directly impact the trust leaders can expect from the profession and from their subordinates in particular.[3] The responsibility military leaders have toward the nation is rather complex since it involves respect for both the people with their expectations, and the state with its agendas. Finally, international responsibility is the outcome of the rise of cosmopolitanism in international relations and foreign policy. Cosmopolitanism stresses that all human beings are citizens of the world, and as Kwame Anthony Appiah explained, cosmopolitanism is thus made of "two strands that intertwine." In Appiah's view, "one is the idea that we have obligations to others, obligations that stretch beyond those to whom we are related by the ties of kith and kin, or even the more formal ties of shared citizenship. The other is that we take seriously the value not just of human life but of particular human lives."[4] Thus, cosmopolitanism advocates a moral duty for every human being toward every other human being—not only toward fellow citizens, friends, family members, or a particular community.

The complexity of military leadership is determined by both the number of actors military leaders are responsible for and

2 Maximilian Weber, *Le savant et le politique*, electronic version by Jean-Marie Tremblay made from the book, *Le savant et le politique* (Paris: Union Générale d'Éditions, 1963), 33–34, http://classiques.uqac.ca/classiques/Weber/savant_politique/Le_savant.html.
3 Emmanuel Goffi, *Les armées françaises face à la morale: Une réflexion au cœur des conflits modernes* (Paris: L'Harmattan, 2011), 160–63.
4 Kwame Anthony Appiah, *Cosmopolitanism: Ethics in a World of Strangers* (New York: W.W. Norton, 2007), xv.

the interaction among those actors. The conflict in Afghanistan, where French troops, as many others, have been deployed with the primary purpose of defending French interests, is a perfect example of this complexity and interaction. Although troops have been deployed to protect French interests, the argument used by the government to explain and justify France's involvement in Afghanistan has been mainly humanitarian. French military leaders in Afghanistan are not always clear regarding the real objective of their mission. Are they supposed to promote French interests, or are they supposed to protect the Afghan people? Or is the intervention in Afghanistan a multipurpose mission?

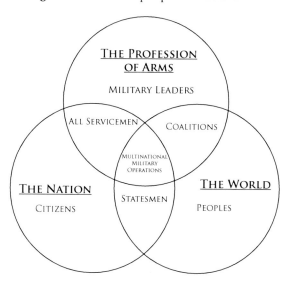

FIGURE 2. **The Circles of Responsibility.**

Figure 2 illustrates the complexity of this situation and how different levels of responsibility interact: in the nation, military leaders are responsible toward fellow citizens; in the profession, they have a double responsibility toward their superiors and their subordinates; and in the world, military leaders are responsible to all human beings. These responsibilities become particularly complex at the intersection of the three circles and particularly when military leaders are deployed in multinational military operations. In these operations responsibilities are blurred, and it is difficult, if not impossible, to determine clearly who is responsible for what.

Before exploring these three levels of responsibility, it is important to explain the meaning of "responsibility."

RESPONSIBILITY: A COMPLEX CONCEPT

George Ambler, executive partner with Gartner Executive Programs, notes that "as leaders we can be given accountability and we can be given authority, but we cannot be given responsibility. We have to take responsibility. Leadership is a choice we make. The attitude of responsibility is a leadership mind-set. We do not become leaders because we have authority and are therefore accountable. We are leaders because of how we choose to respond. Leadership rests on our responsibility, not our authority."[5] In Ambler's view, responsibility is the pillar of leadership. Thus, military leaders should be aware that leadership is deeply rooted in the exercise of responsibility. Michael Walzer provides more clarity on this point. According to Walzer, "Officers take on immense responsibilities . . . for they have in their control the means of death and destruction. The higher their rank, the greater the reach of their command, the larger their responsibilities."[6]

Indeed, responsibility is at the core of the profession of arms. As managers of violence, soldiers and their leaders bear the greatest responsibility. Therefore, they must be held accountable for

---

5 George Ambler, "Responsibility and Its Role in Leadership," The Practice of Leadership (blog), 31 August 2008, http://www.thepracticeofleadership.net/responsibility-and-its-role-in-leadership.
6 Michael Walzer, *Just and Unjust Wars* (New York: Basic Books, 2006), 316.

their actions. This accountability can be divided into two parts: on the one hand, they are responsible for following and applying the rules of law; on the other hand, they are responsible for doing what is morally right.

Currently there is a strong, but necessary, constraint on military actions, especially on the use of force, which is tightly disciplined by the law of armed conflict (LOAC) and the rules of engagement (ROE). Liability of military members can hardly be avoided when a violation of the LOAC has been committed.

Personal liability at all levels of the hierarchy has been the cornerstone of international criminal justice since the Nuremberg trials. As a defendant at Nuremberg, one of the most prominent Nazi leaders, Hermann Göring, argued that he could not be held accountable for complying with orders issued by his superior, Adolf Hitler. He tried to evade his personal liability for the crimes Nazi Germany committed during the Second World War by deflecting the responsibility toward his superiors.

Yet in 1945 the Charter of the International Military Tribunal of Nuremberg determined that "[t]he fact that the Defendant acted pursuant to order of his Government or of a superior shall not free him from responsibility."[7] In 1950, the so-called Nuremberg Principles postulated that "[t]he fact that a person acted pursuant to order of his Government or of a superior does not relieve him from responsibility under international law, provided a moral choice was in fact possible to him."[8]

This point was later included in the Rome Statute that established the International Criminal Court. Article 33 states,

---

7 Agreement for the Prosecution and Punishment of the Major War Criminals of the European Axis, and Charter of the International Military Tribunal, London, 8 August 1945, Art 8, http://www.icrc.org/ihl.nsf/FULL/350?OpenDocument.
8 Principles of International Law Recognized in the Charter of the Nuremberg Tribunal and in the Judgment of the Tribunal, Principle IV, http://untreaty.un.org/ilc/texts/instruments/english/draft%20articles/7_1_1950.pdf.

The fact that a crime within the jurisdiction of the Court has been committed by a person pursuant to an order of a Government or of a superior, whether military or civilian, shall not relieve that person of criminal responsibility unless (a) The person was under a legal obligation to obey orders of the Government or the superior in question; (b) The person did not know that the order was unlawful; and (c) The order was not manifestly unlawful.[9]

Therefore, it is extremely difficult to evade one's responsibilities toward international law. However, when dealing with national jurisdictions, things can be rather confusing. In France, for example, the penal code states that "a person is not criminally liable who performs an action commanded by a lawful authority, unless the action is manifestly unlawful."[10] This statement does not contradict international laws; however, it does not formally forbid the use of the "obedience to authority" argument. The main issue with this article and Article 33 of the Rome Statute is that they imply all military personnel must have a deep knowledge of what is lawful and what is not. However, this is something difficult to achieve. As a result military commanders on the battlefield normally rely on the advice of legal advisors (LEGAD) and on documents such as the rules of engagement. Yet, this is not enough and mistakes can be made. In some circumstances LEGAD might not be available. Often the ROE are complex and unclear. In some cases, there might be a perception that the interest at stake is so high that violating the rules might be a necessary and acceptable compromise. This point is at the center of the debate on torture. Abu Ghraib, al Habbaniyah, and several other detention centers were condemned by the International Commit-

---

9 Rome Statute of the International Criminal Court, Art. 33, http://untreaty. un.org/cod/icc/statute/romefra.htm.
10 French Penal Code, Art. 122.4 § 2, English version, http://195.83.177.9/upl/ pdf/code_33.pdf.

tee of the Red Cross for the abuses that took place. Resorting to torture, even if clearly unlawful, can be wrongly perceived as a necessary means of last resort.

Liability for war crimes includes the perpetrators and also those who were in charge. As Stéphane Bourgon puts it, "[i]t is important, and it goes without saying, that commanders are vested with significant powers. It is necessary to allow them to fulfill the missions they are entrusted with. These powers are, notwithstanding, accompanied by heavy responsibilities based on commanders' duty to control their subordinates and to ensure the respect of international humanitarian law."[11] Leaders' responsibility is obviously not limited to legal considerations. The moral and social consequences caused by the violation of certain rules, such as the prohibition on torture, can weigh heavier than the legal ramifications.

Stanley Milgram, in his famous study on obedience,[12] provided evidence that a significant number of individuals will obey an immoral order issued by an authority perceived as legitimate.[13] It is easy to give up one's responsibility, as had been done by Nazi leaders in Nuremberg or as shown by the Stanford Prison experiment[14] in 1971 and more recently by the French TV program, *Le Jeu de la Mort*, broadcast in March 2010.[15] In this program, which

11 Stéphane Bourgon, "Les tribunaux pénaux internationaux et le droit international humanitaire: La responsabilité du commandement" in, *Les conflits et le droit*, ed. Emmanuel Goffi and Grégory Boutherin (Paris: Choiseul, 2011).
12 In 1961 Stanley Milgram conducted an experiment aimed at measuring the willingness of individuals to obey orders given by an authority figure. A subject was ordered to ask questions to a "learner" and to deliver an electric shock to this "learner" each time he made a mistake. The shock was increased for each mistake to reach highly painful and, eventually, deadly levels. The so-called "learner" was an actor and shocks were not actually delivered.
13 Stanley Milgram, *Obedience to Authority* (New York: Perennial Classics, 2004).
14 Philip Zimbardo, *The Lucifer Effect: How Good People Turn Evil* (London: Rider, 2009).
15 *Le jeu de la mort (Jusqu'où va la télé?)*, broadcast on the French TV channel "France 2," 17 March 2010.

was based on Milgram's experiment, 81 percent of the subjects pushed the lever to the ultimate 440-volt shock just because they were asked to.[16] Leaders must be prepared to avoid the "Lucifer effect"[17] and prevent their subordinates, or even their superiors, from perpetrating evil acts.

Lieutenant General Denis Mercier noted that "[n]o frame, whether legal or moral, will release the soldier—or all the more, the officer—from his responsibility when it comes to the use of force, whatever the complexity of the situation in which he operates."[18] Indeed, as explained by Anthony Hartle, "To be prepared for his responsibility, a commander must be proficient in a variety of areas. He must be a 'tactician, strategist, warrior, ethicist, leader, manager, and technician.'"[19] Often leaders are expected to be knowledgeable about everything and to make good decisions always. Obviously this kind of expectation can be unrealistic, and military leaders can often just choose the "lesser of two evils"; this course of action is the outcome of limited knowledge of the environment where they operate and the interests at stake.

RESPONSIBILITY TOWARD THE PROFESSION

In his classic book, *The Soldier and the State,* Samuel Huntington wrote that military ethics "exalts obedience as the highest virtue of military men."[20] This point of view is widely shared among French field officers. Leaders have a duty to obey and to advise their superiors. Advising superiors, however, can be challenging, particularly when a disagreement might arise and the opinion

16 In the Milgram experiment conducted in 1961, 65 percent of the subjects delivered the final massive shock of 440 volts.

17 Zimbardo, *The Lucifer Effect.*

18 Lieutenant General Denis Mercier, préface in Goffi, *Les armées françaises face à la morale.*

19 Anthony Hartle, *Moral Issues in Military Decision Making* (Lawrence: University Press of Kansas, 2004), 14.

20 Samuel Huntington, *The Soldier and the State* (Cambridge: Belknap, 1957), 79.

offered could be perceived as adversarial. This is indeed an important responsibility for any leader, but even more so for military officers.

In the case of obedience, leaders consider themselves merely professionals who have been given a mission they must execute without questioning.[21] This mind-set is germane to the military culture of obedience and also to the traditional subordination of the military to the political power. In 44 BC, Cicero wrote *"Cedant arma togæ"* or "Let arms yield to the toga."[22] In this statement lies the idea that the military is always expected to submit to civilians' rules. Yet this relation is not always easy. Decisions made by political leaders are sometimes difficult for military leaders to understand. The French-led military *Opération Turquoise*[23] is a good example of the difficulty faced by leaders in fulfilling a political mission with serious consequences in the field, for which "the political responsibility of the French government remains a substantive issue which deserves close scrutiny."[24] It might be easier to blame the military rather than blame the political institution responsible for ordering the use of force.

The conflict in Afghanistan is an interesting and important case. In France the real motivation of the intervention remains unclear to both French military leaders and their troops deployed there. Three factors are useful to explain such a situation. First, officials'

---

21 Jean-Claude Barreau, Jean Dufourcq, and Frédéric Teulon, *Paroles d'officiers* (Paris: Fayard, 2010), 172.
22 Quintus Tullius Cicero, *De officiis* (*On Duties*), I, 77, Stoics.com, http://stoics.com/cicero_book.html.
23 *Opération Turquoise* was a mission in Rwanda proposed and led by France in 1994 under Resolution 929 of the UN aiming at "achieving the objectives of the United Nations in Rwanda through the establishment of a temporary operation under national command and control aimed at contributing, in an impartial way, to the security and protection of displaced persons, refugees, and civilians at risk in Rwanda." France was highly criticized for its controversial role in the conflict. The French army was charged for "complicity of genocide and/or complicity of crimes against humanity."
24 Barreau et al., *Paroles d'officiers*, 179.

discourse, such as that from French political leaders, has stressed the humanitarian dimension of the intervention. Second, the political agendas of Coalition members differ significantly from one another. The reasons that have determined many countries' decisions to intervene in Afghanistan have changed often along with the means used during the intervention. Third, these changes have made a significant impact on the overall environment on the ground and on the perception Afghan people and Coalition troops have of the intervention. As a result, military leaders have often struggled to understand and explain why they are fighting in Afghanistan. In France for instance, exploring why French troops are fighting in Afghanistan can lead to a wide range of explanations. The intervention is justified by humanitarian and strategic arguments, as well as an emphasis on the role France plays within NATO and its relationship with the United States.

It should be noted that making decisions to deploy troops into combat zones and risk their lives for unclear reasons is dangerous. Colonel Eric Maïni, deputy commander of the French Air Force Academy, stated during an informal discussion, "It is a major responsibility to put the name of a subordinate on the list of those who will be sent into combat."[25] This must be taught to cadets and reminded to senior leaders.

Audrone Petrauskaite, at the General Jonas Zemaitis Military Academy of Lithuania, found that 69.1 percent of Lithuanian cadets considered responsibility as the most important quality of a professional officer (responsibility ranked higher than duty–60 percent–and loyalty–50.9 percent). At the same time, 69.1 percent of the respondents agreed with the idea that "militaries should carry out all orders of their commander" and only 25 percent considered "that militaries should carry out only the order that is

_____

25 Quoted with the kind agreement of Colonel Eric Maïni.

in agreement with their consciousness."[26] Cadets overall should be taught that leadership is intrinsically linked to the exercise of responsibility, and a proper exercise of responsibility requires autonomous thinking. This means that they should develop the moral courage required to refuse the execution of illegal and/or immoral orders. According to Hannah Arendt, it is this thinking, considered as free will, that will allow the individual to develop the ability to apply sound judgment and then take responsibility, and if necessary to say "no."[27] Thus, responsibility toward superiors' orders rests in the ability to refuse the execution of those orders that are clearly unlawful and/or immoral. Due to the subjective nature of the perception of morality, the act of refusing the execution of "immoral" orders poses serious challenges to the individual.

On some occasions political and military agendas can collide. According to Robert McNamara, the question often is "how much evil must we do in order to do good? We have certain ideals, certain responsibilities. Recognize that at times you will have to engage in evil, but minimize it."[28] Thus, under what circumstances should military leaders oppose the execution of missions that would place the lives of their troops at risk? This question was raised in Afghanistan when a Dutch battalion refused to carry out a reconnaissance mission as they were not adequately equipped and troops' lives could have been compromised.[29] A similar case occurred in Iraq when U.S. soldiers refused to execute what they

26 Audrone Petrauskaite, "Ethics in Lithuanian Professional Military Education" in *Civil-Military Aspects of Military Ethics*, vol 2, ed. Edwin R. Micewski and Dietmar Pfarr (Vienna: Publication Series of the National Defense Academy, 2005), 90.
27 Hannah Arendt, *Responsibility and Judgment* (New York: Schoken Books, 2005).
28 *The Fog of War: Eleven Lessons from the Life of Robert S. McNamara*, motion picture, Sony Pictures Classics, 2003.
29 "La fronde d'un bataillon néerlandais en Afghanistan," *Le Figaro* (Paris), 1 October 2008, http://www.lefigaro.fr/international/2008/10/01/01003-20081001ART-FIG00433-la-fronde-d-un-bataillon-neerlandais-en-afghanistan-.php.

called a "suicide mission" consisting of operating a fuel convoy in a dangerous area.[30]

In combat commanders are placed under a great deal of pressure generated by the responsibility to execute the orders they have received, while trying to minimize the loss of troops' lives. Leaders deal with multiple responsibilities as subordinates and advisors to civilian authority and toward their subordinates. They are morally and legally accountable for the consequences of the decisions they make. Yet what does accountability mean when responsibility and its sharing are blurred?

One French example provides a good illustration of this concern. In August 2008, 10 troops were killed in an ambush in the Uzbin Valley in Afghanistan.[31] This tragic event paved the way to a strong debate in France about whether France should continue its military commitment in Afghanistan. The most pressing issue was about identifying who was responsible for those deaths. The ambush was widely reported by the media. French public opinion blamed military leaders for the losses. In October 2009, the families of the soldiers killed in action decided to sue military leaders for "deliberately endangering the lives of others [i.e. soldiers]." The legal action was aimed against "individuals who . . . had not assumed their responsibility. . . " Clearly, such a course of action generated a concern about whether military leaders will continue to accept their responsibility in future combat operations.

The French military found that the initiative taken by the families of those killed in action was unfair. The issue is whether military leaders on the battlefield are the only ones who should be held responsible for the casualties. French public opinion per-

---

30 Dan Glaister, "Doubts about U.S. Morale in Iraq as Troops Refuse 'Suicide Mission'," *The Guardian*, 16 October 2004, http://www.guardian.co.uk/world/2004/oct/16/iraq.usa.

31 For further details about the ambush see Bill Roggio, "Taliban Kill 10 French Troops in Kabul Province Ambush," *The Long War Journal*, 19 August 2008, http://www.longwarjournal.org/archives/2008/08/taliban_kill_ten_fre.php.

ceives that French troops are not properly equipped to conduct counterinsurgency campaigns or humanitarian intervention in a highly hostile environment. Indeed, should this be the case, then identifying responsibilities is significantly more complex. Many among the French military stress that the intervention in Afghanistan has been launched without the French parliament's approval as required by Article 35 of the constitution. Therefore, military leaders share no responsibility for such a political decision.

In a situation lacking clarity, military leaders feel a strong responsibility toward their subordinates. This opinion is widely shared within the French military. In addition, before dedication to the nation or fellow citizens, French soldiers' main motivation in combat is to risk their lives for their comrades-in-arms.[32]

When the motivation to enter a conflict is not clear, and as a result, military leaders might find it difficult to explain why troops should kill or might be killed, the very sense of military commitment may shift from the defense of the nation to the defense of comrades. Nonetheless, the defense of the nation remains the traditional—and a strong—responsibility of military leaders.

Responsibility toward the Nation

In the first century BC, Horace wrote, "*Dulce et decorum est pro patria mori*," or "It is sweet and fitting to die for one's country." Today, to serve France with the "ultimate sacrifice" is, as stressed by former French President Jacques Chirac, the "raison d'être" and the "honor of the professional officer."[33] There is a close re-

---

32 Leonard Wong, Thomas A. Kolditz, Raymond A. Millen, and Terrence M. Potter., *Why They Fight: Combat Motivation in the Iraq War* (Carlisle, PA: Strategic Studies Institute, 2003), http://www.strategicstudiesinstitute.army.mil/pubs/display.cfm?pubID=179; and Kelly Kennedy, *They Fought for Each Other: The Triumph and Tragedy of the Hardest Hit Unit in Iraq* (New York: St. Martin's Press, 2010).

33 Jacques Chirac, *Discours de M. Jacques Chirac, Président de la République, à l'occasion du bicentenaire de l'Ecole spéciale militaire de Saint-Cyr*, École Militaire, Paris, 25 January 2002, http://rpr.infos.27.pagesperso-orange.fr/00000003.htm#ANCRE%2025/01/02.

lationship between military personnel and the nation clearly captured by General Georgelin as he notes that "[t]he serviceman is the one who, today as yesterday, thinks about the values which founded the nation, because it is on their behalf that he will, if necessary, use the force he has been entrusted with by his fellow countrymen."[34] Emile Durkheim notes that there is always a tension between patriotism and cosmopolitanism at the moral level. However, he emphasizes the fact that "in contrast with the nation, mankind as source and object of morality suffers this deficiency: there is no constituted society."[35] In addition, the French Defense Code states that "[t]he Armed Forces of the Republic are dedicated to the nation. The mission of the armed forces is to prepare and to insure by force of arms the defense of the homeland and the higher interests of the nation."[36] The defense of the nation has long been the main, if not the only, reason worth dying for.[37] Dying pro patria was, in the First and Second World Wars, a strong motivation for French soldiers.[38]

This seems to be the most important responsibility of any military member. In his study on the U.S. military, Roger H. Nye writes,

> The true American soldier has always cherished a vision of himself as a *servant of the nation*, an identity that runs counter to the idea of professional isolation from the great politic body. In this sense, soldiers feel deeply about patriotism, a patriotism that extols the American nation as man's best hope for guaran-

34 Général d'armée Jean-Louis Georgelin, former Joint Chief of Staff, "L'identité militaire vue par le général Georgelin," http://www.defense.gouv.fr/ema/le-chef-d-etat-major/interventions/articles/12-10-07-l-identite-militaire-vue-par-le-general-georgelin.
35 Emile Durkheim, *Moral Education*, 76.
36 *Code de la Défense*, Article L3211-2, http://droit-finances.commentcamarche.net/legifrance/50-code-de-la-defense.
37 Ernst Kantorowicz, *Mourir pour la patrie et autres textes* (Paris: PUF, 1984), 105–41.
38 Guillaume Cuchet, "Mourir pour la patrie: le poilu entre gloire terrestre et gloire céleste" in *Le sacrifice du soldat: Corps martyrisé, corps mythifié*, Christian Benoit et al. (Paris: ECPAD/CNRS Editions, 2009),74–78

teeing the freedom and peace necessary to man's achieving his great potential. Defense of the nation becomes a great calling. With it comes the duty of protecting the people, their value system, and their material well-being.[39]

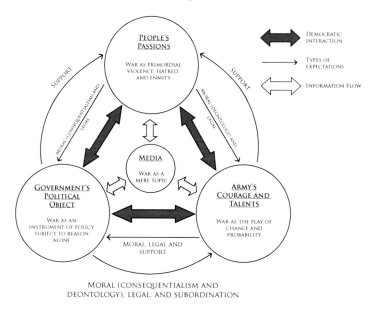

FIGURE 3. **The Remarkable Trinity +1.**

Military leaders have significant responsibilities toward the nation and fellow citizens. Thus, they are placed in a difficult position when the political decision to use force is not supported or when it is even opposed by the "nation." If one accepts Clausewitz's point of view that the military is a tool of politics, and that politics in a democracy is the expression of the nation's will, then military leaders are accountable to their fellow citizens or at least to their representatives. In the case of the French intervention in Afghanistan, the French parliament was excluded from the political decision-making process. More recently the French parlia-

39 Roger H. Nye, *The Challenge of Command: Reading for Military Excellence* (New York: Perigee Trade, 2002), 14.

ment was also excluded from the decision to intervene in Libya. In these situations, military leaders deal with the Clausewitzian "remarkable trinity,"[40] (see figure 3) trying to balance the people, the military, and the government, as well as the media.

In Clausewitz's trinity, military leaders deal with conflicting and often confusing priorities. A balance should be found between the French government's agenda to maintain, and indeed also strengthen, the number of troops in Afghanistan, and the French public opinion's desire to withdraw. In such a blurred situation, military leaders are asked by the government to execute a mission that will risk the lives of their troops while French public opinion is opposed to such a mission. According to Hartle, "[i]n general, military forces and military leaders have always been an obedient arm of the state and strictly subordinate to civilian authority."[41] Therefore, the mission decided by the government must be accomplished even when public opinion is opposed to it. However, such a division will have a major impact on military leaders.

Winning hearts and minds is not only about the people of the countries where troops are deployed, but it's also about their respective public opinions. General Benoit Royal stresses that "today more than yesterday, a military force which does not benefit from the support of the public opinion loses its legitimacy."[42] The people no longer feel responsible for political decisions, and political leaders do not feel responsible since they consider themselves acting on behalf of the people. In this confusing situation military leaders should be able to balance obedience to orders with expectations from their fellow citizens.

---

40 Carl von Clausewitz, *On War* (Ware: Wordsworth Classics, 1997), 24.

41 Hartle, *Moral Issues in Military Decision Making*, 16.

42 Benoît Royal, "La guerre se gagne avec l'opinion publique," *Le Figaro*, 13 February 2009, http://www.lefigaro.fr/debats/ 2009/02/27/01005-20090227ART-FIG00001-la-guerre-se-gagne-avec-l-opinion-publique-.php. .

Responsibility toward the International Community

Since the early 1990s, the international community and its most important actors have decided to play a more robust role to create a safer global environment. Western states have looked at their national security through the prism of international stability. A shift has occurred from a nationalist to a cosmopolitan approach to security, indeed, from a nationalist to a "nationalist-through-cosmopolitan" approach. The French national strategy policy also stresses that "in virtue of its international responsibilities and on the basis of a collective vision of its own security interests, France may occasionally have to take part in an intervention even though its own direct interests are not at stake."[43] French moral altruism comes together with a more pragmatic consideration of France's "indirect interests."

Together with an increased use of force to deal with internal or international disputes in some parts of the world, such as in Africa, the need for justice has increased in a considerable way. This is true both at the national and international levels.

The idea of being responsible for the well-being and the security of other peoples has been largely used by the United Nations to promote and justify humanitarian and peacekeeping operations. As Kwame Anthony Appiah writes,

> Accepting the nation-state means accepting that we have a special responsibility for the life and justice of our own; but we still have to play our part in ensuring that all states respect the rights and meet the needs of their citizens. If they cannot, then all of us—through our nations, if they will do it, and in spite of them, if they won't—share the collective obligation to change them, and if the reason they fail their citizens is that they

---

43 *French White Paper on Defence and National Security* (Paris: Odile Jacob/La Documentation Française, 2008), 69, http://merln.ndu.edu/whitepapers/France_English2008.pdf.

lack resources, providing resources can be part of that collective obligation. That is an equally fundamental cosmopolitan commitment.[44]

Cosmopolitanism is now widely spread throughout the world in the name of a duty to aid others, or "the responsibility to protect." According to Jeremy Bentham, "[e]very man is bound to assist those who have need of assistance, if he can do it without exposing himself to sensible inconvenience."[45] Therefore, military leaders' responsibility has broadened beyond the profession and the nation to include also the international community. Are French soldiers and UN peacekeepers expected to risk their lives to protect other peoples? Should military leaders accept and risk their troops' lives to protect foreign citizens when deployed in peace-support operations? The cosmopolitan answer to these questions is yes. However, it is not clear whether French public opinion would always support such a view. French public opinion might have a limited tolerance for casualties, and its resilience might decrease as French troops' deployment drags on and there might be a perception of diminishing chances of success.[46]

In addition, it is reasonable to explore what cost French public opinion is ready to pay in support of these missions. Michael Gross argues that "[a] nation in a position to help must determine whether costs are reasonable and whether it is willing to sacrifice some of its members to rescue others."[47] It is important to determine what would be considered "reasonable." According to Gross, "[o]nly national defense can obligate individuals to risk

44 Appiah, *Cosmopolitanism*, 163–64.
45 Jeremy Bentham, "Specimen of a Penal Code" in *Works* vol I, ed. J. Bowring (Edinburg: Tait, 1843), 164, cited in Michael L. Gross, *Moral Dilemmas of Modern War: Torture, Assassination, and Blackmail in an Age of Asymmetric Conflict* (Cambridge: Cambridge University Press, 2010), 215.
46 Christopher Gelpi, Peter D. Feaver, and Jason Reifler, *Paying the Human Costs of War: American Public Opinion and Casualties in Military Conflicts* (Princeton: Princeton University Press, 2009), 13.
47 Gross, *Moral Dilemmas of Modern War*, 223.

their lives."[48] Yet in Gross' view, if the nation decides to engage troops in humanitarian interventions, it must provide the best conditions possible to succeed and thus minimize, as much as possible, risks for its service members.

French military leaders involved in the Uzbin ambush were charged as "individuals who . . . had not assumed their responsibilities . . . " by "deliberately endangering the lives of others."[49] The fact that the intervention in Afghanistan was presented to the French public as a humanitarian mission led them to believe that French troops would face a negligible amount of hostility and violence. Humanitarian interventions are not "real wars," and thus there is a mistaken perception that they will be missions without casualties. Political leaders' responsibility and credibility are at stake. Very likely they use a moral argument to justify the use of military force. The humanitarian argument was used by former French President Nicolas Sarkozy to justify the intervention in Afghanistan. A translation of his speech following the Uzbin ambush reads,

> Because fighting here, in the Uzbin Valley in Afghanistan, the French must know it, [we are] protecting our democracies from terrorism. You defend here human rights, and in particular women['s] rights. Thanks to you, millions of Afghan children are attending school. Thanks to you, infant mortality has been reduced by a factor of three. . . . Even if the death toll is very heavy, be proud of what you are doing here.[50]

48 Ibid., 228.
49 Eight complaints, based on article 223-1 of the French Penal Code, have been registered at the *Tribunal aux Armées de Paris* (Paris Military Tribunal). See *Agence France Presse*, "Embuscade d'Uzbin: 6 nouvelles pliantes," *Le Figaro*, 30 November 2009; and Jean-Dominique Merchet, "Plaintes d'Uzbin: le point de vue d'un magistrat et d'un avocat," Secret Défense (blog), 29 October 2009, http://secret-defense.blogs.liberation.fr/defense/2009/10/plaintes-duzbin-le-point-de-vue-dun-magistrat.html.
50 Nicolas Sarkozy, President of the French Republic (speech, Kabul, Afghanistan, 20 August 2008, http://www.elysee.fr/president/root/bank/pdf/president-2073.pdf. (Author's translation.)

This approach emphasizes the moral responsibility that French people are supposed to have toward other nations. The so-called responsibility to protect (R2P) was "first articulated by the independent International Commission on Intervention and State Sovereignty (ICISS) and refined and adopted at the 2005 World Summit."[51] R2P aims at preventing and stopping crimes such as genocide, war crimes, ethnic cleansing, and crimes against humanity.[52] However, several commentators have noted that R2P can be used as an excuse for violating state sovereignty. Seuma Milne considered the intervention in Libya, launched to "protect civilians and civilian-populated areas under threat of attack in the Libyan Arab Jamahiriya,"[53] as a strong evidence of R2P abuse.[54]

Despite the fact that "Western armed forces have historically focused on and attempted to instill, nationalist sentiments—mainly the protection and self-defense of physical integrity and rights of family, friends, and fellow nationals—as ideological incentives to fight,"[55] they must also be prepared to accept the shift to cosmopolitan interventions. Indeed, humanitarian operations are now part of European Union missions, as stated in the Petersberg Tasks.[56] These missions will continue to play an important

51 Edward C. Luck, "The Responsibility to Protect: Growing Pains or Early Promise?" *Ethics & International Affairs* 24 (2010), http://www.carnegiecouncil.org/resources/journal/24_4/response/001.html?sourceDoc=002023.
52 Ibid.
53 *United Nations, Security Council, Resolution 1973* (2011), 17 March 2011, http://daccess-dds-ny.un.org/doc/UNDOC/GEN/N11/268/39/PDF/N1126839.pdf?OpenElement.
54 See Seumas Milne, "There's Nothing Moral about NATO's Intervention in Libya," *Guardian.co.uk*, 23 March 2011, http://www.guardian.co.uk/commentisfree/2011/mar/23/nothing-moral-nato-intervention-libya; and Hany Besada, "Libya and R2P," *Thedailynewsegypt.com*, 26 May 2011, http://thedailynewsegypt.com/global-views/libya-and-r2p.html.
55 Daniel Blocq, "Western Soldiers and the Protection of Local Civilians in UN Peacekeeping Operations: Is a Nationalist Orientation in the Armed Forces Hindering Our Preparedness to Fight?" in *Armed Forces & Society* 36 (2010): 291.
56 *Petersberg Declaration*, Western European Union, Council of Ministers, Bonn, 19 June 1992, http://www.weu.int/documents/920619peten.pdf.

role in foreign and security policy and will very likely grow in number. As a result, the risk that service members might die to protect foreigners will be high. Military leaders should be aware of these tasks and their consequences, so that they can explain them to their subordinates and be better prepared to advise their political leaders. The French white paper (the national strategy) stresses that,

> Apart from the legality of military intervention, the question of its legitimacy has already been raised by the Secretary General of the United Nations and by the Security Council itself. This development concerns not only cases of *"genocide, ethnic cleansing, or other grave violations of international humanitarian law,"* where the "responsibility to protect" falls in the last resort on the international community. It also applies to *"threats to international peace and security caused by terrorist acts."*[57]

France, like the other countries, has a significant responsibility toward the international community. This is due to the fact that France is a permanent member of the UN Security Council and other important international organizations.

Conclusion

The role of leadership is multifaceted and complex particularly when responsibilities are confused and poorly defined. The new international environment, which leads to new expectations, modern conflicts, citizens' lack of interest in politics, and the manipulation of moral arguments to justify the use of force, have made military leadership significantly more complex.

In addition, military leaders often take responsibilities that go beyond their traditional role of defending the nation and national

---

57 *French White Paper on Defence and National Security*, 107, http://merln.ndu.edu/whitepapers/France_English2008.pdf.

interests. The call for cosmopolitanism has generated confusion and doubts about the role the armed forces are supposed to play. Used to defending their nation's interests, military leaders are now facing risks associated with the protection of people all over the world. Meanwhile, public opinion has expectations that lives will be saved, not only among their own country's troops, but also among enemy troops and particularly among civilians.

Commanders and their troops are asked to fight ethically in wars that might be morally questionable. They are expected to save lives in violent conflicts presented as humanitarian, which erroneously implies a low level of violence, for the most part, to the population. In order to generate public opinion support, political leaders do not hesitate to resort to a rhetoric that might lead people to believe that casualties are just a minor occurrence in modern conflicts. Military leaders are thus supposed to wage and win wars without casualties.

The reality of war, however, is significantly different from this portrayal. Wars, even those conducted for humanitarian reasons, are uncertain and often costly for human life. Canadian Army Lieutenant Colonel Richard Walker notes that "within command responsibility resides the three-way command-harm dilemma: the mission vs. risk to our soldiers vs. risk to innocent civilians. The decision process may start with is it legal? And if legal, is it moral or ethical?"[58]

Military leaders are aware that in order to accomplish their missions they will often have to choose among risking their troops, the enemy troops, and in several cases innocent civilians. The choice they will face will rarely be an easy one. Then they will be held responsible, at least morally, for the casualties they might cause or suffer.

58 Richard J. Walker, *Duty with Discernment: CLS Guidance on Ethics in Operations* (Ottawa: National Defence Strategic Edition, 2009), http://www.army.forces. gc.ca/land-terre/downloads-telechargements/aep-peat/duty-servrir/duty-servrir-eng.pdf.

Indeed, the political/strategic decision to use military force must remain under the exclusive control of the nation through its representatives, yet the outcomes—and thus the responsibility of commanders' decisions to use force at the tactical and operational level—must be shared between military and political leaders. In France, the parliament is the legitimate authority to decide when to use military force. Citizens must then be reminded that through their participation in political life, they play a decisive role in the nation's policy making; as a result, they hold a collective responsibility for the use of military force decided by their representatives.

In addition, the objectives of military interventions must be clearly and honestly explained to citizens and to service members. This would reduce the moral gap between public opinion and the military and strengthen support for the troops that are often deployed in a hostile environment. Political leaders should be committed to properly informing people about the context, the stakes, and the risks associated with each intervention, especially when they are conducted for humanitarian purposes.

Political leaders will continue to defend their nation's direct interests and their own population; however, this should not prevent them from assuming even greater responsibility. They might take a cosmopolitan approach to foreign policy. Yet cosmopolitanism should not be used to cover the pursuit of national interests. The real intent, when using force, should never be hidden behind moral arguments, such as humanitarian motives, as they blur the reason given to service members' commitment and create strong and/or false expectations among the public.

Political leaders' responsibility toward the world should be seen as the outcome of two factors: the defense of indirect national interests and the protection of foreign peoples. These two factors must meet two requirements. First, they should not add an unnecessary and heavy burden on the country's military, and second, the nation's public opinion must be supportive of such a commitment.

Indeed, the nation and its people must be placed at the core of any decision that might consider military action and the use of force. In France, this key element of a functioning democracy has been neglected, and today the parliament has no weight in the decision to deploy troops, which is exclusively under the control of the executive branch.

In conclusion, responsibility is indeed a key element of military leadership. Military leaders must take responsibility toward the profession, the nation, and the world. However, their ability to do so might be seriously compromised if a significant distance exists between them, the country's political leadership, and public opinion. Now, more than ever, it is necessary to move the three circles of the remarkable trinity closer to each other.

WHEN LOYALTY TO COMRADES CONFLICTS WITH MILITARY DUTY

J. PETER BRADLEY

On 19 October 2008, a Canadian army captain working with the Afghan National Army fired two rounds into a gravely wounded Taliban fighter. Although shooting the man was a violation of the law of armed conflict, Canadian criminal law, and the Canadian military's code of conduct, the captain's misconduct was not reported by the three Canadian soldiers who were present that day and only became public when an Afghan interpreter reported the incident weeks later.[1] Like most modern militaries, the Canadian forces have regulations that require members to report unlawful behavior, but in this instance, the soldiers ignored these rules and kept quiet, perhaps to protect their captain, to protect themselves from reprisals, or for some other reason. This episode is similar to other incidents of military misconduct in which personnel witness transgressions but keep quiet. For example, a generation earlier, in 1992, soldiers of the Canadian airborne regiment set ablaze the personal vehicles of two unit leaders, a captain and a sergeant, and then formed a wall of silence when authorities tried to iden-

---

1 The captain was later convicted of "disgraceful conduct" by a military court and dismissed from the Canadian Forces. Michael Friscolanti, "A Soldier's Choice," *Maclean's Magazine* 24 May 2010, 20–25, http://www2.macleans.ca/2010/05/18/a-soldiers-choice/; Michael Friscolanti and John Geddes, "A Stern Message about Battlefield Ethics and the 'Soldier's Pact'," *Maclean's Magazine*, 2 August 2010, 28–30, http://www2.macleans.ca/2010/07/26/a-stern-message/.

tify the culprits.[2] More recently in 2003 and 2004, plenty of U.S. soldiers knew of the mistreatment of detainees at Abu Ghraib, but only a few reported the abuse.[3] Likewise, when soldiers of the U.S. Army's 101st Airborne Division raped a 14-year-old Iraqi girl in 2006 and then executed her and her family, several months passed before other unit members reported those responsible for the atrocities.[4]

Incidents like these are examples of "military duty-personal loyalty dilemmas," situations in which the demands of military duty collide with personal loyalty to unit mates. In such cases, military personnel witness the misconduct of their comrades or learn about it later and are then faced with the problem of deciding whether to report the incident as required by their military duty or to protect their comrades by keeping quiet. Such dilemmas happen more often than we realize. Some cases become well-known and can have vast implications, such as U.S. Army Specialist Joe Darby's decision to report the abuse of Iraqi detainees at Abu Ghraib.[5] Other cases are more banal, like a soldier who witnesses a comrade accidently fire his weapon and then must decide whether to report the misconduct or not.

Given the significance the military places on good order and discipline, it is important that military men and women have the moral competence to intervene when wrongdoing is occurring or to report the misconduct if they are unable to intervene. When military personnel ignore the misdeeds of others, the abuses continue, cover-ups occur, and the military profession is diminished. Very little has been written on this issue, so the purpose of this chapter

2 David Bercuson, *Significant Incident: Canada's Army, the Airborne, and the Murder in Somalia* (Toronto: McClelland and Stewart, 1996), 212, 224.

3 Philip Zimbardo, *The Lucifer Effect: Understanding How Good People Turn Evil* (New York: Random House, 2007), 330, 360.

4 Jim Frederick, *Black Hearts: One Platoon's Descent into Madness in Iraq's Triangle of Death* (New York: Harmony Books, 2010), 323.

5 Seymour M. Hersh, "Torture at Abu Ghraib," *The New Yorker*, 10 May 2004.

is to show how relevant elements of social science theory and research can help us understand why military personnel ignore the illegal, unprofessional, or immoral conduct of their unit mates. As much as possible, this analysis draws on published research to explain the forces involved in these duty-loyalty dilemmas. There is still much to be learned in this area, so suggestions for additional research are also included. The chapter has three objectives: (1) to highlight what is currently known about the factors involved in the decision to report (or not) the misconduct of military comrades; (2) to identify areas of further research so we can better understand the forces involved; and (3) to suggest actions that military forces can take to encourage more reporting of military misconduct.

THE MILITARY DUTY-PERSONAL LOYALTY DILEMMA

The duty-loyalty dilemma can occur in an instant or unfold slowly over time. In the rapid version of the dilemma, service members witness a comrade doing something illegal, unprofessional, or immoral, and then find themselves with a difficult decision to make: What do I do? Do I intervene to stop this behavior from happening or do I keep quiet? Because the event occurs in a matter of minutes, or even seconds, the observer has little time to think or act. If the observer has not been prepared for such challenges, he or she may succumb to the influences of the moment and act on impulse without much thought. In the more protracted version of the dilemma, the witnessing military member either saw the violation and did not take action at the time or learned of the misconduct afterwards. Even though the observer did not, for one reason or another, intervene at the time of the misconduct, he or she is still faced with a dilemma after the fact—to report the violation or to keep quiet.

When viewed from a professional or a moral perspective, the dilemma does not seem very complicated. It basically involves two obligations that are in conflict. First, military personnel have a duty to stop illegal actions from happening or to report them if they did not stop them earlier. Second, the same military personnel may feel obliged to help their comrades—or at least to avoid harming them—and to report an infraction would likely lead to harm of some sort, perhaps in the form of disciplinary action or a reprimand from a supervisor.

Some readers of this chapter might suggest that the question to report or not in such a situation does not qualify as a moral dilemma per se because a moral dilemma requires conflicting moral obligations; in this instance, there is only one moral obligation, the professional duty to report the unlawful activity. Strictly speaking, an individual has no moral obligation to protect a colleague who has committed a violation, so the military duty-personal loyalty dilemmas I have described here may not qualify as bona fide moral dilemmas. Instead, they may be closer to what Coleman calls "tests of integrity," situations "where it is reasonably obvious, or even perfectly obvious, what the right thing to do is, but for whatever reason, it is difficult for the person involved to actually do the right thing."[6]

Coleman's distinction between tests of ethics (i.e., moral dilemmas) and tests of integrity provides a degree of conceptual clarity that is useful at a theoretical level (and is also valuable in the ethics classroom), but this distinction may be less helpful to individuals who have to make these choices in the heat of the moment. Indeed, some service members may believe that supporting their comrades is the right thing to do in these instances. The emotional and social pressures involved in tests of integrity can be so

6 Stephen Coleman, "The Problems of Duty and Loyalty," *Journal of Military Ethics* 8 (2009): 106.

overwhelming that an individual facing a duty-loyalty dilemma can easily misinterpret the situation as an ethical obligation, demanding that his or her unit mates be supported. Of course, this is speculation, but there is a vast body of research showing that individuals often think they are making sound decisions when they are actually making poor choices.[7]

Research shows that individuals employ two types of decision making to solve problems: (1) rapid, intuitive reasoning and (2) slower, more deliberate reasoning.[8] Studies also show that the rapid approach is most often used in solving moral dilemmas because individuals typically respond to moral questions very quickly with the first idea that comes to their mind.[9] Clearly, there is a role here for ethics training and education to ensure that an ethical response comes quickly to a service member's mind. Ethics training that emphasizes professional obligations (so personnel will know what is expected of them) and decision-making procedures (so they will know how to reason through moral dilemmas) would be helpful not only with more deliberate decision-making tasks, but the same procedures can be trained to the point that they become almost instinctual in order for military personnel to respond quickly and correctly when unanticipated dilemmas occur.

THE POLICE CODE OF SILENCE

The military is not the only institution in which individuals must wrestle with conflicts between professional duty and loyalty to

7 Carol Tavris and Elliot Aronson, *Mistakes Were Made (But Not By Me): Why We Justify Foolish Beliefs, Bad Decisions, and Hurtful Acts* (Orlando, FL: Harcourt Books, 2007).

8 Jonah Lehrer, *How We Decide* (New York: Houghton Mifflin Harcourt, 2009), xvi.

9 Jonathan Haidt, "The Emotional Dog and Its Rational Tail: A Social Intuitionist Approach to Moral Judgment," *Psychological Review* 108 (2001): 814–34.

comrades. Police officers have been known to look the other way, falsify police reports, and lie in court to cover the misconduct of other police officers.[10] Of course, the military and police communities are not identical,[11] but they are similar enough to warrant comparison: each has a clear chain of command, uniforms, rules for employing force, and members of each community operate in dangerous environments which, in turn, promote reliance on comrades and little mixing with outsiders. Therefore, research showing how and why police officers overlook collegial misconduct may provide some insights to the military.

The practice of police officers protecting other police is so common that there are terms for it: "The Code," "The Code of Silence," "The Blue Curtain"[12] and the "Blue Wall of Silence."[13] The code is not a myth; it "is well documented in court opinions, scholarly literature, news reports, and police investigatory mission reports."[14] Moreover, it is a worldwide problem.[15] Within the United States, several large-scale investigations have confirmed that the code is a significant part of police corruption.[16] The prevalence of the code is well illustrated in two surveys of American police officers. In the first, 52 percent of the respondents agreed that "it is not unusual for a police officer to turn a blind eye to improper conduct by other officers" and 61 percent disagreed with the state-

10 Gabriel Chin and Scott Wells, "The 'Blue Wall of Silence' as Evidence of Bias and Motive to Lie: A New Approach to Police Perjury," *University of Pittsburgh Law Review* 59 (1997–98): 234.

11 Thomas J. Cowper, "The Myth of the 'Military Model' of Leadership in Law Enforcement," *Police Quarterly* 3 (2000): 228–46.

12 Carl B. Klockers, Sanja K. Ivkovich, William E. Harver, and Maria R. Haberfeld., *The Measurement of Police Integrity* (Washington, DC: U.S. Department of Justice 2000), 1, www.ncjrs.gov/pdffiles1/nij/181465.pdf.

13 Chin and Wells, "Blue Wall of Silence," 233.

14 Ibid, 238–40.

15 Rick Sarre, Dilip K. Das, and H.J. Albrecht, *Policing Corruption: International Perspectives* (Lanham, MD: Lexington Books, 2005).

16 Jerome H. Skolnick, "Corruption and the Blue Code of Silence," in *Policing Corruption: International Perspectives*, 303.

ment "police officers always report serious criminal violations involving abuse of authority by fellow officers."[17] In another study of 1,116 full-time police officers in the United States, researchers found that 46 percent "had witnessed misconduct by another employee but took no action."[18]

The code persists because of police loyalty and fear of retaliation.[19] Police officers submit to the code of silence for many practical reasons such as to avoid "being shunned, losing friends, having no one to work with, losing backup support, harassment, physical threats, permanent stigmatization, and exposure of one's own misconduct."[20] The code of silence is also promoted by a subculture of loyalty and protecting colleagues, which in turn, is created by the inherent danger of the job, the closed nature of the police community, prying media, and unsympathetic outsiders.[21] Police officers are first exposed to this subculture at the beginning of their careers, in the academy.

Not all police officers, however, live by the code all the time. Research shows that police officers are more likely to report the violations of other police if they are employed in supervisory roles or serve in larger organizations.[22] These results may be due to the fact that senior personnel usually have more knowledge about what is required by the organization, more commitment to the or-

17 David Weisburd and Rosann Greenspan, "Police Attitudes Toward Abuse of Authority: Findings from a National Study," *The National Institute of Justice Research in Brief*, May 2000, 5, www.ncjrs.gov/pdffiles1/nij/181312.pdf.

18 Neal Trautman, "Police Code of Silence Facts Revealed" (paper presented at the International Association of Chiefs of Police: Legal Officers' Section, 2000), http://www.aele.org/loscode2000.html.

19 Gary R. Rothwell and J. Norman Baldwin, "Whistle-Blowing and the Code of Silence in Police Agencies: Policy and Structural Predictors," *Crime & Delinquency* 53 (2007): 607–8.

20 J.H. Skolnick, "Corruption and the Blue Code of Silence," in *Contemporary Issues in Law Enforcement and Policing*, eds. Andrew Millie and Dilip K. Das (Boca Raton FL: CRC Press, 2008), 306.

21 Rothwell and Baldwin, "Whistle-Blowing," 612.

22 Chin and Wells, "The Blue Wall," 250; Skolnik, "Corruption," 46.

ganization, and more to lose if they don't take correct action. Regardless, the research suggests that the issue of police silence may be more serious among less experienced officers in smaller units. Because smaller units may be more cohesive, they may also have stronger norms for not reporting the infractions of others. Moreover, smaller units provide fewer places for officers who might be contemplating reporting to find refuge from the retaliation of colleagues.

While police officers have a reputation for remaining silent about the misconduct of colleagues, they are not alone in this regard. A study of 197 police officers and 168 public employees working in the state of Georgia revealed that the civilian employees were less likely to report violations of misconduct than the police officers surveyed in the same study.[23] Unsatisfactory colleagues are also tolerated in other occupations. For example, "doctors often overlook the deadly faults of the most marginal members of their profession . . . and just about any teacher knows colleagues who should not be in a classroom, but have been, for years, without a meaningful professional objection raised."[24]

Given the parallels between the military and police communities, it seems logical to find some military members behaving like police officers by keeping silent about the misconduct of comrades. What is perhaps more surprising is that so little is known about military silence in this area when so much is known about the police code of silence. Clearly, the military code of silence is an issue that merits further research.

---

23 Gary R. Rothwell and J. Norman Baldwin, "Ethical Climates and Contextual Predictors of Whistleblowing," *Review of Public Personnel Administration* 26 (September 2006): 216–44.
24 Patrick O'Hara, *Why Law Enforcement Organizations Fail* (Durham, NC: Carolina Academic Press, 2005), 141.

Whistle-Blowing in the Military

Individuals who report the misconduct of others are called whistle-blowers,[25] but sometimes they are given more disparaging labels like "snitch," "squealer," or "rat," terms that convey the contempt others can have for whistle-blowers. Unfortunately, while there is very little empirical research on whistle-blowing in the military, one study—the only published study of its kind—showed shocking results. Employing a variety of research methods (e.g., surveys, interviews, focus groups), the Mental Health Advisory Team IV (MHAT IV) studied the attitudes of U.S. soldiers and Marines serving in Operation Iraqi Freedom in 2006. Of primary interest is the reluctance of the soldiers and Marines in the study to report battlefield misconduct.[26] Shown in figure 4, the results indicate that only 55 percent of the soldiers and 40 percent of the Marines would report a unit member for injuring or killing an innocent noncombatant. These results are startling when one considers that injuring and killing noncombatants are violations of the laws of war and contravene the moral codes of the U.S. armed forces.

When asked in the comfort and safety of the MHAT IV research setting, slightly less than half of the respondents said they would not report the wrongdoings of a unit comrade. But what would they actually do if faced with a situation in which a unit comrade had just injured a noncombatant or committed the other infractions listed in figure 4? Would they take the proper professional action of reporting the misconduct or would they keep silent? There is no way of knowing for certain, but one can expect, from

25 An internal whistle-blower is someone who reports a violation through existing (i.e., internal) organizational channels, while an external whistle-blower reports the wrongdoing to authorities outside the organization, like the media, government, or police.

26 Mental Health Advisory Team (MHAT) IV Operation Iraqi Freedom 05-07, Final Report, 17 November 2006, http://i.a.cnn.net/cnn/2007/images/05/04/mhat.iv.report.pdf.

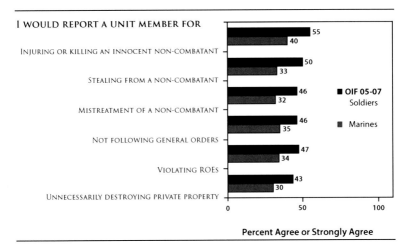

FIGURE 4. **OIF 2005-07 Soldier and Marine Reporting of Battlefield Ethics Violations (from MHAT IV Final Report).**

research mentioned earlier, that many of the respondents would make the choice very quickly, based on intuition, gut feeling, or first impression.[27] When viewed this way, the MHAT IV results may be helpful in predicting what the respondents would actually do, for the answer that came to the minds of almost 50 percent of the respondents was not to report the incident. If not reporting is the default decision for almost half the respondents in a research setting, it is unlikely that larger numbers would report the wrongdoings when presented with the choice in real life.

Another question that emerges from the MHAT IV study is whether these results are unique to U.S. military members. How would the military men and women of other countries respond to the same questions? Would British, French, or Canadian service members be more (or less) inclined than their American counterparts to report the misconduct of their comrades? Again, there is no way of knowing for certain without conducting the same study in these countries, but it is possible that the results would be similar. To the extent that military culture has common characteristics shared across different nations—conservatism, focus

27 Haidt, "The Emotional Dog."

on order, discipline, and control come to mind—and to the extent that these factors influence the attitudes of military members, we could see similar results. However, there may be differences among the military cultures of nations as well, caused by factors such as the intensity of the operations the nation's forces are currently involved in (i.e., military personnel who are fighting wars are likely to be more aggressive than those in peacekeeping operations), the leadership styles promoted within the nation's forces, and the nation's cultural tolerance for violence and diversity.

Perhaps the most important question posed by the results in figure 4 is why the respondents reported as they did. The ethical violations in figure 4 are listed in descending order of harm to others, and the rates of reporting are lower as we move down the list. This suggests that respondents were using some sort of internal scale for deciding whether an offence should be reported or not. If this is the case, where did this scale come from? Are the responses based on individuals' own personal criteria or norms that are promoted within their section, platoon, or company? We see that the lowest rate of reporting, 43 percent of soldiers and 30 percent of Marines, was obtained for unnecessarily destroying property, the least harmful act on the list. Clearly, most of the respondents felt that violations of private property were not serious enough to report.

The MHAT IV research on attitudes about battlefield ethics is a novel and important study that reveals how little we understand about the perceptions and motivations of military personnel on the issue of reporting professional misconduct.

WHAT HAPPENS TO WHISTLE-BLOWERS?

The obvious reluctance to report the violations listed in figure 4 might have less to do with the type of misconduct and more to do with the personal consequences of reporting. A perusal of the research on civilian whistle-blowers shows that the personal cost of reporting the misdeeds of others can be high. One study of 394 civilian internal and external whistle-blowers found that approximately two-thirds of those surveyed "lost their job or were forced to retire (69 percent), received negative job performance evaluations (64 percent), had work more closely monitored by supervisors (68 percent), were criticized or avoided by coworkers (69 percent), [or] were blacklisted from getting another job in their field (64 percent)."[28] In addition to the retaliation they experienced on the job, many whistle-blowers in this study also suffered mental health problems like "severe depression or anxiety (84 percent), feelings of isolation and powerlessness (84 percent), distrust of others (78 percent), declining physical health (69 percent), severe financial decline (66 percent), and problems with family relations (53 percent)."[29] Although there are no equivalent studies in the military sector, one might expect that military whistle-blowers would suffer similar consequences.

Like the respondents in the police research described earlier, some military personnel may keep quiet about misconduct in their units because they are afraid of possible retaliation. This was the case with U.S. Army Specialist Adam Winfield, a soldier in the 5th Stryker Brigade, who was reluctant to go along with his comrades and squad leader in their scheme to kill innocent Afghan noncombatants. Adam's sergeant and a few squad members had already killed several innocent Afghan men and then fabricated evidence to make it appear as though the shootings were lawful killings of insurgents. When word spread through the platoon that he was thinking about reporting the killings to authorities, Winfield was ostracized by his fellow soldiers and threatened by the

---

28 Joyce Rothschild and Terance Miethe, "Whistle-Blower Disclosures and Management Retaliation: The Battle to Control Information about Organization Corruption," *Work and Occupations* 26 (1999): 120.
29 Ibid., 121.

squad leader. The atmosphere in the platoon became so bad that "he was hoping he'd get blown up and just end this mess."[30] In a similar case, Private Justin Watt was worried about the possibility of retaliation when he was thinking about turning in his 101st Airborne Division mates for the rape and murder of a 14-year-old Iraqi girl and her family. He was convinced that the misconduct should be reported, but he "was terrified for his safety. . . . If these dudes would kill a kid, he thought, why wouldn't they kill the soldier who snitched?"[31]

MILITARY CULTURE

There are a number of influences within military culture that may make military personnel reluctant to report the misconduct of their comrades. Many of these influences flow from the military's need for members to cooperate with one another, a key component of unit efficiency. Unfortunately, some of the social processes employed to strengthen unit effectiveness also weaken individual responsibility.

Unit cohesion is a central element of military culture. It is built upon peer bonding that is actively encouraged in the military. A military member learns early in his or her career that success or failure—indeed life or death—depend in large measure on the willingness of comrades to come to his or her aid when needed. As a result, the intense bonding that occurs in military units can lead service members to value peer relations over most other considerations. In fact, the lateral bonds holding peers together can be stronger than the vertical bonds between military personnel and their leaders. This, in turn, can lead to problems, as the path for passing "the professional ethic" to junior personnel is from leader to follower, not peer to peer.

---

30 Luke Mogelson, "A Beast in the Heart of Every Fighting Man," *The New York Times Magazine*, 1 May 2011, 41.
31 Frederick, *Black Hearts*, 322.

When comparing the responses in figure 4, there is an indication that social cohesion may have had some impact on the views of the respondents. It is impossible to know for certain why the U.S. Marines who completed the MHAT IV survey were less inclined than the soldiers to report violations committed by unit comrades, but it may have been due to the influence of unit cohesion. Members of cohesive units stick together more, both in their actions and attitudes, and the Marine Corps has a reputation as a particularly cohesive force.

The decision to report or not may also be influenced to some extent by careerism. Even though military culture emphasizes teamwork, it is individuals who are rewarded with promotions and medals, rarely teams, so a self-serving ethic is also fostered in the military. Most personnel want to advance in rank and they quickly learn that the way to progress is to do one's work well, get along with everyone, and conform to unit norms. Few military personnel would view reporting on comrades, which might jeopardize their colleagues or their units, as the kind of behavior that leads to personal success.

Military life can have a transformative effect on the mind-set of the service member because of social processes like conformity, deindividuation, the bystander effect, and groupthink, which abound in the military environment. Service members who are susceptible to these group pressures are particularly vulnerable when presented with military duty-personal loyalty dilemmas because these processes can erode their personal agency to the point that they readily accept the norms and influences generated by their unit mates.

People are naturally inclined to conform to the actions of others when placed together in groups,[32] and military socialization is

32 Eliot Aronson, *The Social Animal*, 9th ed. (New York: Worth Publishers, 2004), 20.

particularly effective at getting service members to conform to unit norms.[33] The strength of unit cohesion is well illustrated in the following quote from a Canadian soldier:

> You have a bond. You have a bond that is so thick that it is unbelievable! It's the pull, it's the team, the work as a team, the team spirit! I don't think that ever leaves a guy. . . . And that's the whole motivation, that when somebody says we want you to do something, then you'll do it. You'll do it because of the team, for the team and because the team has the same focus.[34]

Through the process of deindividuation, some military members become so absorbed by their small unit that they cast aside their own sense of right and wrong for the norms embraced by the group. Some of the junior soldiers who participated in the abuses at Abu Ghraib, and were later prosecuted for their actions (e.g., Private Lynndie England), undoubtedly fall into this category.[35]

The bystander effect, a concept that accounts for how people can observe harmful events without intervening, helps explain the reluctance of service members to take proper action, particularly in novel situations for which they feel unprepared. Many individuals are unwilling to take action that deviates from the group, and military personnel, like most people, will observe others for cues as to how to act in ambiguous situations. If no one is taking action, they won't either. The bystander effect has been invoked to explain the misconduct of soldiers of the elite Canadian Airborne Regiment during operations in Somalia in 1993. Two paratroopers of the regiment were in a bunker torturing and beating a Somali teenager to death while other members of the regiment, within earshot of the abuse, continued on with their daily activities. It is

---

33 Gwynne Dyer, *War: The New Edition* (Toronto: Vintage Canada, 2005), 31–53.

34 D. Harrison and L. Laliberte, *No Life Like It: Military Wives in Canada* (Toronto: James Lorimer, 1994), 28.

35 Zimbardo, *The Lucifer Effect*, 367.

estimated that up to 17 members of the unit entered the bunker and saw what was happening, but no one tried to stop it.[36]

Finally, "groupthink" emerges when members of a group are more concerned about maintaining harmonious relations within the group than doing the right thing.[37] Groupthink is typically associated with the failures of higher-level leadership teams like the decision makers behind the failed Bay of Pigs operation[38] and the Challenger space shuttle disaster,[39] but it can operate in military squads and platoons as well. Groupthink thrives when people keep their dissenting views to themselves and go along with the prevailing mood of the group. Therefore, any service member who observes his teammates engaged in misconduct and does not speak up is contributing to groupthink.

Taken together, the elements of military culture and social processes outlined above are potent forces that can induce military personnel to go along with their peers, for good or bad. What makes these influences even more powerful is their elusive nature, for many military personnel caught in their grasp are simply unaware that their attitudes and behavior are being shaped by these subtle forces.

IS MILITARY LOYALTY MISPLACED?

One way to view service members' choices to protect comrades over fulfilling their military duty is to characterize these choices

36 George Shorey, "Bystander Non-Intervention and the Somalia Incident," *Canadian Military Journal* 1 (Winter 2000–01), 24, http://www.journal.forces.gc.ca/vo1/no4/index-eng.asp.

37 Irving L. Janis, *Victims of Groupthink* (Boston: Houghton Mifflin, 1972).

38 Roderick M. Kramer, "Revisiting the Bay of Pigs and Vietnam Decisions 25 Years Later: How Well Has the Groupthink Hypothesis Stood the Test of Time?" *Organizational Behavior and Human Decision Processes* 73 (1998): 236–71.

39 James K. Esser and Joanne S. Lindoerfer, "Groupthink and the Space Shuttle Challenger Accident: Toward a Quantitative Case Analysis," *Journal of Behavioral Decision Making* 2 (1989): 167–77.

as misplaced loyalty. But according to an insightful article by Peter Olsthoorn, this type of choice should be expected because it is consistent with the model of loyalty practiced within the military. Olsthoorn argues that military personnel tend to value loyalty to comrades and their military unit over allegiance to the professional ideals espoused in military manuals.[40] Of course, this is an assertion that stands to be empirically validated, but those who have observed military affairs up close would likely endorse Olsthoorn's hypothesis. It certainly explains behavior we see in some theaters of operation, such as the tendency of Western military personnel to place greater emphasis on the safety of their comrades (i.e., force protection) than the safety of indigenous noncombatants, even though the military profession espouses legal and ethical protections for noncombatants. This view of the military model of loyalty is apparent in an article by Thomas Smith showing how a unit of American soldiers interpreted their rules of engagement during checkpoint operations in order to maximize force protection at the expense of noncombatant immunity.[41]

It should come as no surprise that service members have more regard for their comrades than for outsiders. Both in the general population and in social science circles, it is well known that our self-esteem and sense of well-being are related to the quality of relationships we have with others we value.[42] People naturally strive to get along with others and fear of social rejection motivates us to cooperate with our colleagues, friends, and other people who are important to us. This behavior has evolutionary origins, for it served our ancestors well in the past when the group's survival depended on the collective effort of comrades. Because humans

---

40 Peter Olsthoorn, "Loyalty and Professionalization in the Military," in *New Wars and New Soldiers: Military Ethics in the Contemporary World*, ed. P. Tripodi and J. Wolfendale (Farnham, UK: Ashgate Publishing Ltd, 2011), 262.
41 Thomas W. Smith, "Protecting Civilians . . . or Soldiers? Humanitarian Law and the Economy of Risk in Iraq," *International Studies Perspectives* 9 (2008): 154.
42 Aronson, *The Social Animal*, see the chapter on conformity.

tend to favor the members of their own group over nonmembers, it is an easy transition for military recruits to cooperate with unit mates, and military personnel soon learn to value the lives of their comrades over the lives of others, even though (as pointed out by Olsthoorn) ethicists like Michael Walzer maintain that all people have equal worth.[43] The result is that unit loyalty often trumps professional ideals in military matters.

Group work has been hardwired into us over the millennia because of its social utility, but it now undermines our desire to be ethical toward outsiders. How do we then overcome our natural inclination to treat insiders better than outsiders? A possible remedy is to expand our notion of group identity to a broader, more inclusive group—call it humanity—but this is difficult to do, particularly when members of the out-group are shooting at us or supporting those who are shooting. We will return to this topic later.

RESEARCH

So far we have seen that (1) there are some cases of military personnel not reporting the misconduct of their comrades; (2) the issue is a major problem in police culture; (3) civilians who report the misdeeds of their coworkers usually suffer harmful consequences; and (4) some aspects of military culture likely contribute to nonreporting. That said, the analysis in this chapter has been speculative, based on my personal impressions—formed over 33 years of military service—one empirical study in the military, and findings from police research whose results may not apply entirely to the military environment. While there are indications that military personnel turn a blind eye to the misconduct of comrades on occasion, it is not clear how widespread this is, so we need to investigate further to see if there is indeed a problem before seeking potential solutions. Any actions the military institution takes on this issue should be based on evidence, not opinion, so I

43 Michael Walzer, *Just and Unjust Wars* (New York: Basic Books, 1992), 156–58.

suggest a program of empirical research that focuses on two levels of analysis: military members (e.g., attributes and perceptions) and the characteristics of the environments in which they work.

At the individual level, the research should investigate the extent to which nonreporting occurs in order to determine the extent of the problem, and, if there is a problem, to determine whether the issue is one of understanding (not knowing what should be reported or how to report) or will (lacking the motivation to report misconduct). This line of inquiry should be conducted with military personnel of different ranks from both the enlisted and officer corps. The research could employ surveys, interviews, focus groups, or experiments, but the results would be more conclusive if several of these methods were used. Here are some questions that could guide this research:

- Have military personnel observed violations (of the law and military codes of conduct) in the past? Have they reported these violations? Why or why not? What were the consequences for those who reported? Do those who have reported (and those who did not) have any regrets about their decision?

- How do reporters of wrongdoing differ from nonreporters? What are the personal characteristics of those who report and those who do not (i.e., age, rank, education, training, etc.)?

- Do military personnel know which misconduct should be reported? For example, should x, y, z violations of your nation's military code be reported up the chain of command? What types of misconduct do they think should be reported? Which misdeeds should not be reported? Why do some types of misconduct warrant reporting and others do not?

- How likely are military personnel to report misconduct?

- Do military personnel know how to report (i.e., who to report to, when, and how)?

- Why would military personnel want (or not want) to report misconduct? What are the factors and influences that might lead them to report? Are there circumstances that might alter what or when they report? Are military personnel not reporting the violations of others because of fear of retaliation? Are other motivations involved?

- How do military personnel think their unit mates would view someone who reported misconduct? What actions might other military personnel take against someone who reported misconduct?

- Are military personnel more vulnerable to the bystander effect and groupthink than others?

- To what extent do unit characteristics like cohesion, climate, leadership, and such play a role? For example, (A) are military personnel in cohesive units more or less inclined to report misconduct?; (B) how does the ethical climate of a military unit affect whether its members will report misconduct?; (C) how do the impressions of military personnel about their unit leaders affect whether unit members are disposed to report misconduct?; and (D) how do other unit characteristics (unit type, size, mission, etc.) relate to a member's likelihood to report misconduct?

- Is the code of silence encouraged by some aspects of military culture? What characteristics of military culture support nonreporting? What characteristics of military culture encourage reporting?

This list of research questions is not exhaustive. Other questions will surface as researchers think more about the issues and still others will emerge when researchers begin discussing the subject with military members (e.g., in focus groups and/or interview sessions).

At the environmental level, research should explore the impact that military units, training schools, and mission-specific factors have on the inclination of personnel to report misconduct. Studies should be conducted in different types of units (combat, support, administrative) to see what effect, if any, unit type has as well. Given the research evidence showing that the seeds of the police code of silence are planted in the academy, studies should also be conducted with candidates on recruit and occupation training courses to determine if attitudes about reporting (or not) are formed early in one's exposure to military culture. It is also possible that attitudes about reporting may vary according to the type of operations service members are engaged in, therefore the effect of mission-specific factors (e.g., combat intensity, stressors, etc.) should be explored as well.

INTERVENTIONS

A research program based on questions like those listed above will show if there is actually a problem within the military of personnel not reporting the misconduct of comrades. If a problem is identified at the individual level, the results will show if it is due to lack of knowledge (what to do and how to do it) or insufficient motivation (commitment to taking proper action). This information will help in the development of potential interventions. For example, gaps in knowledge can be addressed with further training on regulations and procedures, as well as training on how to analyze a duty-loyalty dilemma to select the correct course of action. Gaps in motivation can be treated with more emphasis on professional identity and obligations, as well as training on the social and situational influences that can prevent people from taking proper action. Similarly, if a problem is identified at the environmental level, special training and policies can be designed for those types of units that need it. While the research outlined above is critical for determining what needs to be done, we can be confident that instruction in three areas—professionalism, psychosocial influences, and military loyalty—will be helpful.

Professionalism

Police research found that senior officers in supervisory roles were more likely to report misconduct, probably because it was part of their role and therefore expected of them. Enhancing the professional identity of junior military personnel could have a similar effect. Impressing on service members that they have a responsibility to intervene when their colleagues misbehave could make them less inclined to protect any transgressing comrades. While this responsibility departs from the traditional view of junior personnel, which has them attending more to following orders than making what are essentially leadership decisions, many of us now subscribe to General Krulak's view of the "strategic corporal," which recognizes that junior personnel have a broader role than that of passive follower.[44] In order to become strategic corporals however, service members need training that emphasizes their professional obligations, and, as military manuals insist, the training must be realistic so as to prepare them for challenges they will actually face on the job.[45] The training should convey what it means to be a professional military member in the nation's forces, stress national values, and highlight international law that is relevant to the unit's current mission. The utility of such training is illustrated in a recent evaluation of ethics training conducted

44 Charles Krulak, "The Strategic Corporal: Leadership in the Three Block War," *Marines Magazine*, January 1999, http://www.au.af.mil/au/awc/awcgate/usmc/strategic_corporal.htm.

45 Paul Robinson, "The Fall of the Warrior King: Situational Ethics in Iraq," in *Ethics Education for Irregular Warfare*, ed. Don Carrick, James Connelly, and Paul Robinson (Farnham, UK: Ashgate, 2009), 75–86.

within a U.S. Army brigade in the aftermath of the MHAT IV research mentioned earlier. Built around realistic battlefield scenarios and conducted by unit leaders, the training began with senior commanders leading their mid-level commanders in ethics discussions, who then did the same with their junior commanders, who in turn, guided their soldiers in discussions of the scenarios. Based on the responses of brigade members to pre- and posttraining surveys, the authors of the study noted that the training led to lower rates of unethical conduct and increased willingness on the part of soldiers to report the misconduct of fellow soldiers.[46] This is an important study that illustrates how effective ethics training can be when it is conducted by a unit's chain of command. In fact, training of this sort that highlights a unit's ethical standards is an excellent opportunity for leaders to establish their professional and moral authority.

Psychosocial Influences

We saw in the research presented earlier that police officers are often reluctant to report the misconduct of colleagues because of cultural norms and peer pressure. These influences are difficult to resist, but unit leaders can help their personnel muster the moral motivation to report comrade misconduct through a mix of stiff penalties for not reporting, increased emphasis on the responsibility to report, and instruction on how social influences can pressure individuals to make unprofessional choices. Behavior that is illegal, unprofessional, or immoral does not happen by chance. It usually follows from earlier actions that weaken the individual's resolve to withstand harmful influences. Accordingly, leaders must be constantly reminded that they can shape the climates of the units their subordinates work in. Leaders can be taught about

---

46 Cristopher Warner, et al., "Effectiveness of Battlefield-Ethics Training During Combat Deployment: A Programme Assessment," *The Lancet* 378 (2011): 915–24.

the kinds of situational forces that encourage misconduct and can learn how to mitigate these corrosive influences.

Military Loyalty

One of the reasons for not reporting misconduct might be that the military places more value on group cohesion and unit loyalty than commitment to professional values. To the extent that this is an accurate assessment (another research question), Nicholas Rescher's model of military obligations can provide some guidance on how the military community could expand its understanding of professional loyalty.[47] The model consists of five broad targets of military loyalty: the chain of command, the service, the nation, civilization, and humanity at large. Most military personnel already grasp the concepts of loyalty to chain of command, service, and nation, as they are well established in the military model of institutional loyalty. However, the obligations to civilization and humanity might be novel ideas for some and would therefore require quality training to ensure that these concepts are accepted by junior personnel. Training on military loyalty could be incorporated into the professionalism training mentioned above, and, like professionalism, would be most effective if delivered by credible leaders from within the unit.

CONCLUSION

Although the dilemma of choosing between protecting a comrade who has committed professional misconduct and reporting the violation as required by one's military duty is not a moral dilemma, it is a very difficult decision, and one that some military personnel are not able to make correctly. As we saw with the

---

47 Nicholas Rescher, "In the Line of Duty: The Complexity of Military Obligation," in *The Leader's Imperative: Ethics, Integrity, and Responsibility*, ed. Carl Ficarrotta (West Lafayette, IN: Purdue University Press, 2001), 245.

MHAT IV study, many military personnel are reluctant to report the violations of their unit mates. Research also shows that the problem of not reporting collegial misconduct is widespread in the police community. It endures at the individual level because police officers are afraid of reprisals and at the institutional level because police culture encourages the code of silence. Similar influences may operate in the military environment, but we will not definitively know unless studies like those conducted in the police community are replicated in military units. This is an important line of research that will tell the military institution if it needs to act and how.

LEADERSHIP AND THE ETHICS OF DISSENT
REFLECTIONS FROM THE HOLOCAUST
PAOLO TRIPODI

The Nazi extermination of European Jews is the most extreme in-stance of abhorrent immoral acts carried out by thousands of people in the name of obedience. Yet in lesser degree this type of thing is constantly recurring: ordinary citizens are ordered to destroy other people, and they do so because they consider it their duty to obey orders. Thus obedience to authority, long praised as a virtue, takes on a new aspect when it serves a malevolent cause; far from appear-ing as a virtue, it is transformed into a heinous sin. Or is it?

-Stanley Milgram[1]

The victims were shot by the firing squad with carbines, mostly by shots in the back of the head, from a distance of one metre on my command. . . . Meanwhile Rottenführer Abraham shot the children with a pistol. . . . The way Abraham killed the children was brutal. He got hold of some of the children by the hair, lifted them up from the ground, shot them through the back of their heads and then threw them in the grave. After a while I just could not watch this any more and I told him to stop. What I meant was he should not lift the children up by the hair, he should kill them in a more decent way.

-SS-Mann Ernst Göbel[2]

---

1 Stanley Milgram, *Obedience to Authority: An Experimental View* (New York: Perennial Classics, 2004), 2.
2 SS-Mann Ernst Göbel quoted in Ernst Klee, Willi Dressen, Volker Riess, eds., *The Good Old Days* (Old Saybrook, CT: Konecky & Konecky, 1991), 197.

There is a strong and distinctive sickening reaction generated by the description of such atrocities and by the many photos[3] of the slaughters perpetrated during the days of the Holocaust. The piles of extremely malnourished bodies and the ashes of human beings who were gassed and incinerated by the thousands make us wonder how this could have happened.

There were many photos taken before these slaughters took place; these photos often depicted women getting undressed in front of a crowd of German SS or army officers and soldiers, just seconds before being killed. On the way to their deaths, in the final moments of their lives, they had to suffer the humiliation of being stripped of their clothes and dignity. In some photos, there are children who are very likely five or six years old, their hands up walking toward the execution place. In the background, a few Germans in uniform watched to make sure that even these young children would be part of the "final solution."

Another set of photographs gives us the same intense sickening feeling, yet the presence of individuals in uniform, the perpetrators or the bystanders, generate a different set of questions: Why did they do it? How could they do what they did? We struggle to make sense of how people who look like us managed to lose their sense of humanity and kill their next door neighbor's children, wife, mother, and then the neighbor himself. According to Michael Geyer,

> These soldiers are people quite unlike anything movies, television, and quite a few books would like to make us believe. They look in uniform much like what they would become in postwar life—your average Fritz, Franz, or Otto. They look perfectly normal but committed extraordinary atrocities. We would

3 These photos can be seen on permanent exhibition at the U.S. Holocaust Memorial Museum (USHMM) in Washington, DC. They can also be viewed on the USHMM website at http://www.ushmm.org/museum/exhibit/exhibit/.

not recognize them for what they did—were it not for the photographs that depict what they did, but did not see for themselves, until years later.[4]

How could these individuals simply "obey" and pull the trigger many times? At what point did they develop a mental callus that allowed them to view what they did as nothing more than a necessary "job"?

The list of questions that are generated by reflecting upon the Holocaust is endless. In this chapter, I will address two: first, should those German officers, NCOs, and soldiers have disobeyed the orders they were given? Clearly, yes. The second question, however, is significantly more complicated and deserves much attention: could they have disobeyed? In this paper, I will deal with this difficult question and try to provide some tentative answers.

THE INDIVIDUAL'S INABILITY TO COMPLY WITH INHUMANE ORDERS: SUBJECTIVE DISOBEDIENCE

The killing of defenseless people began in the days that followed the German invasion of Poland. Hundreds of Polish POWs, Jewish people, and civilians were killed in the beginning of September 1939. In the following months and years, the German approach on the Eastern Front was ruthless. Jürgen Förster noted that "the brutalization of the soldiers began in Poland; the barbarization of warfare itself would begin in Soviet territory."[5] German soldiers did not hesitate to crush resistance or any form of opposition. The search and elimination of Jewish civilians was well organized and, regretfully, very effective.

---

4 Michael Geyer, introduction to *The German Army and Genocide: Crimes Against War Prisoners, Jews, and Other Civilians in the East, 1939-1944*, ed. Hamburg Institute for Social Research, (New York: The New Press, 1999), 9.
5 Jürgen Förster, "Complicity or Entanglement? Wermacht, War, and Holocaust," in *The Holocaust and History*, ed. Michael Berenbaum and Abraham J. Peck (Bloomington: Indiana University Press, 1998), 271.

By the time the Wehrmacht crossed the line of departure to invade the Soviet Union in June 1941, the army had become highly ideological. German soldiers no longer saw themselves as warriors, but rather as the executors of an ideological mission whose end was the elimination of Germany's most deadly enemies. They believed that they were fighting a war for the supremacy and survival of German civilization in which annihilation of the enemy, both combatants and civilians, was a vital and necessary task.[6]

It is difficult, if not impossible, to differentiate with a strong degree of reliability the number of individuals who "happily" executed the orders of killing innocent civilians from those who executed the orders going against their inner nature and soul, and those who actually asked to be exempted from killing. Edward Westermann in *Hitler's Police Battalions* noted that "the range of behavior exhibited by gendarmes in the East extended from direct opposition to the conduct of atrocity, to dutiful obedience, and into the realm of enthusiastic, if not sadistic, support for the most brutal treatment of the local population."[7]

Oswald Rufeisen, a Jew who became an interpreter with a German police unit in Mir, Poland, which today is Belarus, had the unwelcome "opportunity" to get to know the members of the unit quite well. Rufeisen remembers that the second in command, Karl Schultz, seemed "a beast in the form of a man." Nechama Tec, a biographer of Oswald Rufeisen wrote, "Schultz, a brutal man, a sadist, took great pleasure in torturing people in general, and Jews in particular. . . . [W]hen faced with two prospective victims, a mother and child, he would kill the child in front of the mother and only after a day or two execute the mother."[8]

---

6 For an insightful portrayal of the German soldiers' perception of the war in the East see Stephen Fritz, *Frontsoldaten* (Lexington: The University Press of Kentucky, 1995).

7 Edward B. Westermann, *Hitler's Police Battalions: Enforcing Racial War in the East* (Lawrence: University Press of Kansas, 2005), 210.

8 Nechama Tec, *In the Lion's Den: The Life of Oswald Rufeisen* (Oxford: Oxford University Press, 1990), 103.

Yet not all Germans "enjoyed" such a gruesome task. In *Ordinary Men*, Christopher Browning offered an enlightening portrayal of how the policemen of Reserve Battalion 101 dealt with the killing of hundreds of people in a small Polish city. Several of these men's stories provide an opportunity to reflect about the terrible nature of what they were ordered to do. Many among them, however, tried to execute the orders even if they did not know whether they would be able to deal with the killing of harmless civilians in cold blood. The following are a few cases of policemen who struggled to perform such a task.

Georg Kageler realized what he was participating in after he assisted with the first round of killing. He then learned that many of the victims were from Kassell, Germany, the same place he was from. He labeled what was happening as "repugnant" and asked his platoon leader to be released, after which he was assigned to guard the marketplace.[9] August Zorn's first victim was an old man who could not keep pace with the rest of the group headed to the execution site. Zorn and the old man arrived when everybody else had been killed already. The scene must have been devastating for both men, but surely for the old man who threw himself on the ground and refused to move. Zorn shot him on the spot "because I was already very upset from the cruel treatment of the Jews during the clearing of the town . . . I shot too high." For Zorn the view of the damage caused by the bullet on the body of the old man was awful and he asked to be relieved of such a "duty."[10]

Franz Kastenbaum hesitated and then missed the fourth man he was supposed to kill. It was so repugnant to him that it became impossible to shoot accurately. He then ran into the woods, vomited, and sat for a while against a tree.[11] Even the battalion commander, Major Wilhelm Trapp, a member of the Nazi party since 1932, was

---

9 Christopher R. Browning, *Ordinary Men: Reserve Police Battalion 101 and the Final Solution in Poland* (New York: Harper Collins, 1992), 67.
10 Ibid., 66–67.
11 Ibid., 67–68.

appalled by the terrible orders he and his battalion were asked to execute. Yet, with tears in his eyes, he just did what he was told to do.

In "Military Violence and the National Socialist Consensus: The Wehrmacht in Greece, 1941–44," Mark Mazower provides an interesting insight on how some of the men of the 98th Regiment of the First Mountain Division (Gebirgs Division) reacted when they had to execute large numbers of civilians in the Greek village of Komino.[12] In the middle of August 1943, soldiers from the 98th Regiment surrounded and assaulted the small village of Komino. In the attack against the defenseless population, they killed more than 300 people, about 50 percent of the entire village population. In executing the carnage, they made no distinction of age or sex. The division's reputation was that of a unit that would engage in any operation that would help them defeat the enemy, even if this meant killing civilians and enemy POWs. Even so, several soldiers among those who participated in the punishing operation against the Greek civilians resented what they were asked to do.

Mark Mazower provides the reaction from a few soldiers. Karl D. recalls that the soldiers of 12th Company had much discussion about what they had done. According to Karl D., "few thought it [was] right." He was so sickened by the massacre that it took him "weeks to recover my [his] peace of mind." August S. remembers that after the shooting was over, several soldiers were "very depressed."[13] Otto G. was disgusted and determined to do something about it, yet "in the end we lacked the courage to desert. Not a single man deserted."[14] Mazower's research provides strong ev-

---

12 Mark Mazower, "Military Violence and the National Socialist Consensus: The Wehrmacht in Greece, 1941–44" in *War of Extermination: The German Military in World War II, 1941-1944*, ed. Hannes Heer, Klaus Naumann (New York: Berghahn Books, 2004).
13 Ibid.,147.
14 Ibid.,148.

idence of a difficult situation for many of the soldiers in the division. The official report filed by the division chaplain stated, "The mass killing of women and children during operations against the bands is producing a difficult inner burden on the conscience of many men."[15]

Both Browning and Mazower provide evidence that, in the execution of cold and calculated massacres, there were individuals who found it dreadful to murder another human being. Such inhumane behavior made a serious emotional and psychological impact on several of those policemen and soldiers. Yet, despite this inner resistance, they did as they were ordered. Several were able to step back from the very beginning and were assigned to different duties, while others could not continue and were allowed to stop. Yet another group executed the horrible task.

Indeed, although those who objected to the execution of killing tasks technically disobeyed, they did so on the ground that the task was something they found difficult to perform. They provided several reasons for their refusal. Some did not want to kill women and children, as they had families back home; others felt very close to those civilians who came from the same place they came from; still others simply could not stand the gruesome consequences of killing at a close range, the stain of their victims' blood mixed with grey brain matter on their uniforms, the screams of women and children, the implorations, the begging for mercy. They did not disobey on the ground that killing harmless civilians or enemy POWs was morally wrong and therefore unacceptable.

In order to understand individual soldiers' resistance to such terrible orders, it's necessary to explain two forms of disobedience. The first form of disobedience, *subjective*, takes place when an individual is unable to perform or execute the orders. The second form, *objective*, occurs when an individual is in irreconcilable

---

15 Ibid.

disagreement with the order. In the first case, had the individual been able to execute the order, he would probably have done so; in the second case, the individual's decision is firm and probably unchangeable.

The two forms of disobedience can occasionally happen simultaneously; however, it should be noted that there is a significant difference between disobeying in the former case and in the latter. Indeed, it is questionable whether the action of those who objected to the orders of killing in the first case is real disobedience. They knew that the worst consequence they might have suffered was to be ridiculed by their colleagues; identified by their officers as wimps; or in the worst case, they might have compromised their careers. The testimony of an SS sergeant is helpful to illustrate this point. He said, "The reason I did not say to Leideritz [probably the sergeant's immediate superior] that I could not take part in these things was that I was afraid that Leideritz and others would think that I was a coward. I was worried that I would be affected adversely in some way in the future." He did not want to give the impression of being too weak, thus he explained, "I carried out the orders not because I was afraid I would be punished by death if I didn't. I knew of no case and still know of no case today where one of us was sentenced to death because he did not want to take part in the execution of Jews. . . . I thought that I ought not to say anything to Leideritz because I did not want to be seen in a bad light, and I thought that if I asked him to release me from having to take part in the executions, it would be over for me as far as he was concerned and my chances of promotion would be spoilt or I would not be promoted at all."[16]

Indeed, Oswald Rufeisen stressed the fact that there was a significant difference within the German small police unit regarding participation in actions against the Jews and the partisans:

---

16 SS-Scharführer Leideritz, quoted in Klee, *The Good Old Days*, 78.

A select few Germans, three out of thirteen, consistently abstained from becoming a part of all anti-Jewish expeditions. Conspicuously absent from such anti-Jewish expeditions was Meister Hein [the commander of the unit]. Neither he nor the other two were reprimanded for it. No one seemed to bother them. No one talked about their absence. It was as if they had a right to abstain.[17]

Helmut Langerbein in *Hitler's Death Squads* provided an insightful account of how some of the executors dealt with the experience of killing and how some resisted or objected to the task of killing women and children. Harm Willms Harms, a police first lieutenant, is an interesting case. The police lieutenant had obeyed some execution orders and resisted carrying out others. Langerbein noted that "his case is important because it shows that his refusal to kill Jews had no negative repercussions. Harms was not an SS member, but he had already participated in the early massacres of Einsatzkommando Tilsit without hesitation." When he was ordered to supervise the execution of women and children, however, Harms told his commander that he could not do it. Despite a short-lived negative reaction from his direct superior, he was allowed not to take part in the execution. Although Helmut Langerbein used Harms's case to show that a refusal to obey orders to execute Jews had no consequences, he did not provide the right context and consideration. Harms did not refuse to kill Jews, indeed, he had already participated in the execution of Jews, but he could not bring himself to kill women and children. His was not an act of real disobedience; he did not have the "ability" to perform such a terrible task. His commander, after threatening him, told him, "That's all right then. . . . You do not have to do that. You have a wife and children."[18]

17 Tec, *In the Lion's Den*, 102.
18 Helmut Langerbein, *Hitler's Death Squads: The Logic of Mass Murder* (College Station: Texas A&M University Press, 2003), 169.

The case of another police officer, Captain Hans Karl Schumacher, provides even better insight into what motivated certain individuals to obey terrible orders rather than overtly defy them. Captain Schumacher was a policeman who did not disguise his dislike for the Nazis' policies. He was so determined in his objection to the Nazi regime that he had refused to join the Gestapo. In 1941, however, Schumacher was sent to Kiev to join Einsatzgruppe C. His immediate task in Ukraine was to organize a local detective force. However, in a few months, the German forces began a tough antipartisan campaign, which was clearly a pretext to exterminate the local Jewish community. Captain Schumacher, after an initial resistance to the methods used to kill, just accepted what "needed to be done." He overcame his initial resistance and, although he continued to express his disagreement and repugnance for the implementation of the Nazi extermination policy, he participated in eight and maybe ten massacres. Langerbein explained that Captain Schumacher was a firm believer of the German officer's honor code "to obey one's superiors and to set an example for one's men." Therefore, to set an example to his men, he personally killed some of the victims.[19]

It is realistic to consider that those who could not "stomach" killing women and children in cold blood did not really disobey, as long as they knew that such an option was available to them. In a few cases, the most they had to do was to assess how much their refusal might have affected their professional future and the relations with their comrades. True disobedience is the act of challenging the ordering authority on the ground that what was asked was objectively wrong and morally unacceptable rather than subjectively difficult to perform. Clearly this course of action carried consequences much harsher than being assigned to different duties or compromising a possible promotion. Indeed, these acts of true disobedience were met in a rather different way.

---

19 Ibid., 172.

Some reflections offered by Stanley Milgram following his famous obedience experiment are important to this argument.[20] Milgram provided evidence that a significant number of individuals are ready to inflict harm on another individual if they are ordered to do so by someone in a position of authority. In his experiment Milgram explored many key areas, an important one for the objective of this chapter being moral responsibility. He noted that when someone in a position of authority issues an order, he might relieve those who will execute the order of their moral responsibility. Soldiers who have been asked to perform a certain action might see themselves as instruments of a commander's will. From their point of view, they share no moral responsibility for the action.

Thus, special attention should be placed on understanding moral confusion and its impact on soldiers. When a commander issues an order, soldiers' initial moral concern might be about how well they perform that action and live up to the commander's expectation. The confusion occurs when the order they have been asked to execute is illegal and/or immoral. Wolfram Wette noted that "in wartime conditions that imposed both physical and psychological burdens on soldiers, the military's demand of absolute, unquestioning obedience—even to criminal orders—caused many of them to lose most of their sense of individual responsibility and personal guilt. Soldiers' sense of humanity and justice became dramatically deformed."[21]

This sense of "unquestioning obedience" becomes even stronger when the execution of immoral, criminal orders is perceived as a necessary part of the mission that must be accomplished both at the tactical and strategic level. A strong element of the military profession is to accomplish the assigned mission. Officers and

---

20 See section in this essay titled "Could They Have Disobeyed?" for an explanation of Milgram's experiment.
21 Wolfram Wette, *The Wehrmacht: History, Myth, Reality* (Harvard: Harvard University Press, 2006), 158–59.

their soldiers might be inclined to accept some bending of their sense of morality in order to accomplish their mission. They will be even more determined if they perceive that what they have been asked to do, although immoral, is in line with the country's overall mission. Troops might develop a common sense of what is "right," which in the case of the German military was completely immoral.

APPLIED ETHICS: OBJECTIVE DISOBEDIENCE.

The form of disobedience that is probably more powerful is objective disobedience. Rather than stressing their inability to perform a terrible task, a few individuals made a point about how wrong it was to execute such inhumane and immoral orders. They actually challenged the authorities in charge and responsible for issuing the orders. For those individuals, it was not only subjectively impossible to kill another human being in cold blood, but also ethically unacceptable that the institution they belonged to was engaged in such an immoral project. Their refusal to execute the orders had a greater impact as their standing could potentially become an obstacle to the implementation of the Nazi policy. It appears that the number of individuals who took this course of action was even smaller than those who overall opposed, for personal reasons, the killing of defenseless individuals. In addition, I differ from other scholars in this field as I believe that those who objectively refused to execute the orders on legal and/or moral grounds indeed paid, or would have paid, a much higher price for their decision, much different from the first group, as happened in the cases of Feldwebell Schmid and Lieutenant Battell. Their motivations and beliefs—rather than a nauseating reaction to blood, the fragment of skull or human brain, or the sight of executed or about to be executed children—gave them enough strength to defy the orders.

It is important to recognize such a difference because it allows us to understand how distorted military training and discipline can compromise an individual's moral autonomy and thus create the conditions for soldiers to execute inhumane orders. The motivations of those who did not comply because of their inability to execute the task were not at odds with what was happening. They understood these actions as elements of an overall vision in an ideological war against a deadly enemy. Those soldiers and officers had lost the ability to morally assess, in an autonomous way, right from wrong. Under normal conditions, a good number of them might have argued that killing women and children was wrong. However, when immersed in a strong military system, run by "corrupt" leadership with strong discipline, those same individuals probably pulled the trigger many times. Indeed, several individuals among them found a perverted pleasure in killing defenseless people. Yet, here, the purpose of this paper is to understand the motivation of those who might have defied such an order, rather than the motivation of those who received a sick satisfaction by committing these evil deeds. For those who had the potential to defy the orders and did not, it is important to understand moral autonomy and how it works in military institutions.

The coming pages will focus on individuals who defied the orders and decided to disobey in an objective way. This analysis explores their motivations and the consequences of their actions.

Sergeant Anton Schmid

Anton Schmid was drafted in the Wehrmacht, although he was an Austrian who was born in 1900 in Vienna. In 1941, Schmid's unit was stationed in Vilna (today the capital of Lithuania), supporting the frontline troops. Vilna was also the place that hosted one of the oldest Jewish communities in the region. Sergeant Schmid had many chances to interact with members of the Jewish community, as several among them worked in the same military camp. The Austrian sergeant not only became aware of the terrible conditions Jewish people were subjected to in Vilna, he also learned of mass killings and was severely shocked. He learned of the systematic killing of thousands of civilians in the beginning of 1941 in Ponar, just a few miles from Vilna. For Schmid, it was unacceptable to remain a passive bystander, and he took action to protect the lives of innocent and defenseless people. He secretly hosted several of them inside the building he was responsible for, hid many in military vehicles and took them out of Vilna, and distributed "yellow" permits that identified specialized workers deemed essential to the Wehrmacht. From October 1941 to February 1942, when he was arrested for his activity, he probably saved more than 300 people. It took the war court just a few days of trial to decide that Sergeant Schmid should receive the toughest punishment.

In April 1942, Schmid was executed. In one of the last letters to his wife, he wrote of the terrible atrocities he had learned about—the killing of thousands of people and the brutalization of women and children. He simply and powerfully wrote, "I acted as a human being." Wolfram Wette noted that Schmid "was able to preserve the humane orientation he had acquired before his induction into the Wehrmacht and act on it."[22]

Lieutenant Albert Battel

German army Lieutenant Albert Battel, a veteran of the First World War, was a mature man in his fifties and an old-time member of the Nazi party; he had joined the party as early as 1933. Yet the more he learned about the anti-Jewish Nazi policy and the killing of Jews, the more he became committed to protect as many as possible. In 1942, he successfully convinced his commander, Major Liedtke, to instruct the local commanders to protect all the Jews working for the German army in their areas of responsibility. He

---

22 Wette, *The Wehrmacht*, 290.

clearly took an attitude that seriously defied the SS in the region. He went so far as to stop the SS from rounding up the Jews in the Przemysl ghetto in July 1942. The unusual confrontation between the SS and the army units stopped just short of escalating to a fire-fight. In the words of Modecai Paldiel, "It was an event unheard of [sic] in the annals of the Third Reich. A German military officer had dared to raise his weapon against the SS to prevent them from carrying out a fateful deportation action against Jews."[23] Yet, after the SS leaders complained to the army headquarters about the behavior of the two officers, the SS were able to force both Battel and Liedtke to give up the Jews they were protecting.

In the Gestapo report redacted after the incident, Lieutenant Battel was identified as the instigator of the army action against the SS in Przemysl. In October, the commander in chief of the SS, Heinrich Himmler, wrote to Martin Bormann, Hitler's chief adjutant, that he intended to arrest Battel after the end of the war. In the meantime, Battel was reprimanded and sent to a unit on the Eastern Front. To be sent to a frontline unit in Russia at the end of 1942 meant a significant increase in the possibility of being killed compared to serving in the military administrative task Battel had behind the front. David Kitterman wrote that Battel, despite his action, "suffered no serious consequences."[24] Arguably, this only happened because he was lucky enough not to be killed on the Russian Front and because the end of the war did not allow Himmler to arrest Battel. Had Battel been arrested, Himmler likely would have done everything he could to have executed the lieutenant.

Lieutenant Klaus Hornig

Klaus Hornig began his career as a police officer in the early 1930s. It was only because of a series of events over which he had

---

23 Mordecai Paldiel, *The Righteous Among the Nations* (New York: Harper Collins, 2007), 21.
24 David Kitterman, "Those Who Said 'No!': Germans Who Refused to Execute Civilians during World War II," *German Studies Review* 11 (1988): 243.

no control that he received command of a police company deployed in the Eastern Front. In the portrayal of Horning provided by David Kitterman, the German officer came across as a determined man, committed to the values he inherited from his family and critical of several initiatives adopted by the Nazi regime years before the beginning of the conflict. Hornig was an officer who clearly distinguished right from wrong and, as a lawyer, had a good knowledge of the German Military Code of Justice, specifically of Paragraph 47, which clearly established that a subordinate had the "right to refuse an order which he recognized as illegal."

In the weeks after Lieutenant Hornig took command of his company, he was ordered by the battalion commander to eliminate 780 Russian POWs who had been labeled as "commissars."[25] The lieutenant made clear to his commander that he had no intention of executing an order he considered illegal and immoral. He then proceeded to instruct his troops about the German Military Code of Justice and the provisions of Paragraph 47.

None of Hornig's troops participated in the execution. Yet for the Lieutenant it was the beginning of a series of major problems. He had overtly challenged an order and instructed his men on how to avoid executing such orders; it was a case of insubordination. In addition, over the course of the following days he overtly criticized what the SS were doing; his criticisms were categorized as insults. Hornig's life soon became a nightmare. He was put on trial a number of times and jailed in different penitentiaries until he was sent to the infamous concentration camp of Buchenwald where he spent nearly ten months as a "political" prisoner. It was clear that the most senior SS leaders wanted Hornig sentenced to death. It was only the lack of witnesses and evidence that stopped

25 Several weeks before the beginning of Operation Barbarossa, orders and directives were issued to provide the German military with guidelines for the behavior of troops in occupied territories. Among them was the "Guidelines for the Treatment of Political Commissars," better known as the Commissar Order. Troops were instructed to execute immediately any identified Soviet political commissar that might have been captured during military operations.

SS prosecutor Paulmann from sentencing Hornig to capital pun-ishment.[26] Hornig was indeed extremely lucky to survive the hardship of internment, months in a concentration camp, a few deadly marches during which thousands of his fellow inmates were killed, and the SS determination to execute him to set an example.

Captain Paul Grueninger

Perhaps the most significant case of objective disobedience based solely on moral grounds is that of Swiss Border Police Captain Paul Grueninger. In the late 1930s, the flux of Jews leaving Germany and Austria grew dramatically in response to the Nazis' anti-Jewish policies and activities. Switzerland had become a des-tination for many German and Austrian Jews. Yet by August 1938 the Swiss government decided to close the borders; only individ-uals with a valid visa were allowed in country. All border police officers were ordered to comply with the new policy. Clearly, this meant denying access to large numbers of Jews and abandoning them to a terrible fate. All police stations applied the new policy with the exception of the one in St. Gallen. For months, Captain Grueninger helped Jews in St. Gallen find shelter, either some-where in Switzerland or in a hosting country through the Swiss Association of Jewish Refugees. Grueninger was then ordered to stop his activity and to ignore the cry for help coming from the refugees. The implementation of such a policy was in clear conflict with Grueninger's beliefs. He told his family, "I [would] rather break the rules than send these poor, miserable people back to Germany."[27] The Swiss captain not only defied the order, but he also falsified the date of the entry visas on a number of passports.

---

26 David Kitterman, "Refusing to Kill in the Midst of the Holocaust: The Case of Klaus Horning," in *Remembrance, Repentance, Reconciliation,* ed. Douglas Tobler (Lanham, MD: University Press of America, 1998), 107–25.
27 Meir Wagner, Moshe Meisels, Andreas C. Fischer, and Graham Buik, *The Righ-teous of Switzerland: Heroes of the Holocaust* (Hoboken: Ktav Publishing House, 2001), 37.

In April 1939, Grueninger's commitment to save the lives of defenseless people was exposed. The Swiss government took a tough approach: Grueninger was suspended, lost his salary and pension rights, and had to wait for a couple of years to be processed. The sentence he received was only monetary—he had to pay the cost of the investigation and an additional fine of 300 Swiss francs.

The following years and decades became extremely difficult for Grueninger. Without salary, he took whatever job was offered to him. Despite the hardship he had to face for the rest of his life, Grueninger never felt that he was a victim. Rochat and Modigliani stressed that he wished the Swiss authorities would acknowledge what he had done and give him "credit for having upheld one of his country's finest traditions. What is striking about Grueninger's deeds is that he never altered his values to suit the government. The refugees whose lives were in danger took priority over the legitimate orders of the administration."[28]

Indeed, it is difficult to believe that even after the end of the war, the Swiss government refused to rehabilitate Grueninger when it became clear that individuals like him had acted out of humanitarian concerns and deserved great respect. Although his actions were morally sound, Grueninger had committed the sin of disobedience at a time when the government became morally blind to the cry for help. It was in November 1993 when Grueninger was finally rehabilitated by the Swiss government—21 years after he died. The message from the Swiss government was that clear obedience is significantly more important than morality.[29]

COULD THEY HAVE DISOBEYED?

Milgram's obedience experiment is best known in its main

---

28 Francois Rochat and Andre Modigliani, "Captain Paul Grueninger: The Chief of Police Who Saved Jewish Refugees by Refusing to Do His Duties," in *Obedience to Authority: Current Perspectives on the Milgram Paradigm*, ed. Thomas Blass (Mahwah: LEA Associates, 2000), 100.
29 Wagner, *The Righteous of Switzerland*, 32–56.

variant, the one in which the 40 subjects—"the teachers"—were placed in a situation in which they were "ordered" to punish a "learner" who they could not see. They could hear that the learner was suffering because of the electric shock they inflicted on him. In this situation, two-thirds (65 percent) of the teachers inflicted what they believed was a harsh punishment that caused significant harm and possible death to another individual, the learner. Among the other variants of the experiment, Milgram tested how the teacher would behave if the learner was extremely close to him and thus the teacher would be able to hear, but also and more importantly, to experience directly, the consequences of the punishment he inflicted. The outcome of this variant was significantly different. In what Milgram called the "touch-proximity" variation, the outcome was reversed, 70 percent of the teachers defied the order to punish the learner.[30] Through this experiment, Milgram demonstrated that a significant number of individuals are willing to inflict significant pain to another individual, basically because they are ordered to do so. However, many among these individuals would resist if they had visual contact with the victim.

Milgram's findings are extremely interesting; we should bear in mind, however, that in the Yale University laboratory where the initial experiment took place, the conditions under which the teachers were observed did not even come close to the conditions an average German soldier was placed in. It should also be noted that out of 40 individuals assembled from different social backgrounds, living in a free country without any specific training or indoctrination, a disturbing percentage of them decided to comply with what was clearly an immoral order. They were free to do as they wanted—just stand up and walk away—and a few did, but not the majority. Clearly in Milgram's experiment there was not a fear of punishment, as there was not a fear of punishment among those who selectively decided not to kill women and children but went ahead and participated in other types of execu-

---

30 Milgram, *Obedience to Authority,* 36.

tions. Therefore, fear of punishment does not help to explain why thousands of individuals lost their sense of humanity and killed large numbers of defenseless civilians.

Military organizations emphasize the importance of virtues. They need brave, dedicated soldiers who are ready to face the most difficult challenges: kill and being killed. Samuel Huntington in *The Soldier and the State* wrote that "loyalty and obedience are the highest military virtues."[31] Clearly, if these are the highest virtues, disloyalty and disobedience are the greatest vices or sins. Yet it is quite clear that in the case of the use of the military to achieve an evil objective, such as the extermination of a race, disobedience is the highest virtue. The issue for soldiers is the difficulty of defying orders, and this is not because they might fear punishment, since this would be a rather easy but reductive explanation.

Soldiers are placed in a system that compresses, or to use Peter Kilner's definition, "by-passes" moral autonomy.[32] I am not suggesting here that soldiers become automatons. What I am stressing is that the military system, with its training programs and values, tends to stress compliance rather than defiance. In a military operation, soldiers become part of an effort in which their behavior becomes nearly mechanical. Their ability to react to the orders they receive in a quick and efficient manner is fundamental to succeed. However, one wishes that when asked to do something as horrible as the German Armed Forces did during the Holocaust, they would seriously question the orders and actually defy them. That disobedience should not only be subjective, but more importantly, it should be objective. We would expect that soldiers asked to kill women and children would defy the order, not because they do not have the "callus" to execute it but because they disagree with the execution of the order.

---

31 Samuel P. Huntington, *The Soldier and the State: The Theory and Politics of Civil-Military Relations* (Harvard: Harvard University Press, 1981), 73.
32 Peter Kilner, "Military Leaders' Obligation to Justify Killing in War," *Military Review* (2002): 24–31.

Yet the complexity of such a course of action, as rightly noted by Christopher Penny, is that although soldiers have a responsibility to disobey illegal orders, " . . . the lower a soldier's position in a military hierarchy, the less ability he or she will have to effectively question orders."[33] In addition, even if faced with manifest illegitimacy, soldiers will place great trust in their leaders regarding the legality and morality of what they have been asked to do. U.S. Army Lieutenant Colonel Michael L. Smidt provided an excellent explanation of the relationship between leaders and soldiers and the important role leaders play in influencing soldiers' behavior. Smidt stated that

> It is through effective military leadership that a soldier can be influenced to perform acts that transcend the norms of human nature. Only a successful and skilled motivator of troops can inspire a combatant to charge a machine gun position, contrary to the most powerful of human instincts, that of self-preservation, in order to acquire a small and seemingly insignificant piece of turf. Powerful and persuasive leaders are required to build and maintain the degree of commitment necessary to successfully execute an armed conflict.

Even more important for the objective of this paper is what Smidt stressed in relation to leadership and atrocities:

> Just as dynamic military commanders can induce their subordinates to accomplish heroic acts beyond the pale of traditional human limitations, they also, unfortunately, possess the power and means of ordering, encouraging, or acquiescing to acts that are inhumane in the extreme. Through an abuse of legitimate military leadership and authority, a commander may condone or even direct conduct that goes far beyond even the relaxed standard of acceptable violence associated with warfare. Under

---

33 Christopher Penny, "Amoral Automatons: A Moral Critique of Superior Orders as a Defence to War Crimes Charges Before the International Criminal Court," in *The War on Terror: Ethical Considerations*, ed. Daniel Lagacé-Roy and Bernd Horn (Kingston: Canadian Defence Academy Press, 2008), 17.

the direction of persuasive leadership, soldiers have committed acts so atrocious as to exceed any possible rational application of military force. . . . It is to the leader that a young soldier looks for guidance in terms of distinguishing appropriate and inappropriate uses of force during military operations.[34]

The German Armed Forces officers do indeed bear the greatest burden and responsibility for giving up the obligation of dissenting from accomplishing an evil and clearly immoral project. While soldiers' ability to disobey in an objective way was greatly limited, military commanders, and particularly senior leaders, would have been in a much stronger position to oppose Hitler's evil project and the conduct of a war that broke all fundamental rules of decency. Jürgen Förster noted that "the military leaders did not simply comply with Hitler's dogmatic views, they were not mere victims of an all-exonerating principle of obedience. The military leaders, too, believed that the threats of Russia and Bolshevism should be completely eliminated."[35]

Indeed, it is difficult and rare to find cases of senior military leaders who opposed or resisted the execution of Hitler's immoral orders. Michael Walzer in *Just and Unjust Wars* uses General Erwin Rommel's actions as an example of good leadership in war. Believing it was unacceptable to kill POWs, General Erwin Rommel decided to burn Hitler's 1942 commando order, which required German soldiers to kill enemy soldiers found behind German lines, even after these soldiers had surrendered.[36] General Siegfried Westphal, an operations officer in General Rommel's staff, said that they destroyed such orders "as we did not want anything to do with such methods."[37] Other distinguished and charismatic leaders such as

---

34 Michael Smidt, "Yamashita, Medina, and Beyond: Command Responsibility in Contemporary Military Operations," *Military Law Review* 164 (2000): 157–58.
35 Förster, "Complicity or Entanglement?," 273.
36 Michael Walzer, *Just and Unjust Wars: A Moral Argument with Historical Illustrations* (New York: Basic Books, 2006), 38.
37 Siegfried Westphal quoted in Gerald Fleming, *Hitler and the Final Solution* (Berkley: University of California Press, 1984), 38.

Heinz Guderian and Erich von Manstein tried, after the war, to present themselves as reputable leaders who had opposed Hitler's brutal and criminal conduct of the war. The evidence proved that they were also accomplices in allowing the troops under their command to perpetrate the killing of civilians and POWs. Probably the most notable exception was General Johannes Blaskowitz. According to Richard Giziowski, Blaskowitz's biographer, the German army general deeply despised the atrocities conducted by the SS and some army units since the invasion of Poland in 1939.[38]

Yet none of the senior leaders directly defied Hitler's criminal orders, even if this meant compromising the value and reputation of Germany. They gave up the responsibility to disobey in an objective way and to make clear to Hitler and senior Nazi leaders that Germany's values could not be compromised by such an evil project. They would likely have faced a punishment much different from the one Anton Schmid faced.

What should be noted is that those individuals who decided to defy the authorities who ordered the execution of such an evil project had a clear understanding that the values they stood for were much greater than their own lives. They felt they had a moral responsibility, indeed an obligation, to stand for those values, even when everybody else headed in a different direction. These were individuals of strong character who did not allow the situation to overwhelm them. Schimd made a clear case for humanity, while Hornig, a religious man, made a compelling case not only for legality, but also for morality, and more importantly, for Germany itself. Grueninger felt that closing the border to the refugees was a betrayal of what Switzerland stood for, and he did not care if the authorities had taken a different approach. Indeed, he

---

38 Richard Giziowski, *The Enigma of General Blaskowitz* (New York, Hippocrene Books, 1996).

felt that they were taking the country in a direction that negated Swiss history and tradition. Rochat and Modigliani wrote that

> Grueninger believed that the refugees' lives ought to be placed above federal decrees regarding who should or should not be allowed into Switzerland. . . . Grueninger fully believed in Switzerland's long-standing humanitarian tradition—a tradition that called for taking care of people whose lives were in danger . . . . He seemed convinced that the Swiss people felt bad for the refugees—that they felt a sense of responsibility toward them, as they had previously when other persecuted peoples had needed assistance.[39]

In all these cases, these courageous individuals had an extremely clear understanding of what their real duty was and acted accordingly. The case of Grueninger, however, has one additional factor that differs from all others, making it an important case for both practitioners and ethicists. While Schmid, Battel, and Horning defied illegal and immoral orders, Grueninger's orders were absolutely lawful, yet the consequences were immoral. He judged that, under the circumstances, the execution of lawful orders was unacceptable. It is still today an open discussion whether disobedience in a case similar to Grueninger would be praised or punished. A focus on subjective and objective disobedience, rather than on the strict parameters of legality and the abstract boundaries of morality, would pave the way to a more constructive discussion on dissent.

---

39 Rochat and Modigliani, "Captain Paul Grueninger," 106.

ENACTING A CULTURE OF ETHICAL LEADERSHIP

COMMAND AND CONTROL AS UNIFYING MIND

CLYDE CROSWELL AND DAN YAROSLASKI

I envisioned large, sweeping formations; coordinating and synchronizing the battlefield functions to create that "point of penetration," and rapidly exploiting the initiative of that penetration to achieve a decisive maneuver against the armies that threatened the sovereignty of my country. . . . We witnessed in Baghdad that it was no longer adequate as a military force to accept classic military modes of thought.

-Major General Peter W. Chiarelli, U.S. Army[1]

[A] few successful individual leaders in charge aren't enough. They must build a culture of leadership that becomes the identity of the organization rather than just that of top leaders.

-General Anthony C. Zinni, U.S. Marine Corps[2]

The topic of this article is derived from a presentation to the NATO Senior Officer Policy Course at NATO School made by Mr. Tom Randall, Legal Advisor to the NATO Supreme Allied Commander Operations. Thanks also to Mr. Sherrod Lewis Bumgardner for his generous contributions to this article.
1 Major General Peter Chiarelli and Major Patrick Michaelis, "Winning the Peace: The Requirement for Full-Spectrum Operations," *Military Review* (2005): 4.
2 Tony Zinni and Tony Koltz, *Leading the Charge: Leadership Lessons from the Battlefield to the Boardroom* (New York: Palgrave Macmillan, 2009), 58.

"Enaction" is the new paradigm emerging in cognitive science about how the mind works. In the past, "cognition" was defined simply as the processing of information about the world "out there" as if it already existed. In contrast, "enaction" means that the way the human species thinks and behaves is actually cocreating the world in which we live. This chapter introduces this emerging, dynamic, multidisciplinary paradigm.

In living biological and organizational systems,[3] leadership emerges from many levels at the same time: the individual level that includes cells, organs, and organisms; the organizational level that includes teams, military units, and groups; and even at the community, society, and supranational/global levels. Leadership's complexity changes the global and local environment day-by-day, moment-by-moment for each living self. While the term "global" means geographical to some, in the context of enaction, it pertains to a *biologically adaptive, living process through which parts connect to create a meaningful whole*. Global is a unifying principle of self-organizing minds, a comprehensive concept that explains the self-organizing process between global and local mind[4] based on top-down *and* bottom-up dynamic coemergence or reciprocal causality. For the military, this process may be thought of as command and control where the commander (global/top down) provides commands and intent while subordinates (local/bottom up) provide control and specificity in the form of action and feedback. Global mindfulness is a comprehensive concept—a means for imagining how transformation actually works in collective human experience—a primal way of thinking about enacting a culture of ethical leadership and creating a globally unified mind.

3 James Grier Miller and Jessie Miller, "A Living Systems Analysis of Organizational Pathology," *Behavioral Science* 36 (1991), 239–52.
4 Francisco Varela and Natalie Depraz, "Imagining: Embodiment, Phenomenology, and Transformation," in *Buddhism and Science: Breaking New Ground*, ed. B. Alan Wallace (New York: Columbia University Press, 2003), 195–230.

The authors' military service includes operations in Vietnam and the Middle East, and in both enlisted (bottom-up specifics) and field grade officer (top-down constraint) service. Drawing upon our enlisted experience, we quickly agreed that blind obedience from the bottom up is seldom ideal. Further, we now contend that a less mindful acceptance of the classic strategic mind, which focuses on simply attaining goals with little thought to the global impact, often generates complexities and emerging problems heretofore unimaginable. Focusing simply on ends without a global, unifying approach creates extremely complex problems, i.e., the My Lai massacre during the Vietnam War, Abu Ghraib during Operation Iraqi Freedom, and countless other instances of civilian atrocities.[5]

This chapter is firmly grounded on two vital documents that serve as the philosophical foundation of the Marine Corps: Marine Corps Doctrinal Publication (MCDP) 1, *Warfighting*, and 6, *Command and Control*. Within MCDP 1, the concept of viewing the adversary as a "complex system of individual parts" is balanced with the recognition that individual human will is a central component in war.[6] MCDP 6 "sees command as the exercise of authority and control as feedback about the effects of the action taken" based on "an interactive process involving all the parts of the system and working in all directions."[7] The goals of this chapter are threefold: arm the reader with a new vocabulary; create an expanded understanding of the ethical nature of leadership complexity, sense making, and exercise of authority; and ground tactical, operational, and strategic thinking all within the biological, natural principles of living systems.

---

5 Paolo Tripodi, "Understanding Atrocities: What Commanders Should Do to Prevent Them," in *Ethics, Law and Military Operations*, ed. David Whetham (New York: Palgrave Macmillan, 2010), 173–88.
6 U.S. Marine Corps, Doctrinal Publication 1, *Warfighting* (Washington, DC: Department of the Navy, 1997), 12–13.
7 U.S. Marine Corps, Doctrinal Publication 6, *Command and Control* (Washington, DC: Department of the Navy, 1996), 40.

Our premise is that to succeed in future complex engagements, the U.S. military must create a culture of ethical leadership that enacts creative and ethical thinking at all organizational levels, both inside and outside the bonds of social or organizational culture. A culture of ethical leadership can be the major source of energy for a unifying mind. Moreover, ethical leadership by and in an organization sees people in other countries as legitimate parts of civilization, not as separate entities. Recognizing another's legitimacy—even the enemy's—is vital because of what can be learned from the other since any "other" can provide information, knowledge, or experience that can help create increased understanding of a complex problem and inform decision-making processes.

Recognizing "the other" as legitimate is not a new idea. General Alfred M. Gray, former Marine Corps Commandant, often said that Marines are not too proud to learn *anything* from *anybody*, since *every* person can be a source of legitimate information or a learning resource. Much can be learned from the civilian populace during combat, in complex political and social situations, and in rugged foreign terrain, particularly in order to generate peace after conflict. A unifying culture of ethical leadership must pervade every level of leadership, beginning with strategic guidance appropriate for complex engagements in both war and peace and including ethical conduct and moral autonomy for every member in the organization. The purpose of this chapter then is not only to describe "what" a culture of ethical leadership's key values are, but also to explain "how" and "why" a culture of ethical leadership is created and required.

PRINCIPLES OF ADAPTATION: MINDFULNESS, ETHICS, AND EMOTIONS

Nobel physicist Max Planck said when we change the way we look at things, the things we look at change. Fitness of mind is vital to success and effectiveness, as is fitness of body and spirit, and innovative behavior and action are closely related to creative living and the practice of mindfulness.[8] "An *innovation* is an idea, practice, or object that is perceived as *new* by an individual or other unit of adoption."[9] Mindfulness can be simply seen as (1) creation of new categories; (2) openness to new information; and (3) awareness of more than one perspective.[10]

Enaction is "movement into context,"[11] where the movements of mind and body are simultaneously prereflective, nonverbal reflective, or verbal reflective.[12] The embodied mind operates prereflectively as mindfulness, and when developed and practiced can be, as anthropologist Gregory Bateson said,[13] the difference that makes the difference.

Our new science of mind, ideas of ethics, and affective neuroscience are radically reconceptualizing leadership's complexity and revealing why many conventional leadership strategies often fail. That is precisely why research on the practice of mindfulness in the military and Mindfulness-Based Mind Fitness Training (MMFT) is contributing to creating a culture of ethical leadership.[14]

Fitness of mind is just as vital to success as fitness of body, and it is increasingly evident in Iraq and Afghanistan that conventional twentieth-century models, linear leadership theories, and stra-

---

8 Jon Kabat-Zinn, *Guided Mindfulness Meditation* (Boulder, CO: Sounds True, 2002), 1.

9 Everett Rogers, *Diffusion of Innovations*, 4th ed. (New York: The Free Press, 1995), 11.

10 Ellen Langer, *Mindfulness* (Cambridge, MA: Perseus, 1989), 63–70.

11 Clyde Croswell and Scot Holliday, "Generating Organizational Awareness: The Primordial Nature of Language and Emotioning" (working paper, The Center for the Study of Learning, George Washington University, Washington, DC, 2004).

12 Daniel Stern, "Pre-Reflexive Experience and Its Passage to Reflexive Experience: A Developmental View," *Journal of Consciousness Studies* 16 (2009): 307–31.

13 Gregory Bateson, *Mind and Nature* (Cresskill, NJ: Hampton Press, 2002), 64–65.

14 Elizabeth Stanley and John Schaldach, "Resources," Mindfulness-Based Mind Fitness Training (MMFT) Institute webpage, http://www.mind-fitness-training.org/resources.html.

tegic minds often constrain effectiveness. Conventional strategy can become the cause of "wicked" problems in foreign cultures, ever-changing climates, and the rugged mental landscapes and terrains of twenty-first-century military action.

Natural principles can guide, cultivate, and nourish an attitude of wisdom and creative living.[15] Essentially, bio-adaptive principles generate organizational improvisation, innovation, and creativity,[16] not unlike Meyers and Davis' "improvisation-ready" Marines.[17] Life-conserving principles provide a species ethic guided by global maxims and are the source of inter-enactive emergence,[18] or *enacting a culture of ethical leadership.*[19] Principles of ethical leadership include

- Ethics is "a reflection on the legitimacy of the presence of others."[20]

- Value is the expression of relationship between self and other (other may be a person, thing, innovative idea, organization, etc.).[21] In living systems, value is seen as whatever an organism chooses [consciously or subconsciously] to be attracted by or self-propelled toward.[22]

---

15 Karl Weick, *Making Sense of the Organization* (Oxford: Blackwell Publishing, 2001), 361.

16 Karl Weick, *Making Sense of the Organization: The Impermanent Organization* (West Sussex: John Wiley & Sons, 2009), 267.

17 Christopher Meyer and Stan Davis, *It's Alive: The Coming Convergence of Information, Biology, and Buisness* (New York: Crown Business, 2003), 151.

18 Giovanna Colombetti and Steve Torrance, "Emotion and Ethics: An Inter-(en) active Approach," *Phenomenology and the Cognitive Sciences* 8 (2009): 505–26.

19 Douglas Griffin, *The Emergence of Leadership: Linking Self-Organization and Ethics* (New York: Routledge, 2002), 175

20 Humberto Maturana and Francisco Varela, *The Tree of Knowledge: The Biological Roots of Human Understanding* (Boston: Reidel, 1992), 247.

21 Tsunesaburo Makaguchi and Dayle Bethel, *Education for Creative Living*, trans. Alfred Birnbaum (Ames: Iowa State University Press, 1989), 70–74.

22 Elliot Jaques, *The Life and Behavior of Living Organisms: A General Theory* (Westport, CT: Prager, 2002), 245.

- The process of structuring has two aspects: components (individuals) and organization (relationships).[23]

- Mindfulness is "the awareness that comes from systematically paying attention on purpose in the present moment, and nonjudgmentally, to what is closest to home in your experience: namely this very moment in which you are alive, however it is for you—pleasant, difficult, or not even on the radar screen—and to the body sensations, thoughts, and feelings that you may be experiencing in any moment."[24]

- Learning is the transformation of behavior through experience.[25]

- Communication is the coordination of behavior,[26] and communication (implicit or explicit) has no meaning except in the context of the receiving apparatus.

- We human beings are both constitutive of and constituted by our world. Constitution is the process of providing an ever-clearer meaning.[27]

- Enaction is movement into context (whether physical, verbal, mindful, intentional, or not).[28] Any movement affects others and consequently evokes an affective response in others in the environment.

---

23 Humberto Maturana, "The Biology of Language: The Epistemology of Reality," in *Psychology and Biology of Language and Thought: Essays in Honor of Eric Lenneberg*, ed. George Miller and Elizabeth Lenneberg (New York: Academic Press, 1978), 27–63.

24 Jon Kabat-Zinn, *Guided Mindfulness Meditation*, 1.

25 Humberto Maturana and Francisco Varela, *Autopoiesis and Cognition: The Realization of Living* (Boston, MA: Riedel, 1980), 35; Maturana and Varela, *The Tree of Knowledge*, 172.

26 Maturana and Varela, *The Tree of Knowledge*, 193.

27 Stephen Strasser, *The Idea of Dialogal Phenomenology* (Pittsburg, PA: Duquesne University Press, 1969), 69.

28 Croswell and Holliday, "Generating Organizational Awareness," 9.

- Emotions are "dynamic processes in the brain and body that prepare the body for future actions and enable it to carry them out,"[29] which means "emotion is the energy of transformation."[30] It is not trivial that the source of time is the valence and constitutional dynamics of affect.[31]

- Violence is the imposition of a point of view with little or no understanding of the other's point of view.[32]

INNOVATIVE MINDS AND LANGUAGE

The mindful insights found in the opening quotes by Major General Chiarelli and General Zinni illuminate and confirm changes and shifts in both the global and local landscapes. General Chiarelli, who had just arrived in Baghdad as the commanding general of the U.S. Army 1st Armored Cavalry Division, realized something new was emerging simultaneously across all levels—strategic, operational, and tactical. We propose a new mental model (see figure 5) for enacting a culture of ethical leadership that comprehensively accounts for the complexity of global problems, local situations, sense making, living adaptation, and the diffusion of command (exercise of authority) and control (dynamic feedback from numerous sources).

---

29 Walter Freeman, *How Brains Make Up Their Minds* (New York: Columbia University Press, 2000), 91–92.

30 Clyde Croswell and Krishna Gajjar, "Mindfulness, Laying Minds Open, and Leadership Development: An Enactive Approach to Leadership Complexity and Practical Wisdom" (joint paper to the Interdisciplinary Conference on Cognition, University of Central Florida Cognitive Science Program and International Association for Phenomenology and the Cognitive Sciences, Orlando, Florida, 20–23 October 2007, 25).

31 Francisco Varela and Natalie Depraz, "At the Source of Time: Valence and the Constitutional Dynamics of Affect," in *Ipseity and Alterity: Interdisciplinary Approaches to Intersubjectivity*, eds. Shaun Gallagher, Stephen Watson, Phillipe Brun, and Phillipe Romanski (Rouen, France: University of Rouen, 2004), 153–74.

32 William Isaacs, *Dialogue: The Art of Thinking Together* (New York: Random House, 1999), 132.

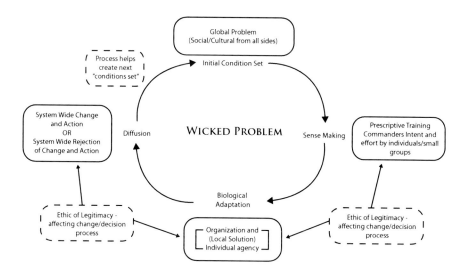

FIGURE 5. **Culture of Ethical Leadership.**

A new mental model is vital to replace outmoded thinking, otherwise one remains imprisoned by the "iron cage of memory," with little to no creativity or possibility of change.[33] By changing the language one uses, or the meaning of the language, one can change the way one thinks. However, innovative, adaptive thinking and solutions to complex problems often meet resistance, so brief descriptions of key components will provide subtle insights as we move through the process, flow, and meaning of the model.

The starting point for creating a culture of ethical leadership is the recognition that individuals and organizations now and in the future face inherently complex problems (see figure 6). Complex problems are generated by the interaction between cultural, societal, paradoxically global (shared), yet locally specified realities. When the interactive nature of complex problems or interactive complexity requires political judgment rather than scientific reason or preconceived strategy alone to resolve, they are consid-

---

33 Croswell and Holliday, "Generating Organizational Awareness," 1.

ered "wicked problems."[34] General Chiarelli quickly realized that the situation in Baghdad was wicked, complex, and composed of enemy, friendly, and neutral cultures, each with unique goals, objectives, and desires. Further, he also realized he needed to change the mind-set and modus operandi of his own organization.

To better understand how humans experience a phenomenon like complexity or wicked problems, two key terms require explanation:

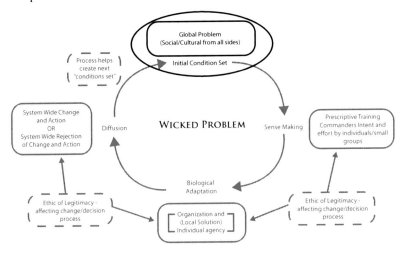

FIGURE 6. **Initial Condition Set.**

- *Phenomenon*: Something perceived or experienced, especially an object as it is apprehended by the human senses as opposed to an object as it intrinsically is in itself. A phenomenon can be other selves, happenings, experiences, and ideas as they are *perceived* by human beings through the use of their senses and imagination.

- *Phenomenology*: The study of human experiences; understanding that the meaning of an object transcends the nature

---

34 Hoirst Rittel and Melvin Webber, "Dilemmas in a General Theory of Planning," *Policy Sciences* 4 (1973): 160.

of the object.[35] When human beings experience a phenomenon, we immediately ascribe meaning, specific to self.
When one person says to another "think about a dog," the
speaker might imagine one type of "dog" (poodle), while
the listener might likely imagine a different "dog" (dachs-
hund). Both images share common characteristics, but the
specific meaning of each person's image may be completely different. The implication is that when human beings

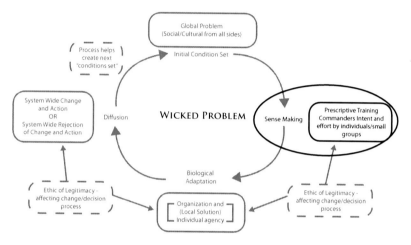

FIGURE 7. **Sense Making.**

gather and organize, a comprehensive unifying mind
emerges as culture as images gain shared meaning. Consequently, culture is "an intersubjective system of meaningful
experiences, institutions, activities, symbolic expressions
of the ritual and art, together with their products, which
are shared by the members of a given society"[36] (i.e., local
people inter-enacting cocreate their society, community, or
culture).

35 David Bidney, "Phenomenological Method and the Anthropological Science
of the Cultural Life-World," in *Phenomenology and the Social Sciences*, ed. Maurice
Natanson (Evanston, IL: Northwestern University Press, 1973), 117–18.
36 David Bidney, "Phenomenological Method," 133.

While experiencing a phenomenon, an individual autonomously begins the process of sense making in order to respond or determine appropriate actions (see figure 7). Making sense of situational (local) problems occurs at both an individual and organizational unit level, taking into account previous experience, plans, training, and education as well as cultural, societal, and personal biases. Yet sense making is far more than mere retrospection. "Mindfulness," "participatory sense making," and "global" are three concepts vital for effective sense making:

- *Mindfulness*: "The awareness that comes from systematically paying attention on purpose in the present moment, and nonjudgmentally, to what is closest to home in your experience; namely, this very moment in which you are alive, however it is for you—pleasant, difficult, or not even on the radar screen—and to the body sensations, thoughts, and feelings that you may be experiencing in any moment."[37]

- *Participatory Sense Making:*[38] A biologically self-grounded natural response to stimulus in the environment whose purpose is to conserve and sustain the life of a living organism (an adapting, autonomous agent). In this sense, self is more than a psychological or social/cultural construct; *self is primarily a living, self-creating, self-conserving, self-organizing unity that adapts to preserve its autonomy/self.* Sense making is the autonomous agency and intentionality of a living system, whether individual, team, organization, community, society, or planet/globe.[39] The self operates with biological autonomy to conserve its own life; biologists call this process *autopoiesis*, meaning self-creating or self-organizing.[40]

---

37 Kabat-Zinn, *Guided Mindfulness Meditation*, 1.
38 Hanne De Jaegher and Ezequiel Di Paolo, "Participatory Sense-Making: An Enactive Approach to Social Cognition," *Phenomenology and the Cognitive Sciences* 6 (2007): 485–507.
39 James Miller, *Living Systems*, 239.
40 Maturana and Varela, *Autopoiesis and Cognition*, 82.

- *Global*: The unifying principles of living, self-organizing systems; not global in the geographical sense, but the comprehensive mind that identifies and unifies the place and mental space of all persons, cultures, and ideas. Global is a species ethic[41] or unified field of mind. Local is exclusively a simple unity point of view (bottom up); global is inclusively the composite unity view (top down). Thus, global mind creatively imagines[42] relationships and potential connections that already exist between components or all peoples. In nature, these two points of view operate with complementarity and dynamic coemergence, not as polar opposites.

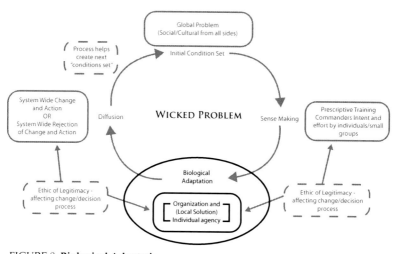

FIGURE 8. **Biological Adaptation.**

The next process in the model, biological adaptation, emerges when individuals and organizations attempt to create solutions to wicked problems after having made sense of both perceptible and cognitive phenomena (see figure 8). After achieving a shared sense, one moves into action (or context) based upon that sense.

---

41 Jonathan Shay, *Achilles in Vietnam: Combat Trauma and the Undoing of Character* (New York: Scribner, 1994), 206.
42 Varela and Depraz, "Imagining: Embodiment, Phenomenology, and Transformation," 195–230.

Key to this process is the idea that prior to and during the process of adapting, individuals and organizations are affected and therefore are experiencing the energy of transformation. Accordingly, affect is vital to understanding biological adaptation.

- *Affect*: A transitive verb essentially meaning "to move somebody emotionally" or to act upon something.[43] Affect is a primordially adaptive impulse or natural form of energy that arises in living beings. The emergence of affective science, affective neuroscience, and neurobiology are revealing that cognition and emotion (mind and body) are *inseparable* and operate with complementarity, not as polarized opposites that can be excluded one from the other. They operate as parallel aspects of the embodied mind. Affect is manifested as six major phenomena in human experience: emotion, feelings, mood, attitude, affective style, and temperament.[44] In this context, affect or emotion can be seen as the energy of transformation.[45] Finally, affect has energic valence[46]—positive (productive), negative (destructive), or neutral (ambivalent/ambi-valent)—as it emerges.

The final process in enacting a culture of ethical leadership is diffusion and whether an idea (innovation) is either adopted or rejected by an individual or organization (see figure 9). Diffusion essentially means that the individuals within an organization have rejected or accepted a proposed innovation. Rejection or acceptance of the innovation includes specific feedback from the individual (local) level to the organizational (global) level about what will or will not work and what is or is not acceptable. In wicked problems, diffusion and implementation of an

43 *Webster's New Collegiate Dictionary*, 1979, s.v. "affect."
44 Richard Davidson, Klaus Scherer, and H. Hill Goldsmith, *Handbook of Affective Science* (Oxford: Oxford University Press, 2003), xiii.
45 Croswell and Gajjar, "Mindfulness, Laying Minds Open, and Leadership Development," 25.
46 Varela and Depraz, "At the Source of Time," 153–74.

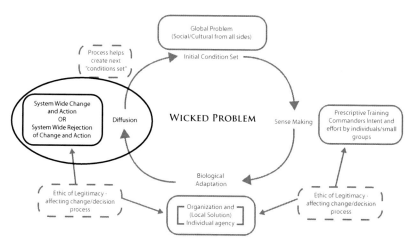

FIGURE 9. **Diffusion.**

innovation create the next condition set that must be resolved. The vital term for understanding diffusion is "enaction" or the "enactive approach."

- *Enaction*: The emerging paradigm of cognition that breaks with modern cognitive theory and the myths of mere information processing, computationalism, and representationalism. The enactive approach to cognition is more comprehensive than classic single-disciplined psychological or social approaches to cognitive science known as first generation Computational Theory of Mind (CTM). Conventional cognitive theory conceptually reduces and represents the things, people, and events observed in a stable external world by seemingly independent consciousness. Enacted meanings are constituted both globally and locally within the embodied mind by an interpretation of the experience and physical history of a culture, organism, unit, organization, person, etc.[47] In enaction, "knowledge depends on being in a world that is inseparable from our bodies,

---

47 Francisco Varela, Evan Thompson, and Eleanor Rosch, *The Embodied Mind: Cognitive Science and Human Experience* (Cambridge, MA: MIT Press, 1991), 9.

our language, and our social history—in short, from our embodiment."[48]

In the new paradigm, enaction is based on dynamic coemergence[49] of self and other and the inseparability of a living organism from its environment. Enaction is "movement into context"[50] by a living organism that is both verbal and nonverbal, prereflective and reflective.[51] When the body of an individual speaks or moves, that enaction or embodied life and mind is a stimulus to which the environment reflexively responds, including the responses of other human beings, organizations, or nation states. What we say or do partly specifies the world we live in; we are our world. At the same time, how the global mind or world we live in responds to our movement often constrains our self or who we are. Thus, a living organism actually specifies or enacts the world it lives in, and with synchronicity, the world constrains or enacts its components or parts. The enactive approach constitutes both maturity[52] and wisdom.[53]

The proposed mental model is designed to assist individuals and organizations in the process of becoming aware.[54] Several operational narratives are debunked by both enaction and leading-edge neuroscience, as technological innovations like brain and positron emission tomography (PET) scans, functional magnetic resonance imaging (fMRIs), consciousness studies, and the emerging science of mind and life unmask these outdated myths. Still, innovative

48 Ibid., 149.
49 Evan Thompson, *Mind in Life: Biology, Phenomenology, and the Sciences of the Mind* (Cambridge, MA: Harvard University Press, 2007), 38.
50 Croswell and Holliday, "Generating Organizational Awareness," 9.
51 Stern, "Pre-Reflexive Experience and Its Passage to Reflexive Experience," 307–31.
52 Norman Fischer, *Taking Our Places: The Buddhist Path to Truly Growing Up* (New York: Harper Collins, 2004), 188.
53 Weick, *Making Sense of the Organization*, 361.
54 Natalie Depraz, Francisco Varela, and Pierre Vermersch, *On Becoming Aware: A Pragmatics of Experiencing* (Philadelphia: Jon Benjamins, 2001), 1.

conclusions and ideas from science meet resistance in the mind and body; and, as we all have learned through experience, resistance to change often creates wicked problems. To begin unifying mind for enacting a culture of ethical leadership, we first must understand the natural complexity of "wicked problems."

WICKED PROBLEMS

Success in complex problems, particularly wicked global problems, requires a combination of mindfulness, creativity, and wisdom, along with a culture of ethical leadership. Heretofore, military training and doctrine were highly focused on penetrating an enemy system and taking decisive action to terminate a local threat. General Chiarelli realized that classic modes of strategy alone and action directed toward linear threat termination are often obsolete. Military actions require a more global mindfulness, innovation, and organizational learning,[55] not less mindfulness[56] or mere routine performance. Mindfulness is particularly relevant since recent history shows that the military is the branch of the United States government normally called upon to lead and coordinate efforts to respond in the global environment.

When contextualized by the two quotes at the beginning of this chapter, it becomes apparent that a new, more dynamic model of ethical leadership and mindfulness of the complexity of moment-by-moment adaptation is vital. In complex adaptive systems,[57]

---

55 David Schwandt and Michael Marquardt, *Organizational Learning: From World-Class Theories to Global Best Practices* (Boca Raton, FL: St. Lucie Press, 2000), 61.

56 Daniel Levinthal and Claus Rerup, "Crossing an Apparent Chasm: Bridging Mindful and Less-Mindful Perspectives on Organizational Learning," *Organization Science* 17 (2006): 502–13.

57 David Schwandt, "Individual and Collective Co-Evolution: Leadership as Emergent Social Structuring," in *Complexity Leadership: Part I: Conceptual Foundations*, ed. M. Uhl-Bien and R. Marion (Charlotte, NC: Information Age Publishing, 2008), 102.

leadership emerges at all levels (local and global, individual and collective). As individuals take local action on behalf of the United States, they do so through immediate sense making of the situation, and then take the corresponding present moment behavior in response to the situation. This behavior then creates the conditions that specify the future action required to deal with a new, emerging situation.

The perfect example of this is what took place at Abu Ghraib. In making sense of their situation and role as prison guards, the soldiers acted in a way they thought was appropriate for the situation. The consequence was a new strategic reality felt throughout the world, particularly in Muslim and Middle Eastern societies. The practice of mindfulness[58] precisely cultivates wise creativity and creative wisdom[59] because an autonomous agent "creates products through creative processes aimed to optimize its own life chances in surroundings of continuous livability."[60] Thus, solving locally emerging problems requires more than global, conventional military strategy, operations, and tactics. Effective problem solving requires a culture of ethical leadership.

Twenty-first-century conflicts and the complexity of wicked problems defy clear, simple, linear solutions, and they are often contextualized by societal norms embedded in cultural phenomena that require resolutions based solely on political judgment. Moreover, political stability is often fleeting in a rapidly evolving world, and these emerging wicked problems are open-ended and affected by factors often well beyond any one leader's or individual's understanding and span of influence. Problems are wicked when the

---

58 Karl Weick and Ted Putnam, "Organizing for Mindfulness: Eastern Wisdom and Western Knowledge," *Journal of Management Inquiry* 15 (2006): 275–87.
59 Hans Knoop, "Wise Creativity and Creative Wisdom," in *Creativity, Wisdom and Trusteeship: Exploring the Role of Education*, ed. Anna Craft, Howard Gardner, and Guy Claxton (Thousand Oaks, CA: Corwin Press, 2008), 119.
60 Ibid., 126.

mission/situation is ambiguous and not likely to become apparent as to whether or not a problem has been resolved.[61] For these reasons and in order to "create, connect, and evolve"—to adapt and overcome—Marine Corps doctrinal publications dispensed with conventional command and control and redefined the organization's command structure.[62] Marine Corps doctrine states that command and control is really "a system that provides the means to adapt to changing conditions . . . we can thus look at command and control as a process of continuous adaptation."[63] Whether manifested in the form of command and control or some other process, *adaptation occurs primarily in self-organizing, living systems as the agency and intentionality of biological autonomy—not solely psychological, political, economic, societal, nor cultural constraints alone.* This vital aphorism reveals the profound complexity of human systems and organizations and is a definitive point explained throughout the chapter.

The profound complexity of wicked problems has 10 distinguishing features. Six of the proposed 10 features quickly rise to the level of greatest value to this chapter. The term "wicked" does not connote evil or cruel; Rittel and Weber used the term to mean extraordinarily complex, murky, and an overt challenge to reason. Thus, when facing a truly wicked problem, any leader must realize that *first*, the problem itself is a symptom of another problem and *second*, every actor is a potential leader precisely because every stakeholder enters the situation with a distinct and personal history and a unique causal narrative,[64] affirming the value of the "enactive" approach to cognition.

Leadership is being reconceptualized and its ethical meaning is rapidly being reconstituted. Scholars, practitioners, nonprofit or-

---

61 Rittel and Webber, "Dilemmas in a General Theory of Planning," 160.
62 Meyer and Davis, *It's Alive*, 152.
63 U.S. Marine Corps, *Command and Control*, 46.
64 Rittel and Webber, "Dilemmas in a General Theory of Planning," 165.

ganizations, governments, businesses, and military units alike are finally beginning to realize that leadership is something generated dynamically and with complementarity. Generated leadership emerges at all levels of an organization from the top down *and* bottom up, globally *and* locally by the actions, participation, agency, and behaviors of traditionally conceived leaders *and* followers/informal leaders alike. Thus, leadership is a complex, collective phenomenon, not created solely by one person at the top of an organization in a single role or position of authority.

Effective leadership emerges (or not) in the immediate space and time where tactics meet commander's intent. In Meyers and Davis' provocative text, *It's Alive*, Lieutenant General Paul K. Van Riper argues effectively that "the power comes from the bottom up, not from the top down. Those involved have some awareness of both what's happening around them [locally] and what's happening on a larger scale [globally], and they will self-organize [locally] to achieve the commander's intent."[65] Lieutenant General Van Riper's use of the word "power" is *not* trivial in this case, as power is an analogue for force, energy, affect, and emotion as explained above, most literally, physical movement. Both power and emotion have valence that can produce, destroy, or become neutral.

Since a wicked problem is symptomatic of another problem, leaders must face the *third* distinguishing feature of a wicked problem, which is that each wicked problem is truly unique.[66] The problem may have varying levels of similarity to other problems, yet the problem itself cannot always be solved by doing exactly what worked for a previous problem or the last time a similar problem was encountered. Albert Einstein clarified this feature of wicked problems when he stated that "the significant problems

---

65 Meyer and Davis, *It's Alive*, 153.
66 Rittel and Webber, "Dilemmas in a General Theory of Planning," 164.

we face cannot be solved at the same level of thinking we were at when we created them."[67]

Consequently, the uniqueness and causality of any problem leads to the *fourth* distinguishing feature that any proposed solution is a "one-shot operation."[68] Thus, dealing with wicked problems requires mindfulness and wisdom of organizational sense making.[69] In a social/political situation, a single operation or action/tactic may now serve as another causal factor in the next uniquely problematic situation. As a result of the experimental, unpredictable, one-shot nature of a proposed solution, a wicked problem will not present an obvious end to an open-ended causal chain. Likewise, the *fifth* distinguishing feature, based on the nonlinearity and individual narrative associated with the problem, is that a proposed solution cannot be seen as either true or false, but rather as good or bad for each of the actors involved.[70]

Finally, the *sixth* and most significant feature of wicked problems for any leader, given the complexity and emerging, adaptive nature of self-organizing and the potential damage a bad solution can create, is that a leader "has no right to be wrong."[71] With all of these characteristics at work, any single leader facing a wicked

---

67 Albert Einstein, "Quotes on Problem Solving," LeadershipNow webpage, http://wwwleadershipnow.com/problemsolvingquotes.html. The authors would like to point out that many scholarly references are made to this quote as written but that there is some debate about the exact wording. Alice Calaprice, in *The New Quotable Einstein* (Princeton, NJ: Princeton University Press, 2005) attributes the quote to a paraphrasing of a 1946 *New York Times* article ("Atomic Education Urged by Einstein," *New York Times*, 25 May 1946, http://www.turnthetide.info/id54.htm), where Einstein writes, "A new kind of thinking is essential if mankind is to survive and move toward higher levels." Either version of the quote would prove the same point being made in this chapter, that a new thinking is necessary if we want to effectively deal with complex systems.
68 Rittel and Webber, "Dilemmas in a General Theory of Planning," 163.
69 Karl Weick, Kathleen Sutcliffe, and David Obstfeld, "Organizing and the Process of Sensemaking," in *Organization Science* 16 (2005): 409–21.
70 Rittel and Webber, "Dilemmas in a General Theory of Planning," 162.
71 Ibid., 166–67.

problem has a very low probability of consistent success. Even when a proposed solution is acceptable and effective, there will be a voice of dissonance stating the solution was not good enough. That is why understanding leadership as the self-organizing process of dynamic coemergence is a more accurate conceptualization than are the previous simple, linear leadership theories.

Complex, wicked problems are all too often the exact types of problems military personnel are called on to solve. The proposed mental model begins with the wicked problem primarily because predeployment or preengagement training attempts to help individuals and units within a larger organization create the rules of engagement, command climate, and commander's intent to help individuals understand what they are seeing, react properly, and report accordingly. These actions within the context of the wicked problem are all directed toward individual and organizational sense making. Participatory sense making is an important second step toward understanding command and control.

PARTICIPATORY SENSE MAKING AND THE ENACTIVE APPROACH

Seven properties that have informed organizational decision making for over four decades have conceptually described sense making in organizations.[72] Weick's early publications describing sense making in organizations revealed seven characteristic properties. Sense making is (1) grounded in identity construction; (2) retrospective; (3) enactive of sensible environments; (4) social; (5) ongoing; (6) focused on and by extracted cues; and (7) driven by plausibility rather than accuracy.[73] What has become clear over the past half-century is that organizations are far less stable than assumed; in fact, they are impermanent and structured in ways

72 Synthesis of the evolution of sense making in Karl Weick's published works in 1969, 1979, 1995, 2001, and 2009.
73 Karl Weick, *Sensemaking in Organizations*, (Thousand Oaks, CA: Sage, 1995), 82.

and relationships that are not always evident.[74] Everything is connected and many organizational relationships change continuously, both internally and externally, which is not always evident to autonomous agents.

Most recently, however, the classic approach to sense making is being expanded. Emerging sense-making research, along with calls for understanding the true nature of participatory sense making, has ushered in this expansion.[75] Understanding participatory sense making may provide insights into why current strategic plans often do not create the intended results; why emerging operations often generate unanticipated responses, unintended consequences, and unintended outcomes; and why conventional military tactics often fail to achieve goals and objectives.

Recent literature from sense-making and affective neuroscience research is reconceptualizing the conventional notion of sense making. Both illuminate the *adaptive* nature of sense making that has not yet been explained. Weick, Sutcliffe, and Obstfeld, in dealing with high-reliability organizations (HROs) including those found in the military, have recently concluded that sense making occurs more in the present moment and is more future-directed than was originally proposed.[76] More importantly, they indicate that sense making is more behaviorally defined, emotion-driven, and action-oriented than previously thought. This means that sense making often operates without one's awareness; it is an experience or sense that actually occurs biologically and psychologically within an individual—an "embodied phenomenon." Furthermore, leading scholars of the sense-making concept recognize the important role of emotions and are calling for accounts and research of emotions during the act of sense making.[77]

---

74 Karl Weick, *The Impermanent Organization*, 3–4.
75 DeJaegher and DiPaolo, "Participatory Sense-Making," 485–507.
76 Weick, Sutcliffe, and Obstfeld, "Organizing and the Process of Sensemaking," 409.
77 Sally Maitlis and Scott Sonenshein, "Sensemaking in Crisis and Change: Inspiration and Insights from Weick," *Journal of Management Studies* 47 (2010): 551–80.

Concurrently, affective neuroscience and research from the National Institute of Health (NIH), the National Institute of Mental Health (NIMH), and other institutions around the world reveal that cognition and emotion are aspects of mind that operate interdependently with dynamic coemergence as inseparable processes.[78] From twenty-first-century science of mind, the affective domain is vital for organizational learning, mindfulness, and critical thinking, as this new science reveals insights about how the mind transforms the brain and how mindfulness emerges as participatory sense making. This new understanding of the mind's adaptive nature has significant implications for leadership, ethics, and self-organizing global systems.

These insights manifest themselves particularly in violent settings (i.e., war/combat), and sense making as adaptation is generated spontaneously in the form of affect or primary emotional systems.[79] These autonomic responses take the form of forceful emotions like *seeking, fear, lust, care, panic, play,* and in particular, *rage.* In *Achilles in Vietnam,* Shay's empirical portrayal of the traumatic effects of combat explain more long-lasting affective and behavioral forms in the mind and body than purely momentary rage.[80] The root cause of such extreme reactions is the betrayal of "what's right," leading to the undoing of character. This betrayal creates a loss of social trust emanating from perceptions of injustice therefore producing effects far greater and longer lasting than a single act of situational rage. Stress-induced phenomena in the climate of a combat environment are a primary source of Posttraumatic Stress Disorder (PTSD), and a unified culture of ethical leadership may help reduce perceived betrayal and the wicked problems it creates.

---

78 Justin Storbeck and Gerald Clore, "On the Interdependence of Cognition and Emotion," *Cognition and Emotion* 21 (2007): 1212–37.
79 Jaak Panksepp, *Affective Neuroscience: The Foundations of Human and Animal Emotions* (Boston, MA: Oxford University Press, 2005), 51.
80 Shay, *Achilles in Vietnam.*

Adopting the enactive approach when making sense of wicked problems may lead to a better understanding of how participatory sense making emerges as the process of continuous adaptation. Essentially, within the dynamic relationship between subjective perception (specifically of the bodily senses) and objective conception (of socially constructed and culturally constrained realities) lies the space and source of timely meaning emergence. Finally, sense making's emotional and affective natures are now understood as having affective valence—positive (productive), negative (destructive), or neutral (ambivalent).[81] Earlier insights revealed that emotion/affect is the energy of transformation.[82] Combined with research that reveals cognition does not occur without emotion, and vice versa, the myth that all emotions are always destructive and should always be avoided is debunked, and the oft-perceived negative emotion of fear may in reality produce life-saving behavior.

Affect and emotions are *natural adaptation*, natural selection that often results in life-saving, life-producing, vital processes. For effective leaders, this means sense making is grounded in the understanding that emotions can no longer be ignored or excluded from sense making and decision making.[83] Therefore, affect and emotions are vital to adaptation and ethics.[84] Given this scenario and recent research across various disciplines and fields, particularly affective neuroscience, the unifying mind of ethical leaders and followers takes into account the affective nature and complexity of relationships between self and other. In the intersubjectivity and reciprocity of communications, the whole body (mutual incorpo-

---

81 Varela and Depraz, "At the Source of Time," 159.
82 Croswell and Gajjar, "Mindfulness, Laying Minds Open, and Leadership Development," 25.
83 Lesley Fellows, "The Cognitive Neuroscience of Human Decision-Making: A Review and Conceptual Framework," *Behavioral and Cognitive Neuroscience Reviews* 3 (2004): 159–72.
84 Colombetti and Torrance, "Emotion and Ethics: an Inter-(en)active Approach," 505–26.

ration) is the source of mind, not the head or one brain alone[85]—the mind is embodied, and emotions are "dynamic processes in the brain and body that prepare the body for future actions and enable it to carry them out."[86] The top down/bottom up reciprocity of communications and the prominent role of emotion in sense making echo the concept of "implicit communications" found in MCDP 6, when describing the relationship between individuals controlling local execution and the commander who is overall in charge of the mission.[87] Thus mindfulness during sense making generates leadership as the outcome and complementarity of both local and global agents and actors (components) inter-acting in relationship.

Dynamic coemergence then is equivalent to command and control in the military. A global *and* local approach is vital for overcoming the shortfalls, failures, and myopia of strategy, operational crises, and tactical misperceptions of the situation at hand. The bottom-up approach is analogous to Colonel John Boyd's famous OODA Loop[88]—observe, orient, decide, act—which is why virtually every organizational member is a leader. Similarly, in the quantum realm, followers cannot be constrained to action as mere observers; *we are all participating observers*. This means that every individual component participates in enacting ethical leadership. Each is far more than a mere observer (follower) awaiting orders or instructions from the top down. Truly effective mindfulness and sense making are both participatory and adaptive, and this type of command and control (moment-by-moment reciprocal influence or relating) is clearly specified in MCDP 6.[89] The classic

---

85 Francisco Varela, "Steps to a Science of Inter-Being: Unfolding the Dharma Implicit in Modern Cognitive Science," *The Psychology of Awakening: Buddhism, Science, and our Day-to-Day Lives*, ed. Gay Watson, Stephen Batchelor, and Guy Claxton (York Beach, ME: Samuel Weiser, 2000), 73–89.
86 Walter Freeman, *How Brains Make Up Their Minds*, 91–92.
87 U.S Marine Corps, *Command and Control*, 79.
88 Ibid., 63.
89 Ibid.

top-down-only approach to command and control was formally abandoned by the Marine Corps over 15 years ago.

Although participatory sense making is necessary in creating knowledge, understanding the situation, and creating a proper response, knowledge and understanding alone are not enough to create viable solutions to complex problems. Wisdom and creativity are vital. Knowledge and understanding (of the commander's intent and the situation studied during operational planning) must be expanded by adding specific meaning to data and information (local control or feedback from subordinate or outside sources). Without feedback and meaning, learning as the transformation of behavior through immediate experience can be severely limited and adaptation curtailed. Learning is far more than mere recall, memory alone, or knowledge acquisition; learning requires a change in behavior. Under these conditions and in the complexity of wicked problems in particular, mere routine memory or classic trained performance can often fail.

Ethical leadership means that *global constraints (command) and local specifics (control) are both vital components for unifying mind.* Failure by an organization to heed both components from the strategic to the tactical level will continue placing U.S. servicemen and women, allies, and U.S. and foreign civilian populations in unimaginably dangerous situations. A lack of this understanding can equate to a perceived betrayal of "what's right," often leading to a breakdown in social and cultural trust. This breakdown in trust further erodes or creates the undoing of moral character and may result in or at least contribute to long-term pathologies like chronic anger and PTSD in veterans and civilians alike.[90]

The enactive approach to participatory sense making and "implicit communications" are grounded in the embodied mind.[91]

90 Shay, *Achilles in Vietnam*, 165–81.
91 Varela, Thompson, and Rosch, *The Embodied Mind*, 72–77.

This means adaptation in living systems occurs naturally as self-creation or *autopoiesis*;[92] and as a living system, the human body's primordial nature is reflexively, subconsciously, and biologically structured for adaptivity, purposeful self-conservation, and agency.[93] Participatory sense making is a naturally creative, generative process that occurs in the present moment, not from retrospective past experience alone, and with a different, innovative future in mind. So, paradoxically, sense making is both temporal and atemporal in nature. Sense making has double intentionality in that it is both immediate, as in the current situation, and longitudinal, as in oriented toward future goal achievement. As such, effective sense making is based on past, present, and future moments, not retrospective memory alone. Therefore, when seen as the transformation of intentional behavior through experience, learning includes cognition, is more behaviorally defined, and is infused with the transforming energy (+, -, o) of emotion. The immediate and longitudinal "time" dimensions of sense making, seen as urgency and participation respectively, are correlated to dialogic leadership and organizational learning.[94]

In sum, as participatory sense making occurs, we act on that sense. Our own enacted sense in turn effects or evokes a response from or enacts the environment, even though the environmental response cannot be predicted or assumed. So, in essence, enaction really is movement into context.[95] Complex adaptive systems (i.e., humans in organizational relationships) experience participatory sense making, affectivity, mindfulness (more or less), and rationalized contexts (less than more) in ways unique to each individual. As one process for dealing with wicked problems and imagining

92 Maturana and Varela, *The Tree of Knowledge*, 48.
93 Ezequiel DiPaolo, "Autopoiesis, Adaptivity, Teleology, Agency." *Phenomenology and the Cognitive Sciences* 4 (2005): 429–52.
94 Schwandt and Marquardt, *Organizational Learning: From World-Class Theories to Global Best Practices*, 179.
95 Croswell and Holliday, "Generating Organizational Awareness," 9.

innovative solutions, sense making's outcome is carried (embodied) into context mentally, physically, and spiritually. This movement into context enacts a response from the environment, and because the diffusion of innovations in the context of crises and change is often naturally resisted, the actual process of diffusion of innovation requires further exploration.

THE DIFFUSION OF INNOVATION

[I]t was no longer adequate as a military force to accept classic military modes of thought.

-Major General Peter Chiarelli[96]

Major General Chiarelli's recognition that the traditional way of thinking about problems was no longer acceptable demonstrates a mindfulness that ultimately allowed his organization to innovate. The general's openness helped him recognize that a change was required for his unit to be successful and brought with it opportunities for innovation. According to the change model proposed by Everett Rogers in *Diffusion of Innovation*,[97] an innovation occurs at and between the local/individual levels. Specifically, in order for a new idea to reach acceptance—or be valued by others—there are four key elements that must be in play: the innovation itself, communication, a social system, and time.[98]

First, the innovation itself must be recognized by the individual as new, and it is not all that important if the idea is objectively new or not.[99] Second, communication refers to the method through which the innovation is spread. The specific method of communication or medium is not always essential, as long as some form of

---

96 Chiarelli and Michaelis, 4.
97 Everett Rogers, *Diffusion of Innovations*, 1st ed. (New York: The Free Press, 1962), 161, 168–71, 306.
98 Ibid., 12.
99 Ibid.

effective communication occurs.[100] Thus, communication does not happen in a vacuum; rather, it occurs in a social system defined as "a population of individuals who are functionally differentiated and engaged in collective problem-solving behavior."[101]

Yet in living systems, communication is far more than the simple linear transfer or passing of information or knowledge. Instead, communication in complex adaptive systems is "the coordination of behavior"[102] and value is defined as the expression of relationship between self and other (i.e., idea, person, thing, organization, etc.).[103] The affective valuation, cognitive evaluation, and transformation of information into goal-referenced knowledge are vital processes of leadership in organizational learning.[104]

Third, a social system is composed of an entire population, yet the source of control arises within an individual because an innovation can only be considered truly diffused once individual members value and accept the idea. Rogers identified a continuum of adoption/action within his third element. Some innovations can be accepted (valued) quickly by an individual and then acted on, while other innovations may be accepted (valued) by the individual, but not translated into action until a large enough group of adopters has been formed. Conversely, some innovations require only a few but powerful adopter/decision makers before a change may be made.[105]

Finally, the fourth element—time—is vital to organizational learning. While time is not fixed, it does include the process that takes an individual from the point of exposure to the idea to full adoption. For the sake of clarity then, adoption of an idea is far

---

100 Ibid.
101 Ibid., 13.
102 Behavior and *autopoiesis* as explored in Maturana and Varela's two previously cited works, "Autopoiesis and Cognition" and *Tree of Knowledge*.
103 Tsunesaburo Makiguchi, *Education for Creative Living*, (Ames: Iowa State University Press, 1989), 70–75.
104 Schwandt and Marquardt, *Organizational Learning*, 43.
105 Rogers, *Diffusion of Innovation*, 1st ed., 15–17.

more than a cognitive psychological process related to rational goal achievement and is clearly not the same thing as adaptation. Adaptation involves the reflexivity of affective valuation (e.g., an emotional value or response) and often prereflective behavioral intention as well (e.g., a feeling that something is either right or wrong).

A decision to adopt an innovation has societal influences, yet decisions are ultimately most effective when the local/individual level adopts the innovation and decides to implement it both internally and externally. A decision cycle to adopt or accept an innovation often *excludes* the plausibility of sense making to determine the feasibility and "how to" of implementing or applying the innovation. To that end, Rogers 5th edition of *Diffusion of Innovations* offers a model that describes the five-step process an individual utilizes to adopt or reject a particular innovation: knowledge, persuasion, decision, implementation, and confirmation.[106] Rogers is clear that his model frames, or contextualizes, the process of innovating as a rational-empirical cognitive model only and intentionally leaves out the affect of emotion and other nonrational elements. Rogers' assumption of rational-empiricism ignores and possibly conceals affective valuation and adaptive behavioral intention otherwise denied by social and cultural taboos, false assumptions, beliefs, and attitudes.[107] However, this chapter obviously expands on and complements his diffusion of innovation model rather than undermines the strength of his argument. Figure 10 graphically represents Roger's five-step process:

Although the process appears causally linear, it does not always result in adoption or acceptance of a proposed innovation due to the nature of Complex Adaptive Systems (CAS). CAS have a mind of their own and cannot be controlled. Rejection or resistance or

---

106 Everett Rogers, *Diffusion of Innovation*, 5th ed. (New York: The Free Press, 2003), 169.
107 Davidson, Scherer, and Goldsmith, *Handbook of Affective Sciences*, xiii.

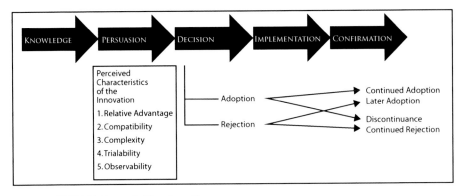

FIGURE 10. **Rogers's Five-Step Process.**

even ambivalence may emerge at any point in the process, and even an initial adopter may eventually reject the change. Thus, five significant criteria make any innovation persuasive: relative advantage, compatibility, complexity, trialability, and observability. Understandably, opinion leaders who demonstrate adoption of the innovation are vital for the organization (relationships) even within a formal and strict social structure like the military. Particularly in the case of the military, leadership is diffused and tasks are often executed with high levels of individual autonomy. Thus if the individual leader/actor/agent finds that the innovation does not provide an advantage over another solution, or is not compatible with an assigned mission and the individual's mind-set, simple to understand, or proven to work, the individual will quickly revert to habits of preinnovation behavior when left unsupervised.

The elements of diffusion/adoption are an important first step to understanding the evaluation of an innovation, yet alone they don't provide direct insight into how organizational change is finally made. To gain this insight, it is vital to understand Rogers' five adopter categories and how each individually and collectively effect (enact) adoption: innovators, early adopters, early majority, late majority, and laggards.[108] The innovator is generally described as someone who is venturesome. He or she does not have to be a formally recognized leader with official authority, position, or role. Generally, an innovator usually has a peer group of other innovators who encourage risk taking and help protect each other when a new idea falls short of expectations or completely fails.[109] The real change agents in the adoption process are the early adopters. Early adopters are usually highly respected by their peers within the social system and quickly become "the *embodiment* of successful and discrete use of new ideas."[110] (italics added.)

Once an innovation has passed through the innovators and early adopters, it is diffused to the majority, which has two categories, early (classified as deliberators) and late (classified as skeptics), each with its own critical opinion leaders. Members of the early majority are rarely in leadership positions and deliberate before adopting the innovation, whereas members of the late majority, with their skeptical approach, hold out until the overwhelming opinion is that the innovation is necessary and the "right way to go."[111] Finally, laggards are the traditionalists with few opinion leaders and a general fear of innovation. Laggards' values derive from what has been done in the past, and, if they do eventually adopt an innovation, it has most likely been superseded by the next innovation.[112] One negative side of adoption is that if an innovation is adopted and then fails, it may be perceived as a betrayal of "what's right." This betrayal, as previously mentioned, could result in the undoing of character in combat, which is one antecedent of PTSD.[113]

---

108 Ibid., 168–71.
109 Ibid., 169.
110 Ibid., 169–70.
111 Ibid., 170–71.
112 Ibid., 171.
113 Shay, *Achilles in Vietnam*, 3.

Rogers' studies identified patterns of adoption within a normal population, demonstrated by figure 11 below:

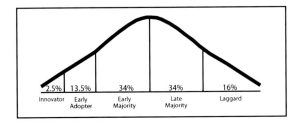

FIGURE 11. **The Diffusion of Innovation.**

In addition to rational-empiricism, there are several other false assumptions concealed by the Rogers model. These false assumptions do not negate the value of the model, but they do create possible variance when military personnel and units adopt innovations. The dynamics of combat action and fog of war are anything but general, nor are combatants the "normal" population. Diffusion of innovations in any crisis and emerging change to routines are often impeded by fear of the unknown; no one can fully predict future outcomes, particularly in twenty-first century military actions.

Yet innovators as complex adaptive living systems are continuously learning in order to transform their behavior through immediate experience. The adaptive capacity of self-organizing systems often appears in the form of resistance to change and even adaptive resistance to stability. In nature, change and stability operate with complementarity and ambivalence rather than as opposing polarizations.[114]

Classic military strategy, theory, and models, innovative as they may be, are often cumbersome and met with resistance, even rendered obsolete by globalization and technology. In fact, many

---

114 Carrie Leana and Bruce Barry, "Stability and Change as Simultaneous Experiences in Organizational Life," *Academy of Management Review* 25 (2000): 753–59.

strategies, operations, tactics, theories, and linear models actually cause wicked problems, albeit unintentionally. The human mind and experience naturally resist control, change, and innovations during the process of participatory sense making. Phenomenology, science of mind, and the biology of the enactive approach provide an alternative, more comprehensive, natural means for understanding and adapting to instability, complex changing conditions, and wicked problems often generated by flawed decision models or simple linear thinking.

Enacting a Global Culture of Ethical Leadership

Evaluation and synthesis of the previous sections reveal two insights. First, any military unit, or individual, is a living, complex adaptive system composed of components and an organization (relationships between the components). In living systems, there are things, people, and ideas/ideals that are both creative and destructive. Value *is expressed and created by relationships generated together with others and with self, both cognitively (conceptions) and affectively (perceptions)*. These values are created, reinforced, and realized by behaviors and actions in organizations through the learning that takes place in the daily interaction of the components/individuals. Without constant reflection, revitalization, evaluation, creativity, and mindfulness, the less mindful habits of blind routine performance can quickly imprison the minds of individuals and organizations in the "iron cage of memory" alone.[115]

Second, every member of an organization is naturally a leader able to coordinate and modify his/her own behavior in the organization moment-by-moment by practicing mindfulness of our own verbal and nonverbal communication.[116] A leader embod-

115 Croswell and Holliday, "Generating Organizational Awareness," 2.
116 Daniel Stern, "Pre-Reflexive Experience and Its Passage to Reflexive Experience," 307–11.

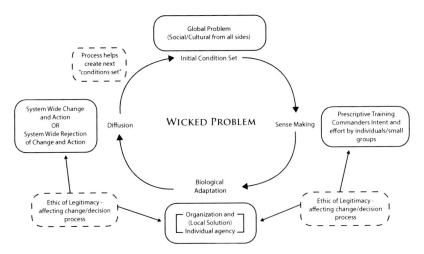

FIGURE 12. **Culture of Ethical Leadership.**

ies mindfulness, enacts policies, and generates emerging ethical know-how for a specific culture—both from bottom-up control and top-down command—by respecting the legitimate presence of all organizational members and other members in external organizations with whom he/she interacts. Each unit member models ethical behavior (or not) for those around them and even for those individuals outside the unit or immediate organization.

The value of ethical leadership is that during times of crisis and change, each member of the unit is able to productively focus the energy of self-regulating emotion in order to cocreatively transform unit potential and its subsequent relationships with other components in the environment. Wholesome transformation is accomplished when each person mindfully discerns a more comprehensive (global) meaning, heeds one's local perceptions and human experience, and includes the insightful sense of one's own embodied mind (self) with the sense of others. Healthy transformation then is learning how to live together.

Mindfulness is self-organizing, or the space and time between perception and conception, where real meaningful living is respected, created, and cultivated in the voice, energy, value, and ethics of learning to live together. When we practice mindfulness, local tactics emerge to meet, inform, and specify the commander's global intent, thereby enacting a global culture of ethical leadership. Command and control is then a process of healthy living, unifying mind, and living to conserve the life of self and others. Every person, component, and relationship has leadership potential. Each leader, wherever in the organization, can act mindfully [more or less] with ethical awareness of others, not from the delusion of "*I* know and have all the answers."

CONCLUSION

The Nobel poet T. S. Eliot reminds us in *The Four Quartets* that every end is a beginning, every beginning an end.[117] This chapter intended to move the reader's mind beyond the confines of memory, linearity, and mere objectivity into the space, time, and complexity of command and control, enaction, and mindfulness. We also introduced an innovative approach for enacting a culture of ethical leadership. The bioadaptive principles were explained and expanded the traditional self-limiting conceptualization of leadership. We saw that enaction, the new paradigm of cognitive science,[118] emerges naturally from the embodied mind,[119] and the enactive approach to participatory sense making can be seen as vital for creative living, diffusing innovations, and realizing the infinite potential and value of human transformation during crisis and change.

Reframing command and control as a dynamic, vital, life-giving process is the ultimate goal of this chapter. Command and control is not a top-down phenomenon only, it operates from the bottom up with complementarity and synchronicity. As members of the

---

117 T. S. Eliot, *Four Quartets* (Orlando, FL: Harcourt Brace & Company, 1943).
118 This subject is fully explored in John Stewart, Oliver Gapenne, and Ezequiel DiPaolo, eds., *Enaction: Toward a New Paradigm for Cognitive Science* (Cambridge, MA: MIT Press, 2010), vii.
119 Varela, Thompson, and Rosch, *The Embodied Mind*, 177.

military, we can change the way we think by practicing mindfulness and by changing the language, or at least the meaning of the language, we use to think. We can transform the way individuals and organizations make sense of wicked problems, how we adapt to emerging situations in a timely manner, and subsequently create and diffuse human value. We cocreate and enact potential resolutions to wicked problems. There is no single road map to follow or guide to cultural transformation; rather, we believe we have provided a solid foundation for enacting a culture of ethical leadership.

Training the Rules of Engagement for the Counterinsurgency Fight

Winston Williams

> *Counterinsurgency is war, and war is inherently violent. Killing the enemy is, and always will be, a key part of guerilla warfare. . . . But successful counterinsurgents discriminate with extreme precision between . . . combatants and noncombatants.*

> -David Kilcullen, *Counterinsurgency*[1]

The application of the Army's Counterinsurgency Doctrine[2] and the rules of engagement (ROE)[3] in modern irregular warfare has become a controversial issue, given Marines', soldiers', military officers', and lawmakers' expressed frustration with the doctrine's application in Afghanistan.[4] In particular, one commander stated his unit felt "they [had] their hands tied behind their backs"

---

1 David Kilcullen, *Counterinsurgency* (Oxford: Oxford University Press, 2010), 4.
2 Department of Army, FM 3-24, *Counterinsurgency* (Washington, DC: Headquarters, Department of the Army, 2006)
3 Chairman of the Joint Chiefs of Staff, Instr. 3121.01B, *Standing Rules of Engagement/Standing Rules for the Use of Force for U.S. Forces*, 13 June 2005.
4 See Colonel David M. Fee, "Rules of Engagement and the Obligations of the Strategic Commander" (unpublished Strategic Research Project, 16–17, 19 February 2010) (on file with author); and Wesley Morgan, "Weighing Threats and Rules of Engagement in Afghanistan," *New York Times*, 23 August 2010, http://atwar. blogs.nytimes.com/2010/08/23/weighing-threats-and-rules-of-engagement-in-afghanistan.

because they complied with the rules of engagement.[5] Furthermore, Congressman Walter Jones (R-NC) avowed that the rules of engagement "have proved too often to be fatal" to Marines and soldiers and "defies our [the American people's] fundamental belief in the right of self-defense . . . "[6] Unfortunately, General Stanley A. McChrystal, USA, a former commander of the Joint Special Operations Command who ran all special operations in Iraq and then became the top commander of American forces in Afghanistan, ignited this frustration and criticism when he issued a tactical directive in July 2009.

Although General McChrystal's directive sought to minimize civilian casualties,[7] many military and political officials saw the directive as a limitation on the right of self-defense.[8] In fact, General McChrystal acknowledged the limitation's potential risks within the directive when he wrote, "I recognize that the carefully controlled and disciplined employment of force entails risks to our troops . . . [b]ut excessive use of force resulting in an alienated population will produce far greater risks."[9] Some soldiers and Marines, however, believe that these rules "give the advantage to the Taliban."[10]

---

5 Sara A. Carter, "U.S. Troops Battle Both Taliban and Their Own Rules," *Washington Times*, 16 November 2009.

6 Dan Lamothe, "Rep.: Hold Rules of Engagement Hearing Now," *Army Times*, 15 April, 2010.

7 Memorandum from General Stanley McChrystal, *Tactical Directive*, Afghanistan, ISAF Headquarters, 6 July 2009, www.nato.int/isaf/docu/official_texts/Tactical_Directive_090706.pdf. This tactical directive is no longer in effect but General David Petraeus incorporated most of the principles into his Counterinsurgency Guidance.

8 Diana West, "Diana West: Afghan War Rules Endanger U. S. Troops," *Washington Examiner*, 12 September 2009, http:// washingtonexaminer.com /op-eds /2009/09/ diana-west-afghan-war-rules-endanger-us-troops.

9 Memorandum from General Stanley McChrystal.

10 "Soldiers Decry Military's Use of Force Rules" CBSnews.com, 25 June 2010, http://www.cbsnews.com/stories /2010/06/25/world/main6618493.shtml. See also Ann Scott Tyson, "Less Peril for Civilians, But More for Troops," *Washington Post*, 23 September 2009, http://www.washingtonpost.com/wp-dyn/content/article/2009/09/22/AR2009092204296.html.

Nonetheless, General McChrystal's guidance has high-level supporters, including General David H. Petreaus (former U.S. Central Command [CENTCOM] and International Security Assistance Forces [ISAF] commander). Specifically, General Petreaus disagrees with the proposition that the current rules of engagement give the advantage to the Taliban. In fact, Petreaus's guidance to ISAF states that "if we kill civilians or damage their property . . . we will create more enemies than our operations eliminate. That's exactly what the Taliban want."[11]

Similarly, civilian casualties in Afghanistan have strained the relationship between ISAF and Afghan President Hamid Karzai,[12] who demanded that ISAF "decrease the civilian casualties as long as they remain [in Afghanistan]."[13] Thus, the challenge in this counterinsurgency, as with other counterinsurgency conflicts, is balancing between the inherent right of self-defense and winning the support of the local populace.

Though the objective of counterinsurgency is to garner the support of the local populace,[14] the standing rules of engagement state that "[u]nit commanders always retain the inherent right and obligation to exercise unit self-defense in response to a hostile act or hostile intent."[15] The counterinsurgency objectives and the right of self-defense may force a commander to choose between protecting his soldiers or protecting the population. Thus, the crucial ethical issue is if units can garner populace support but not undermine their right of self-defense.

---

11 Memorandum from General David Petraeus, *COMISAF's Counterinsurgency Guidance*, Afghanistan, ISAF Headquarters, 1 August 2010.
12 Eltaf Najafizada, "Karzai Says Afghan Army Woes Bring NATO Raids, Civilian Deaths," Bloomberg.com, 23 November 2010, http://www.businessweek.com/news/2010-11-23/karzai-says-afghan-army-woes-bring-nato-raids-civilian-deaths.html.
13 Ibid.
14 FM 3-24, *Counterinsurgency*, para. 1-159.
15 Chairman of the Joint Chiefs of Staff, Instr. 3121.01B, app. A.

Achieving the counterinsurgency objectives while effectively ex-ercisng the right of self-defense requires persistent and deliber-ate training in the rules of engagement.[16] This primer will show that the current ROE training methods do not sufficiently prepare leaders for the complexities of counterinsurgency. Adequate ROE training for the counterinsurgency fight can educate and prepare leaders of Marines and soldiers for the current operational en-vironment, and leaders should tailor ROE training to meet the overall goal of winning the support of the local populace. Thus, to ease the friction between population-centric objectives and the right of self-defense, commanders, with the assistance of their judge advocates, must tailor ROE training to meet the mission re-quirements of the counterinsurgency fight.

To support this proposition, this chapter is divided into three parts. Part I will provide a brief overview of the counterinsur-gency doctrine and the challenges associated with applying the current rules of engagement to counterinsurgency. Part II will provide proposed guidelines for ROE training directed at prepar-ing units for the counterinsurgency fight. Finally, Part III will sum-marize the importance of ROE training to winning the support of the local populace.

OVERVIEW OF COUNTERINSURGENCY DOCTRINE AND THE RULES OF ENGAGEMENT

Counterinsurgency doctrine and the rules of engagement are in-separable at the tactical level. Troops at this level face an indis-tinguishable enemy who attacks them from buildings or other areas populated with noncombatant civilians to provoke the use

16 Mark S. Martins, "Rules of Engagement for Land Forces: A Matter of Training, Not Lawyering," *Military Law Review* 143 (1994). See also Department of the Army, FM 3-07-22, *Command and Control in a Counterinsurgency Environment* (Washington, DC: Headquarters, Department of the Army, 2004) para. 2-72. This manual expired on 1 October 2006.

of force in self-defense.[17] Major General Robert B. Neller wrote, "[t]hough the inherent right of self-defense will always remain paramount in a COIN environment, the default reaction must always be to 'not shoot.'"[18] General Neller's proposition is a harsh reality for Marines and soldiers at the tactical level. Surprisingly, the publication of Field Manual (FM) 3-24, *Counterinsurgency*, did not draw as much criticism as the application of this doctrine along with comparable rules of engagement in Afghanistan.[19] Thus, the source of the friction is the application at the tactical level with seemingly incongruous rules of engagement. The first step to understanding the friction is to understand the foundation of the FM 3-24.

Counterinsurgency Doctrine

COIN is an extremely complex form of warfare. At its core, COIN is a struggle for the population's support.

-FM 3-24, *Counterinsurgency*[20]

The Army published FM 3-24, *Counterinsurgency*, in December 2006 "to fill a doctrinal gap."[21] Field Manual 3-24 defines counterinsurgency as "military, paramilitary, political, economic, psychological, and civic actions taken by a government to defeat insurgency."[22] This definition highlights the military as one of six components of counterinsurgency and shows that there is more to counterinsurgency than combat operations. David Galula was one of the first scholars to recognize the military and nonmilitary aspects of counterinsurgency, and FM 3-24 includes many of Galula's theories.[23]

---

17 Charles J. Dunlap, Jr., "Lawfare: Decisive Elements of 21st-Century Conflicts?" *Joint Forces Quarterly* 54 (2009): 36.
18 Robert Neller, "Lessons Learned," *Marine Corps Gazette*, February 2010, 2.
19 Kilcullen, *Counterinsurgency*, 18. See also Colonel Gian Gentile, "Time for the Deconstruction of Field Manual 3-24," *Joint Forces Quarterly* 58 (2010): 116.
20 FM 3-24, *Counterinsurgency*, para. 1-159.
21 Ibid., "Foreword."
22 Ibid., para. 1-2
23 Ibid., para. 2-42.

Galula wrote extensively on counterinsurgency doctrine, drawing from his experiences in Algeria, Indochina, and China.[24] The crux of his theory was that gaining the support of the population is the objective of a counterinsurgency.[25] The insurgents seek the active or passive support of the local populace against the established government; the counterinsurgents need the active support of the local populace to maintain order and rule of law. Galula describes this conflict between insurgents and counterinsurgents as a "fight between a fly and a lion, the fly cannot deliver a knockout blow and the lion cannot fly."[26] The population's support is vital for the insurgent's success.[27] Thus, Galula's population-centric approach is the logical method for defeating the insurgency—separating the insurgent from the support of the population.

The modern insurgency still thrives on the support of the population, which makes winning the population's support fundamental to overall mission success.[28] Although the objective is clear, the application of the counterinsurgency doctrine is not an easy task because the implementation of this doctrine is "counterintuitive to the traditional U.S. view of war."[29] In a traditional war, each side tries to defeat the other through attrition—killing the enemy through overwhelming force and firepower. This is not the case for counterinsurgency. For this reason, Field Manual 3-24 iden-

---

24 David Galula, *Counterinsurgency Warfare: Theory and Practice* (London: Praeger Security International, 1964), 2.

25 Ibid., 52. Galula's "First Law" states that "the support of the population is as necessary for the counterinsurgent as for the insurgent."

26 Ibid., xii. Galula focuses on the strengths and weaknesses of the insurgent by showing that the insurgent cannot defeat the counterinsurgent in a conventional fight and the counterinsurgent cannot win the unconventional fight with its superior military might. See also Memorandum from General Stanley McChrystal, 6 July 2010.

27 Ibid., 9.

28 Memorandum from General David Petraeus, 1 August 2010. General Petraeus stated, "The decisive terrain is the human terrain. The people are the center of gravity. Only by providing them security and earning their trust and confidence can the Afghan government and ISAF prevail."

29 FM 3-24, *Counterinsurgency*, para. 1-148.

tifies a number of paradoxes that are worth highlighting.[30] One paradox is "[s]ometimes, the more force is used, the less effective it is."[31] Another paradox is "[s]ometimes doing nothing is the best reaction."[32] Also, "[s]ome of the best weapons for counterinsurgents do not shoot."[33] Generals McChrystal and Petraeus incorporated many of these paradoxes in their directives in Afghanistan.[34] In addition, based on United Nations Assistance Mission-Afghanistan (UNAMA) data, the civilian deaths attributable to ISAF have decreased by 30 percent since the issuance of General McChrystal's tactical directive.[35] Thus, understanding and applying these contradictions may in fact contribute to decreasing civilian casualties.

These paradoxes show that counterinsurgency is different from conventional warfare. Counterinsurgency requires the minds of leaders and troops to be adaptive, which impacts not only battle drills and standard operating procedures but also the rules of engagement. Specifically, in counterinsurgency, commanders are required to simultaneously minimize civilian casualties and not limit the inherent right of self-defense.[36] The standing rules of engagement (SROE) for U.S. forces define the inherent right of self-defense.[37] Although there are multiple theater-specific ROE that are more relevant to the current counterinsurgency fight, the SROE are the proper starting point for analyzing the application of the ROE.

---

30 Ibid.
31 Ibid., para. 1-150.
32 Ibid., para. 1-152.
33 Ibid., para. 1-153.
34 Memorandum from General David Petraeus, 1 August 2010.
35 United Nations Assistance Mission Afghanistan, "Afghan Civilian Casualties Rise 31 Percent in First Six Months of 2010," press release, 10 August 2010, http://unama.unmissions.org/ Default.aspx?tabid= 1760&ctl=Details&mid=2002 &ItemID=9958.
36 Colonel Richard C. Gross (Staff Judge Advocate, U.S. Central Command), telephone interview, 13 January 2011. Colonel Gross was the Staff Judge Advocate for ISAF during the implementation of General McChrystal's tactical directive.
37 Chairman of the Joint Chiefs of Staff, Instr. 3121.01B, 2.

The Application of the Rules of Engagement
in a Counterinsurgency

> *A soldier fired upon in conventional war who does not fire back with*
> *every available weapon would be guilty of a dereliction of his duty;*
> *the reverse would be the case in counterinsurgency warfare, where*
> *the rule is to apply the minimum of fire.*

-David Galula, *Counterinsurgency*
Warfare Theory and Practice[38]

The current SROE for U.S. forces "establish fundamental policies and procedures governing the actions to be taken by U.S. commanders and their forces during all military operations."[39] These rules set the foundation for the development of theater-specific ROE in the form of mission-specific ROE or supplemental measures.[40] In fact, because the SROE are fundamentally permissive, all commanders must notify the secretary of defense of any further restrictions placed on the SROE.[41] The SROE provide definitions and procedures for the use of force in self-defense and encompass proportionality.[42] These concepts of self-defense and proportionality are at the heart of the friction between counterinsurgency doctrine and the rules of engagement.

The SROE have included the inherent right of self-defense since the first draft over 10 years ago.[43] The SROE describe the inherent right of defense by stating that

> Unit commanders always retain the inherent right and obligation to exercise unit self-defense in response to a hostile act or demonstrated hostile intent. Unless otherwise directed by a unit

---

38 Galula, *Counterinsurgency Warfare Theory and Practice,* 66.
39 Chairman of the Joint Chiefs of Staff, Instr. 3121.01B , 1.
40 Ibid., 2.
41 Ibid., I-1. "SROE are fundamentally permissive in that a commander may use any lawful weapon or tactic available for mission accomplishment, unless specifically restricted by approved supplemental measures . . . "
42 Ibid., A-2, A-3.
43 Ibid., A-2; see also Chairman of the Joint Chiefs of Staff Instr. 3121.01A, *Standing Rules of Engagement/Standing Rules for the Use of Force for U.S. Forces,* 13 January 2000.

commander as detailed below, military members may exercise individual self-defense in response to a hostile act or demonstrated hostile intent. When individuals are assigned and acting as part of a unit, individual self-defense should be considered a subset of unit self-defense. As such, unit commanders may limit individual self-defense by members of their unit.[44]

Under the current SROE, individual self-defense is not without limits and the unit commander regulates this individual right, which differs from the 2000 SROE. The previous definition separated unit self-defense from individual self-defense and defined individual self-defense as

The inherent right to use all necessary means available and to take all appropriate actions to defend oneself and U.S. forces in one's vicinity from a hostile act or demonstrated hostile intent is a unit of self-defense. Commanders have the obligation to ensure that individuals within their respective units understand and are trained on when and how to use force in self-defense.[45]

The difference between these definitions is important to the counterinsurgency fight for two reasons. First, this change allows the unit commander to control the use of force in self-defense situations. The commander has the discretion to respond, or not respond, to a hostile act by an insurgent that is designed to create civilian casualties. For example, a common tactic of insurgents in Afghanistan is to commit a hostile act "with the primary purpose of enticing counterinsurgents to overreact, or at least to react in a way that insurgents can exploit."[46] On one occasion, the Taliban "held a wedding party hostage" while engaging soldiers to provoke a violent response, which would create civilian casualties.[47] Contrary to the Taliban's intent, the on-scene commander in this instance had the discretion to limit the indi-

---

44 Chairman of the Joint Chiefs of Staff, Instr. 3121.01A, A-4.
45 Ibid., A-3.
46 FM 3-24, *Counterinsurgency*, para. 1-152.
47 Dunlap, "Lawfare: Decisive Elements of 21st Century Conflicts?"

vidual right of self-defense to avoid civilian casualties, which is consistent with General Petraeus' guidance and the SROE.[48] The second difference is the persistent, dated belief that the right of individual self-defense is absolute.[49] Thus, the 2005 SROE clarify the role and limits of individual self-defense as a subset of unit self-defense. Any use of force in self-defense, however, whether individual or unit self-defense, must comply with the principle of proportionality.[50]

The principle of proportionality is an indispensable part of the use of force in self-defense situations. The SROE's definition of proportionality, however, is often confused with the law of war principle of proportionality, which mandates that the "loss of life and damage to property must not be out of proportion to the military advantage to be gained."[51] The practical application of this principle in counterinsurgency is a paradox in and of itself.[52] While the law of war's principle of proportionality universally applies in any targeting decision,[53] the SROE's principle of proportionality only applies in self-defense.[54] The SROE stipulates that

> The use of force in self-defense should be sufficient to respond decisively to hostile acts or demonstrations of hostile intent. Such use of force may exceed the means and intensity of the

---

48 Memorandum from General David Petraeus, 1 August 2010; Chairman of the Joint Chiefs of Staff, Instr. 3121.01B, A2.

49 Rajiv Chandrasekaran, "Petraeus Reviews Directive Meant to Limit Afghan Civilian Deaths," *Washington Post,* 9 July 2010, http://www.washingtonpost.com/wp-dyn/content /article/ 2010/07 /08/ AR 2010070806219_2.html?sid=ST2010070905635; see also John J. Merriam, "Natural Law and Self Defense," *Military Law Review* 206 (2010).

50 Chairman of the Joint Chiefs of Staff, Instr. 3121.01B, A-3.

51 Department of the Army, FM 27-10, *The Law of Land Warfare* (Washington, DC: Headquarters, Department of the Army, 1956) para. 41 (15 July 1976) C1.

52 Matthew L. Beran, "The Proportionality Balancing Test Revisited: How Counterinsurgency Changes Military Advantage," *Army Lawyer,* August 2010, 1, 4.

53 Protocol Additional to the Geneva Conventions of 12 August 1949, and Relating to the Protection of Victims of International Armed Conflicts (Protocol I) annex I, 8 June 1977, 1125 U.N.T.S. 3.

54 Chairman of the Joint Chiefs of Staff, Instr. 3121.01B, A-3.

hostile act or hostile intent, but the nature, duration, and scope of force used should not exceed what is required.[55]

Although unit commanders have control over the proportional response in self-defense situations,[56] higher-level commanders have directed de-escalation in areas where civilians may be present.[57]

One key component of the strategy in Afghanistan to reduce civilian casualties is the focus on an element of de-escalation, withdrawal,[58] prior to employing airstrikes or other uses of force in residential compounds.[59] Consequently, under this strategy, troops are required to consider other courses of action short of the use of force, including withdrawal, in response to a hostile act or demonstration of hostile intent. This type of de-escalation is consistent with General Petraeus's statement that "[e]very Afghan civilian death diminishes our cause. If we use excessive force or operate contrary to our counterinsurgency principles, tactical victories may prove to be strategic setbacks."[60] Although this strategy has achieved some success in controlling the use of force to reduce civilian casualties, some legal scholars have debated the effectiveness of the current ROE and the law of war in a counterinsurgency.[61] Hence, applying the rules of engagement in a popu-

---

55 Ibid., 2
56 Ibid., A-2.
57 Chandrasekaran, "Petraeus Reviews Directive."
58 The term "withdrawal" under the SROE's principal of de-escalation allows the enemy, "when time and circumstance permit," the "opportunity to withdraw or cease threatening actions." Chairman of the Joint Chiefs of Staff, Instr. 3121.01B, A-3. For the purposes of this section, the term is used consistent with the directives in Afghanistan, which require ISAF to consider withdrawing to de-escalate rather than escalate force in residential areas.
59 Brendan Groves, "Civil-Military Cooperation in Civilian Casualty Investigations: Lessons Learned From the Azizabad Attacks," *Air Force Law Review* 65 (2010): 1–5.
60 International Security Assistance Force, "Updated Tactical Directive Emphasizes Disciplined Use of Force," press release, 4 August 2010, http://www.isaf.nato.int/article/isaf-releases/index.php.
61 David E. Graham, "Counterinsurgency, The War on Terror, and the Law of War: A Response," *Virginia Law Review* 95 (2009).

lation-centric operating environment is not an easy task. In order to prepare leaders, Marines, and soldiers for the challenging tasks associated with counterinsurgency, commanders have to use the existing rules of engagement concepts and develop training that complies with not only the rules of engagement but also the counterinsurgency doctrine.

COUNTERINSURGENCY RULES OF ENGAGEMENT TRAINING

Soldiers execute in the manner they train; they will carry out their tasks in compliance with the ROE when trained to do so.

-FM 1-04, *Legal Support to the Operational Army*[62]

The unique challenges of counterinsurgency require leaders to train soldiers and Marines for a decentralized fight in a complex environment.[63] The general trends in Afghanistan "indicate the need for decentralized positions, distributed operations, effective small unit[64] leaders, and well-trained small units that must bear the brunt of close combat."[65] Furthermore, the relative adaptability of the enemy compounds the challenges these small unit leaders face.[66] With these challenges in mind, leaders and judge advocates should focus counterinsurgency ROE training toward empowering small unit leaders to make critical decisions on the application of force. In order to achieve this goal, the training must be rooted in principles[67] and reinforced regularly.[68]

---

62 Department of the Army, FM 1-04, *Legal Support to the Operational Army* (Washington, DC: Headquarters, Department of the Army, 2009), A-44.
63 Michael A. Vane and Robert M. Toguchi, "Achieving Excellence in Small-Unit Performance," *Military Review* (May-June 2010): 73.
64 Ibid., 74.
65 Ibid., 73.
66 FM 3-24, *Counterinsurgency*, para. 1-155.
67 Mark S. Martins, "Rules of Engagement for Land Forces," 83.
68 FM 3-24, *Counterinsurgency*, para. D-8.

Training ROE Principles for the Decentralized Counterinsurgency Fight

> You must train the squad leaders to act intelligently and independently without orders.
>
> -David Kilcullen, *Counterinsurgency*[69]

In counterinsurgency, as in most conflicts, leaders, Marines, and soldiers face situations where principle-based decisions are more effective than adherence to hard and fast rules.[70] When these "principles conform both to tactical wisdom and to the relevant legal constraints on the use of force, then the larger system of ROE governing the ground component in a particular deployment will best serve military objectives and national interest."[71] The two relevant principles to the military objective in counterinsurgency are self-defense and proportionality. Judge advocates must develop training for small unit leaders to educate their troops on these two principles. To be effective, judge advocates should leverage Situational Training Exercise (STX) lanes[72] as the primary forum for company commanders and senior noncommissioned officers to train their troops.

The Situational Training for Self-Defense and Proportionality

The best method for teaching the application of self-defense and proportionality to counterinsurgency is through situational training.[73] This type of training "focuses on one or a small group of tasks—within a particular mission scenario—and requires that soldiers practice until they perform the task to standard."[74] Decentralized ROE training requires commanders and judge advo-

---

69 Kilcullen, *Counterinsurgency*, 34.
70 Martins, "Rules of Engagement for Land Forces," 6.
71 Ibid., 6.
72 FM 1-04, *Legal Support to the Operational Army*, para. A-45.
73 Ibid.
74 Ibid.

cates to establish a "uniform standard."[75] Thus, judge advocates should work with the commanders to develop standards for ROE training far in advance of a deployment to a counterinsurgency fight.[76] After establishing these standards, the judge advocate develops realistic ROE vignettes based on lessons learned in theater.

The most effective ROE vignettes are from the relevant theater of operation. Realistic ROE vignette training provides "a window into how [a] soldier thinks" and gives the leader an "opportunity to train the soldier and teach him a different way of looking at the situation."[77] Hence, ROE vignette training provides the proper forum for judge advocates to assist leaders with training soldiers on the application of self-defense and proportionality in a counterinsurgency. At the completion of vignette development, the judge advocate should identify the proper trainers at the company and platoon levels.

Empowering Small Unit Leaders for Counterinsurgency ROE Training

Training a brigade combat team on the ROE is a difficult task for judge advocates because of limited legal assets at the brigade and battalion level.[78] A recent after action report (AAR) comment from Afghanistan shows that "[i]t is very hard for a brigade legal team to train and educate a 6,000-person BCT on . . . the rules of engagement on a regular basis."[79] To alleviate this burden, judge advocates should conduct ROE training that empowers senior noncommissioned officers and company commanders to train

---

75 Howard H. Hoege III, "ROE . . . Also a Matter of Doctrine," *Army Lawyer*, June 2002, 1, 5.

76 Martins, "Rules of Engagement for Land Forces," 83.

77 "Leading Our Soldiers to Fight with Honor," *Army*, November 2006, 62, *www3.ausa.org/pdfdocs/armymag/nov06/cc_nov.pdf*.

78 Center for Law and Military Operations, The Judge Advocate General's Legal Center and School, *After Action Report: 4th Brigade Combat Team, 4th Infantry Division, 2009-2010*, 10 August 2010, 18.

79 Ibid.

their subordinates on the ROE.[80] Thus, the judge advocate is the primary trainer for the company commanders and the senior non-commissioned officers. These company-level leaders will be the primary trainers for their troops. The efficacy of this approach depends on timing; therefore, the training should start well in advance of a deployment.[81]

Many judge advocates in the field have insufficient time to conduct adequate ROE training for small unit leaders in advance of their deployment.[82] Due to competing requirements, these judge advocates are unable to conduct STX lanes with the unit on the ROE prior to the combat training center rotation.[83] Furthermore, at the combat training centers, ROE training by the units is typically limited to a vignette-driven briefing that is not integrated into STX lanes.[84] These current ROE training trends show that units are not incorporating the ROE into their collective training events during their predeployment timeline.[85]

A unit's timeline for deployment flows from its force generation cycle, which typically includes three distinct phases of training: individual, collective, and mission readiness training.[86] The critical phases for ROE training are the first two phases where the

---

80 Hoege, "ROE . . . also a Matter of Doctrine," 4.
81 Martins, "Rules of Engagement for Land Forces," 82.
82 Captain Matthew Lund (senior observer/controller, Joint Readiness Training Center, Charlottesville, VA), interview, 4 March 2011; Major William Johnson, e-mail message, 24 February 2011 (on file with author). Both of these officers conduct mission readiness exercises for units deploying to Afghanistan. International Operational Law Department, The Judge Advocate General's Legal Center and School, JA 422, *Operational Law Handbook* (2010).
83 Captain Matthew Lund, interveiw.
84 Major William Johnson, e-mail message. All brigade combat teams deploying to Afghanistan execute a mission readiness exercise at one of the three combat training centers.
85 Marine Corps Order 3502.6, "Marine Corp Force Generation Process," 26 April 2010; Department of the Army, FM 7-0, *Training for Full Spectrum Operations* (Washington, DC: Headquarters, Department of the Army, 2008), para. 4-1.
86 FM 7-0, *Training for Full Spectrum Operations,* para. 4-1.

units conduct individual and collective training.[87] Judge advocates should endeavor to conduct all the ROE individual training during the first two phases. The audience for the initial individual and collective training should be the company commanders and senior noncommissioned officers.[88]

Early execution of the situational training for company commanders and senior noncommissioned officers will allow them to incorporate ROE vignettes into the squad-level collective training. The judge advocate plays a supervisory role in the latter phases of this process by getting feedback[89] from the collective training and after action reviews during the combat training center rotation.[90] The benefit of this approach to ROE training is that it produces more trainers at the company level and below, which enables frequent rules of engagement training.[91] More ROE trainers in the small units are invaluable to providing the necessary feedback for the unit to conduct ROE refresher training in theater.

Periodic ROE Reinforcement Training in Theater

Training counterinsurgents in ROE should be reinforced regularly.

-FM 3-24, *Counterinsurgency*[92]

The complex counterinsurgency environment often requires units to update and/or change their initial plans to meet the demands of the dynamic operational environment.[93] Furthermore, this environment entails a "cycle of adaptation . . . between insurgents and counterinsurgents; both sides continually adapt to neutralize

---

87 Ibid., para. 4-3–4-4.
88 Ibid., para 4-20–4-23.
89 Hoege, "ROE . . . Also a Matter of Doctrine," 5.
90 FM 7-0, *Training for Full Spectrum Operations*, para. 4-188.
91 Hoege, "ROE . . . also a Matter of Doctrine," 5.
92 FM 3-24, *Counterinsurgency*, para. D-8. See also Department of the Army, FM 3-0, *Operations* (Washington, DC: Headquarters, Department of the Army, 2008), para. A-24.
93 Ibid., para. 5-114.

existing adversary advantages and develop new (usually short-lived) advantages of their own. Victory is gained through a tempo or rhythm of adaptation that is beyond the other side's ability to achieve or sustain."[94] In order to keep pace with this cycle of adaptation, ROE training should be continuous throughout the deployment.[95] Judge advocates should leverage the unit's update briefs and the small unit leadership to adjust ROE training to enemy tactics and distribute training resources to the lowest level. The small unit leaders, who understand ROE, will not only disseminate these training resources but also provide input on their relevance and effectiveness.

Most units in theater have some form of update brief on a daily or weekly basis, which provides the staff and the commanders with situational awareness.[96] These update briefs will provide the requisite situational awareness[97] to develop new ROE vignettes. The shift change briefing is a briefing conducted by the staff, which includes significant enemy activity over a 24-hour period.[98] During this briefing, the intelligence section briefs "significant enemy actions" and "changes in the most likely enemy courses of action."[99] This portion of the brief gives the legal team a snapshot of enemy activity,[100] which will enable the team to identify trends and update the vignettes for periodic ROE training. While judge advocates use these briefs to gain situational awareness,

---

94 Ibid.
95 See Center for Law and Military Operations, The Judge Advocate General's Legal Center and School, *After Action Report, Special Forces Task Force-81, 2009-2010*, 5 October 2010, 8.
96 Department of the Army, FM-Interim 5-0.1, *The Operations Process* (Washington, DC: Headquarters, Department of the Army, 2006), para. 2-76 (change 1, 14 March 2008). Situational awareness is "knowledge of the immediate present environment including the knowledge of METT-TC." (METT-TC is an acronym for Mission, Enemy, Terrain, Troops, Time, and Civil Considerations.) Ibid.
97 FM 3-24, *Counterinsurgency*, para. 4-22.
98 FM-Interim 5-0.1, *The Operations Process*, para. 2-76.
99 Ibid., para. 2-77.
100 Ibid.

they should also seek feedback from the results of mandatory investigations[101] related to the ROE and their primary trainers—company commanders and senior noncommissioned officers. These investigations often highlight deficiencies in a unit's understanding or application of the ROE.

Since the company leaders "bear the brunt" of the combat operations in counterinsurgency, these leaders are the subject matter experts on enemy tactics and trends. The company commanders rely on their company intelligence support teams to provide them with the updated enemy situation, analysis, and trends.[102] Consequently, prior to developing ROE refresher training in theater, judge advocates should seek input from company commanders and senior noncommissioned officers. The input from these leaders will enhance the relevance and effectiveness of the training. After gathering all the input from the shift change briefs, the investigations, and the company leaders, the judge advocate develops and disseminates updated ROE vignettes for refresher training.[103]

Small unit leaders have multiple methods of conducting refresher training and reinforcing the ROE at their level outside of the standard classroom briefing. These leaders can incorporate the

---

101 Department of the Army, Reg. 15-6, *Procedures for Investigating Officers and Boards Officers* (Washington, DC: Headquarters, Department of the Army, 2006), para.1-6.

102 Rod Morgan, "Company Intelligence Support Teams," *Armor* (July-August 2008): 23–24, http://www.dami.army.pentagon.mil/site/dig/documents/COIST-Armor%20Magazine-JUL-AUG08.pdf.

103 One method of disseminating the updated ROE vignettes is by fragmentary order (FRAGO). A FRAGO is an "abbreviated form of an operation order issued as needed after an operation order to change or modify that order . . . " Department of the Army, FM 5-0, *The Operations Process* (Washington, DC: Headquarters, Department of the Army, 2010), para. I-9.

updated ROE vignettes in the unit's rehearsals.[104] For the battalion-level operations, company commanders can include the updated ROE vignettes in the unit's combined arms rehearsal.[105] For company-level operations and below, squad leaders can update their battle drills and standard operating procedures in accordance with the latest vignettes. The integration of the ROE into these rehearsals provides the leaders with the necessary knowledge to adapt and continue to achieve the counterinsurgency objectives while not undermining the right of self-defense.

CONCLUSION

What is dubbed the war on terror is, in grim reality, a prolonged, worldwide irregular campaign—a struggle between the forces of violent extremism and those of moderation. Direct military force will continue to play a role in the long-term effort against terrorists and other extremists.

- Robert M. Gates[106]

The counterinsurgency fight will likely persist for the near future,[107] especially in Afghanistan with the uncompromising Taliban.[108] As long as this type of warfare continues, leaders at

---

104 Department of the Army, FM 3-24.2, *Tactics in Counterinsurgency* (Washington, DC: Headquarters, Department of the Army, 2010), para. 4-132. There are five types of rehearsals: "confirmation brief, the back brief, the combined arms rehearsal, the support rehearsal, and the battle drill or SOP rehearsal."

105 FM 5-0, *The Operations Process*, para. E-3. This type of rehearsal is a synchronization tool for subordinate units and occurs after these units receive an operation order.

106 Robert M. Gates, "A Balanced Strategy," *Foreign Affairs*, January/February 2009, http://www.jmhinternational.com/news/news/selectednews/files/2009/01/20090201_20090101_ForeignAffairs_ABalancedStrategy.pdf.

107 Ibid.

108 Bobby Gosh, "Obama Afghanistan Plan Breaks Old Ground," *Time*, 28 March 2011, http://www.time.com/time/nation/article/0,8599,1888257,00.html.

all levels will struggle with the challenges of "winning the hearts and minds" of the local populace[109] and exercising the right of unit self-defense. Unit predeployment ROE training and theater refresher training can assist leaders with clearing some of the "fog of war"[110] related to applying the ROE in the counterinsurgency fight. Incorporating and co-opting leaders and noncommissioned officers into ROE training are vital to soldiers' and Marines' understanding of the application of the ROE in the counterinsurgency fight. These leaders will continually train their troops in theater to adapt to the changing enemy situation and garner the support of the local populace.

---

109 FM 3-24, *Counterinsurgency*, para. A-26.
110 Ruth Wedgewood, "Law in the Fog of War," Time.com, 13 May 2002, http://www.time.com /time /magazine /article/0,9171,1002407-1,00.html.

Rules of Engagement
Law, Strategy, and Leadership
Laurie R. Blank

We can't win without fighting, but we also cannot kill or capture our way to victory. Moreover, if we kill civilians or damage their property in the course of our operations, we will create more enemies than our operations eliminate. That's exactly what the Taliban want. Don't fall into their trap. We must continue our efforts to reduce civilian casualties to an absolute minimum.

-General David Petraeus, U.S. Army[1]

You can't fight a war like this.

-Anonymous U.S. Marine officer[2]

Events of the past several years reinforce that United States and Coalition operations in Afghanistan are a constant testing ground for counterinsurgency doctrine. Less frequently discussed, however, until the past year, is the fact that conflict in Afghani-

I would like to thank the participants in the Aspects of Leadership Symposium at Quantico, Virginia, and the participants in the 4th National Security Law Workshop at the Judge Advocate General's Legal Center and School in Charlottesville, Virginia, for their very helpful insights and comments on earlier drafts.
1 General David Petraeus, *COMISAF's Counterinsurgency Guidance*, Afghanistan, ISAF Headquarters, 1 August 2010.
2 U.S. Marine officer, cited in Rajiv Chandrasekaran, "'This Is Not How You Fight a War': Military Reviews Directive that Restricts Troops in Effort to Limit Afghan Civilian Deaths," *Washington Post*, 9 July 2010.

stan is also a critical test for, and demonstration of, the role that rules of engagement (ROE) play during military operations and conflict. The quotes above—one from General Petraeus' August 2010 COMISAF (Commander of International Security Assistance Force) Guidance and the other from a junior officer on the ground in Afghanistan—highlight the essential role that leadership must play in the dissemination, training, and communication of ROE.

ROE distill law, strategy, and policy into tactical instructions for Marines, soldiers, airmen, and sailors regarding when and against whom they can use force.[3] The law of armed conflict governs conduct during wartime and provides the overarching parameters for the conduct of hostilities and the protection of persons and objects. Strategic policy determines the goals of the overall operation and of specific missions. ROE are then the most specific and direct manifestation of both law and policy on the ground. In effect, ROE tell Marines and soldiers how they can accomplish their mission—who they can kill and what they can destroy in the process of mission fulfillment.

In June 2009, General Stanley A. McChrystal, then-commander of the International Security Assistance Force (ISAF), issued a tactical directive setting forth guidelines and rules for the use of force in Afghanistan, as well as the basic principles behind those rules. The tactical directive established guidelines for using air power and other munitions against enemy targets and for providing support to multinational forces engaged with the enemy on the ground. In the strategic and tactical environment of a counterinsurgency—mixed with counterterrorism operations as well—these guidelines focus on the key overarching goal of minimizing civilian casualties, a critical component of counterinsurgency success.

---

3 This chapter will focus on Marines and soldiers, the primary military forces engaged in operations in Afghanistan.

In the past year, however, some have framed the tactical directive and the ROE for U.S. and other multinational soldiers in Afghanistan as handcuffs that restrict the ability of troops to fight the enemy effectively and endanger service members by ostensibly denying the use of air power to protect troops engaged with the enemy. The mainstream media has frequently highlighted this portrayal, as have blogs and other communications from the frontlines.

However, these critiques show, above all else, a failure to understand the purpose and use of ROE. The current debate also—and more importantly—misses the mark. Readers of the news and the complaints believe that there is a fundamental debate about whether the ROE are correct and are left with the impression that perhaps the rules formulated at the top are problematic or even faulty. An examination of the key goals and principles of counterinsurgency strategy show that this is not the case, as will be discussed briefly below.

This chapter will examine and refocus the debate about ROE to analyze the critical intersection of law, strategy, and leadership that ROE represent during armed conflict. When strategic counterinsurgency goals of minimizing civilian casualties are mistaken for legal rules that do not allow for civilian deaths in wartime, we need to delineate the differences between law and policy, between legal parameters governing the use of force and the targeting of persons and tactical considerations driven by strategic policy. Senior commanders see the value of the ROE in their everyday operations and relationships with local military and government officials, but some enlisted soldiers and Marines, as well as officers, complain about what they view as unreasonable restrictions on their ability to use force. As a result, we need to examine the role that leadership plays in communicating the direct relationship between the overall mission, the law of armed conflict, and the tactical needs on the ground. Articulating the mission and the ROE in a manner that is tactically, operationally, and strategically logical is always central to effective mission success. In the complex counterinsurgency environment of Afghanistan and other contemporary conflicts, doing so requires leaders to take an abstract concept—the broader strategic goal of protecting civilians and "winning hearts and minds"—and translate it into an operational concept.

ROE are a key leadership tool and a key leadership opportunity. ROE are also a critical component of military operations, one in which commanders must be proficient, just as they are proficient in weapons systems, tactics, and other aspects of military operations. And yet ROE are rarely, if ever, discussed as a component of leadership. Rather, ROE are generally referenced as instructions, as the manifestation of the legal framework and the mission at hand—which is, of course, wholly accurate. But it also unfortunately ignores the critical function ROE play as a facet of leadership in military operations, both at the level of the commander's capabilities and expertise and the level of how the mission and the means to accomplish that mission are communicated from top to bottom.

This chapter will first provide foundational background about the law of armed conflict, strategic policy—namely counterinsurgency strategy—and ROE to set the stage for the discussion of leadership as a component of ROE and of ROE as a component of leadership. The second section will focus on ROE and leadership, examining the current debate about the ROE in Afghanistan and how leadership is a critical aspect of communicating the mission in Afghanistan to troops on the ground and the U.S. public. ROE form a critical component of everything the military does, and as such, it is essential to ensure that military leaders all along the chain of command understand, craft, and communicate the ROE properly to multiple audiences.

ROE, Strategy, and the Law of Armed Conflict

ROE are "directives issued by competent military authority that delineate the circumstances and limitations under which U.S. [naval, ground, and air] forces will initiate and/or continue combat engagement with other forces encountered."[4] These circumstances and limitations stem from the three components that contribute to the ROE: law, strategy, and policy.

> U.S. rules of engagement are . . . based upon three pillars: national policy, operational requirements, and law. To be truly effective, the rules of engagement that govern the military forces of the United States must be fully consistent with the political objectives of our national policy, the dictates of the law, and the safety and survival of our forces during the prompt and effective accomplishment of their mission.[5]

As a result, a preliminary examination and understanding of the law governing military operations (the law of armed conflict) and military strategy (primarily counterinsurgency strategy) is useful and relevant as background to any discussion of ROE.

Law of Armed Conflict

The law of armed conflict (LOAC)—otherwise known as the law of war or international humanitarian law—governs the conduct of both states and individuals during armed conflict and seeks to minimize suffering in war by protecting persons not partici-

---

4 Joint Pub 1-02, *Dictionary of Military and Associated Terms*, http://www.dtic.mil/doctrine/dod_dictionary/; see also International and Operational Law Department, *U.S. Operational Law Handbook* (Charlottesville, VA: The Judge Advocate General's Legal Center and School, 2010), www.loc.gov/rr/frd/...Law/pdf/operational-law-handbook_2010.pdf.

5 Richard J. Grunawalt, "The JCS Standing Rules of Engagement: A Judge Advocate's Primer," *Air Force Law Review* 245 (1997): 246-47. See also International Institute of Humanitarian Law, *Sanremo Handbook on Rules of Engagement* (Sanremo, Italy: International Institute of Humanitarian Law, 2009), 1, 6 ("In addition to *self-defence*, ROE will therefore generally reflect multiple components, including political guidance from higher authorities, the tactical considerations of the specific mission, and LOAC.").

pating in hostilities and by restricting the means and methods of warfare.[6] LOAC applies during all situations of armed conflict, with the full panoply of the Geneva Conventions and customary law applicable in international armed conflict and a more limited body of conventional and customary law applicable during non-international armed conflict. It is U.S. policy, however, to apply the full body of the law of war whenever the U.S. military is deployed and in any military operations.[7] In all circumstances, therefore, LOAC provides the basic framework for all actions, obligations and privileges; it is, in essence, the outer parameters for all military conduct.

The law of armed conflict has multiple purposes that all stem from or contribute to the regulation of the conduct of hostilities and the protection of persons and objects affected by conflict. The most obvious, perhaps, is the humanitarian purpose, the focus on protecting persons who are caught up in the horrors of war. Equally important, however, is the regulation of the means and methods of warfare for the direct purpose of protecting those who are fighting—soldiers and others—from unnecessary suffering during conflict. Finally, it is crucial to recognize that the law of war does not exist to inhibit military operations or prevent war; rather, the goal of this body of law is to enable effective, moral, and lawful military operations within the parameters of the aforementioned two protective purposes. Each of these purposes plays a critical role in the development of both standing ROE and mission-specific ROE.

---

6 See International Committee of the Red Cross (ICRC), *International Humanitarian Law in Brief*, ICRC.org, http://www.icrc.org/web/eng/siteeng0.nsf/htmlall/section_ihl_in_brief. The law of armed conflict is set forth primarily in the four Geneva Conventions of 14 August 1949 and their Additional Protocols.
7 Chairman of the Joint Chiefs of Staff, Instr. 5810.01D, Implementation of the DOD Law of War Program, 30 April 2010: "Members of the DOD components comply with the law of war during all armed conflicts, however such conflicts are characterized, and in all other military operations."

Four fundamental principles lie at the heart of the law of armed conflict: distinction, proportionality, military necessity, and humanity. Each of these principles helps to carry out the law's goals of protecting civilians and regulating the conduct of hostilities, and together they create a framework that can guide examination of the obligations and actions of parties to conflicts and the rights and privileges of individuals in the conflict zone. When viewed as a whole, these four principles clearly underline the delicate balance the law strikes between military necessity and humanity.

The principle of distinction requires that any party to a conflict distinguish between those who are fighting and those who are not, and direct attacks solely at the former. Similarly, parties must distinguish between civilian objects and military objects and target only the latter. Article 48 of Additional Protocol I sets forth the basic rule:

> . . . [i]n order to ensure respect for and protection of the civilian population and civilian objects, the Parties to the conflict shall at all times distinguish between the civilian population and combatants and between civilian objects and military objectives and accordingly shall direct their operations only against military objectives.[8]

Distinction thus lies at the core of LOAC's seminal goal of protecting innocent civilians and persons who are hors de combat.

The principle of proportionality requires that parties refrain from attacks in which the expected civilian casualties will be excessive in relation to the anticipated military advantage gained.[9] This

---

8 AP I, art. 48, 8 June 1977, 1125 U.N.T.S. 3. Article 48 is considered customary international law. See Jean Marie Henckaerts and Louise Doswald-Beck, *Customary International Humanitarian Law* (Cambridge: Cambridge University Press, 2005), 1.

9 AP I, art 51(5)(b) (prohibiting any "attack which may be expected to cause incidental loss of civilian life, injury to civilians, damage to civilian objects, or a combination thereof, which would be excessive in relation to the concrete and direct military advantage anticipated").

principle balances military necessity and humanity and is based on the confluence of two key ideas. First, the means and methods of attacking the enemy are not unlimited. Rather, the only legitimate objective of war is to weaken the military forces of the enemy. Second, the legal proscription on targeting civilians does not extend to a complete prohibition on all civilian deaths. The law has always tolerated "the incidence of some civilian casualties . . . as a consequence of military action,"[10] although "even a legitimate target may not be attacked if the collateral civilian casualties would be disproportionate to the specific military gain from the attack."[11] That is, the law requires that military commanders and decision makers assess the advantage to be gained from an attack and assess it in light of the likely civilian casualties. Proportionality is not a mathematical concept, but rather a guideline to help ensure that military commanders weigh the consequences of a particular attack and refrain from launching attacks that will cause excessive civilian deaths. The principle of proportionality is well accepted as an element of customary international law applicable in all armed conflicts.

The principle of military necessity recognizes that a military has the right to use any measures not forbidden by the laws of war "that are indispensable for securing the complete submission of the enemy as soon as possible."[12] Critically, military necessity does not justify departures from the law of armed conflict. A doctrine popular among German theorists at the turn of the twentieth century, called *kriegsraison*, suggested that military necessity should override the law and that one could abandon the laws of

---

10 Judith Gardham, "Necessity and Proportionality in Jus ad Bellum and Jus in Bello," in *International Law, the International Court of Justice, and Nuclear Weapons*, ed. Laurence Boisson De Chazournes and Philippe Sands (Cambridge: Cambridge University Press, 1999), 283–84.
11 *Legality of the Threat and Use of Nuclear Weapons in Armed Conflict*, Advisory Opinion, 8 July 1996, 1996 I.C.J. Reports, 936. (Dissenting Opinion of Judge Higgins, dissenting on unrelated grounds.)
12 Department of the Army, FM 27-10, *The Law of Land Warfare* (Washington, DC: Headquarters, Department of the Army,1956), para 3(a).

war in situations of extreme danger. Never accepted, this doctrine remains simply in the archives of legal history. Most important, military necessity is inherent in existing LOAC norms and incorporated into numerous provisions. It does not exist as a norm separate from existing black letter law that can be presented as an alternative approach. Indeed, in this way, military necessity exists in a delicate balance with the fourth and final core principle of LOAC, the principle of humanity.

The principle of humanity—also referred to as the principle of unnecessary suffering—aims to minimize suffering in armed conflict. To that end, the infliction of suffering or destruction that is not necessary for legitimate military purposes is forbidden. Once a military purpose has been achieved, the infliction of further suffering is unnecessary. For example, if an enemy soldier is "out of the fight" by dint of being wounded or captured, continuing to attack him serves no military purpose. Another facet of this core principle is that weapons causing unnecessary suffering, such as dum-dum bullets or asphyxiating gases, are outlawed. Similarly, direct attacks on civilians serve no military purpose; the principle of humanity affirms the immunity of civilians from attack. Originally set forth in the Martens clause,[13] humanity provides a means to fill in potential gaps in LOAC stemming from an erroneous belief that anything not prohibited is permitted in conflict. It is thus "much more than a pious declaration. It is a general clause, making the usages established among civilized nations, the laws of humanity, and the dictates of public conscience into the legal yardstick to be applied if and when the specific provisions of the Convention and the Regulations annexed to it do not cover spe-

---

13 *Convention with Respect to the Laws and Customs of War on Land*, preamble, July 29, 1899, 32 Stat. 1803, 26 Martens Nouveau Recueil (ser. 2), 949. ("Until a more complete code of the laws of war has been issued, the High Contracting Parties deem it expedient to declare that, in cases not included in the Regulations adopted by them, the inhabitants and the belligerents remain under the protection and rule of the principles of the law of nations, as they result from the usages established among civilized peoples, from the laws of humanity, and the dictates of public conscience.")

cific cases occurring in warfare, or concomitant to warfare."[14] The Martens clause also serves as a constant reminder that the principle of humanity remains relevant and retains its primacy even as new developments—whether in the types of conflicts, technology, or weapons—outpace codification. As the International Court of Justice stated in the *Nuclear Weapons Advisory Opinion*, "the Martens Clause . . . has proved to be an effective means of addressing the rapid evolution of military technology."[15]

These four principles and the related concepts underlie LOAC as found in international conventions, such as the Geneva Conventions and the Additional Protocols, and, equally important, in U.S. codification of the law, particularly in the Law of Land Warfare and other U.S. law of war policy and manuals.[16] They therefore form the backbone of the law informing ROE and the foundation of the specific legal principles that set the outer parameters for conduct during armed conflict and other military operations.

Counterinsurgency Strategy

Counterinsurgency (COIN) fits within the broader category of irregular warfare and includes "military, paramilitary, political, economic, psychological, and civic actions taken by a government to defeat insurgency."[17] Thousands of pages have been written about counterinsurgency and the appropriate strategy, tactics, and policy for battling insurgents. This chapter will not attempt to reinvent the wheel or even delve beneath the surface in discussing COIN strategy. Rather, this brief discussion of COIN strategy and mission goals will simply lay the foundation for understanding how they contribute to the development, implementation, and communication of ROE during irregular warfare and COIN, such as in Afghanistan.

---

14 *United States* v. *Krupp*, 9. Trials of War Criminals Before the Nuremberg Military Tribunals Under Control Council Law No. 10, 1341 (1950).

15 *Legality of the Threat and Use of Nuclear Weapons*, para 78.

16 *Convention with Respect to the Laws and Customs of War on Land.*

17 Department of theArmy, FM 3-24, *Counterinsurgency* (Washington, DC: Headquarters, Department of the Army, 2006), 1-1.

COIN focuses primarily on enabling a legitimate government to create effective governance and therefore, "[t]he cornerstone of any COIN effort is establishing security for the civilian populace. Without a secure environment, no permanent reforms can be implemented and disorder spreads."[18] Indeed, "[b]ecause insurgents gain strength from the acquiescence of the population, the focus of counterinsurgency is building the population's trust, confidence, and cooperation with the government."[19] Just like any other central tenet of any military strategy, this fundamental goal drives the implementation of COIN strategy in the form of the ROE. In particular, unlike conventional warfare, in COIN, winning the trust of the civilian population and building the legitimate government's capacity is equal—perhaps even greater—in importance than killing enemy fighters.

> Like politics, all COIN operations are local, and the people are the prize—. . . not terrain, not a body count, not the number of patrols run or civil affairs projects completed. The security and well-being of the people are the only metrics that determine your success.[20]

The application of force becomes measurably more complicated in such operations. As the U.S. Army and Marine *Counterinsurgency Manual* states, therefore, "[i]n a COIN environment, it is vital for commanders to adopt appropriate and measured levels of force and apply that force precisely so that it accomplishes the mission without causing unnecessary loss of life or suffering."[21]

In the current conflict in Afghanistan, U.S. and allied forces face a challenging task of battling Taliban militants, al Qaeda, and as-

---

18 Ibid, 1-23.
19 Ganesh Sitaraman, "Counterinsugency, the War on Terror, and the Laws of War," *Virginia Law Review* 95 (2009): 1745, 1747.
20 Robert Neller, "Lessons Learned," *Marine Corps Gazette*, 2 February 2010.
21 FM 3-24, *Counterinsurgency*, 1-25. See also David Kilcullen, *Counterinsurgency* (Oxford: Oxford University Press, 2010), 4, who notes that while killing the enemy is an essential part of guerilla warfare, "successful counterinsurgents discriminate with extreme precision between . . . combatants and noncombatants.").

sociated terrorist groups while building a secure environment for the government to develop its capacity, rule of law, and security operations. Minimizing civilian casualties has become a key feature of U.S. strategy in Afghanistan in its effort to win "the hearts and minds" of the local population. As General Stanley McChrystal, former ISAF Commander in Afghanistan, stated in testimony before Congress,

> I would emphasize that how we conduct operations is vital to success. . . . This is a struggle for the support of the Afghan people. Our willingness to operate in ways that minimize casualties or damage, even when doing so makes our task more difficult, is essential to our credibility.[22]

Recent analysis of operations in Afghanistan, the incidence of civilian casualties, and the growth or decrease in insurgent operations in particular areas have confirmed the emphasis on minimizing civilian casualties as a key aspect of U.S. strategy. Thus, civilian casualties serve in many ways as the best recruitment tool the insurgents could have: "The data are consistent with the claim that civilian casualties are affecting future violence through increased recruitment into insurgent groups after a civilian casualty incident."[23]

Rules of Engagement

As stated above, ROE are directives to military forces regarding the parameters of the use of force during military operations. One of the most famous examples of ROE is from the Battle of

---

22 *Hearing to Consider the Nominations of Admiral James G. Stavridis, USN for Reappointment to the Grade of Admiral and to be Commander, U.S. European Command and Supreme Allied Commander, Europe; Lieutenant General Douglas M. Fraser, USAF to be General and Commander, U.S. Southern Command; and Lieutenant General Stanley A. McChrystal, USA to be General and Commander, International Security Assistance Force and Commander, U.S. Forces, Afghanistan Before S. Comm. on Armed Services*, 111th Cong. 11 (2009) (statement of LtGen Stanley A. McChrystal).

23 Luke N. Condra, Joseph H. Felter, Radha K. Iyengar, and Jacob N. Shapiro, *The Effect of Civilian Casualties in Afghanistan and Iraq* (National Bureau of Economic Research Working Paper 16152), 3.

Bunker Hill in the Revolutionary War, when an American officer ordered, "Don't one of you fire until you see the whites of their eyes."[24] ROE are based on three key components: law, strategy, and policy—the legal framework of the law of armed conflict, the military needs of strategy and operational goals, and the national command policy of the United States. Equally important, both the standing ROE (SROE) and mission-specific ROE provide for the right of all troops to use force in self-defense. In all military operations, therefore,

> . . . rules of engagement are designed to provide for the safety and survival of U.S. military forces that come into harm's way and to ensure successful accomplishment of any mission that those forces may be tasked to undertake. Our rules of engagement are also the principal mechanism of ensuring that U.S. military forces are at all times in full compliance with our obligations under domestic as well as international law.[25]

The three purposes of ROE provide a useful foundation for exploring the interrelationship between ROE and leadership. As a political purpose, ROE imbue the actions of commanders with the goals and objectives of national policy. "For example, in reflecting national political and diplomatic purposes, ROE may restrict the engagement of certain targets, or the use of particular weapons systems, out of a desire to tilt world opinion in a particular direction, place a positive limit on the escalation of hostilities, or not antagonize the enemy."[26] An example of policy-based ROE

---

24 Mark S. Martins, "Rules of Engagement for Land Forces: A Matter of Training, Not Lawyering," *Military Law Review* 143 (1994): 34, citing John Bartlett, *Familiar Quotations* (New York: Little, Brown & Company, 1968), 446 and note 1 (attributing slight variations of the same statement to Prince Charles of Prussia, Israel Putnam, and Frederick the Great).
25 Grunawalt, "The JCS Standing Rules of Engagement," 246-7. See also International Institute of Humanitarian Law, *Sanremo Handbook on Rules of Engagement*, 1. ("ROE appear in a variety of forms in national military doctrines, including execute orders, deployment orders, operational plans, or standing directives. Whatever their form, they provide authorisation for and/or limits on, among other things, the use of force, the positioning and posturing of forces, and the employment of certain specific capabilities.").
26 *Operational Law Handbook*, 73–74.

can be seen in Executive Order 11850, which prohibits first use of herbicides and riot control agents without prior approval.[27] Military purpose-based ROE give the commander parameters within which he or she should operate to fulfill the designated mission, such as "granting or withholding the authority to use particular weapons systems or tactics."[28] Finally, in fulfilling legal purposes, ROE ensure that commanders and troops act within the framework of the law of armed conflict and other applicable laws. An example of law-based ROE would be the mandate that "hospitals, churches, shrines, schools, museums, and any other historical or cultural sites will not be engaged except in self-defense."[29] In some circumstances, ROE provide greater restrictions than those the law requires, depending on the needs and objectives of the overall mission.

The SROE apply to all military operations and contingencies outside U.S. territory and to air and maritime defense missions within U.S. territory. "They provide implementation guidance on the inherent right of self-defense and the application of force for mission accomplishment [and] are designed to provide a common template for development and implementation of ROE for the full range of operations, from peace to war."[30] The SROE define individual and unit self-defense, distinguish between the use of force in self-defense and in furtherance of the mission, and provide guidance for understanding the concepts of hostile force, hostile act, and hostile intent. Each Combatant Command has specific ROE as well and every military operation, such as Operation Iraqi

---

27 Martins, "Rules of Engagement for Land Forces," 25.
28 *Operational Law Handbook*, 74.
29 Headquarters, Joint Task Force South, *Operations Order 90-2, ROE Card* (Key West, FL: December 1990), para. L (summarizing ROE stated in annex R of the Corps-level operations order for Operation Just Cause in Panama) (on file with CLAMO). This rule fulfills multiple United States LOAC treaty obligations to protect religious and cultural objects, historic monuments, and hospitals, as long as they are not being used at the time for military purposes.
30 *Operational Law Handbook*, 74.

Freedom or Operation Enduring Freedom, has specific ROE tailored to meet the legal, policy, and military parameters and objectives of the particular operations. Furthermore, multinational operations have multinational and combined ROE. These ROE can often present multilayered challenges, as the U.S. Army JAG School's *Operational Law Handbook* explains:

> Each nation's understanding of what triggers the right to self-defense is often different, and will be applied differently across the multinational force. Each nation will have different perspectives on the LOW [law of war], and will be party to different LOW obligations that will affect its ROE. And ultimately, each nation is bound by its own domestic law and policy that will significantly impact its use of force and ROE.[31]

In essence, therefore, ROE represent the intersection of law, policy, operational strategy, and even diplomacy or multinational coordination, the center of four interlocking frameworks.

THE ROLE OF LEADERSHIP

According to one Marine Corps definition, leadership "is the sum of those qualities of intellect, human understanding, and moral character that enables a person to inspire and to control a group of people successfully."[32] The Army's definition of leadership is similar: "Leadership is the process of influencing people by providing purpose, direction, and motivation while operating to accomplish the mission and improving the organization."[33] The Army's leadership framework highlights three levels: direct leadership, organizational leadership, and strategic leadership. All three are relevant to the instant discussion of ROE and leadership.

---

31 Ibid, 77.
32 Lejeune Leadership Institute, *Leadership Guide*, http://www.au.af.mil/au/awc/awcgate/usmc/leadership_guide.pdf.
33 Department of the Army, FM 6-22, *Army Leadership* (Washington, DC: Headquarters, Department of the Army, 2006), 1-2.

Direct leadership is "face-to-face or first-line leadership"[34] and is about direct, in-person communication. Organizational leaders "influence, operate, and improve their outfits through programs, policies, and systems. They must concern themselves with the higher organization's needs, as well as those of their subordinate units and leaders."[35] Strategic leaders are "responsible for large organizations and influence several thousand to hundreds of thousands of people. They establish force structure, allocate resources, communicate strategic vision, and prepare their commands and the Army as a whole for their future roles."[36]

As one of the primary tools for communicating strategy, the legal framework, and other key considerations relevant to a particular mission or broader military operation, ROE are, in a fundamental way, about leadership. This leadership function has multiple audiences, however, which can be divided into two primary categories. The first, of course, is the military, with numerous audiences, from senior commanders to junior commanders to the Marines and soldiers on the ground. While ROE in the most immediate sense provide Marines and soldiers on the ground with the parameters for the use of force, making them in some sense the most direct audience, ROE are equally important at the higher levels of the military. The second category is policy makers and the general public. In today's globally interconnected world of 24-hour news cycles and the Internet, the Marine or soldier on the ground in a faraway conflict has a much louder voice than in the past. The government also does not exercise the same measure of control over information about a conflict as it might have a century, or even several decades, ago. ROE thus play a role—whether intended or not from the beginning—in communicating the mission to

34 Ibid., 3-7.
35 "Army Leadership: Doctrine and the New FM 22-100," Army Study Guide website, http://www.armystudyguide.com/content/army_board_study_guide_top-ics/leadership/army-leadership-doctrine-.shtml.
36 FM 6-22, *Army Leadership*, 3-7.

the general public and policy makers. The manner in which military leaders convey the ROE and mission parameters to these audiences, and the effectiveness of that communication, is therefore another component of the way ROE and leadership interact in a critical manner. Finally, communicating with both sets of audiences in a consistent manner is thus at the heart of the essential interrelationship between ROE and leadership.

The nature of counterinsurgency in particular brings several important aspects of military leadership to the fore. The Marines, for example, highlight the leadership concept of "decentralization." This concept emphasizes that subordinate commanders should make decisions based on their understanding of the commander's intent[37] using their own initiative, rather than passing questions up the chain of command and waiting for an answer. To do so, junior commanders must have a thorough understanding not just of the senior leadership's intent and objectives, but also of how to apply that understanding to the facts on the ground and developments as they arise. As the Marine field manual, *Leading Marines*, states, "In order to generate the tempo of operations we desire and to best cope with the uncertainty, disorder, and fluidity of combat, command and control must be decentralized."[38] Such decentralized operations inherently produce a greater need for increased understanding of ROE and shared vision. In the same way, the permissive structure of the SROE goes hand in hand with leadership considerations based on trusting those at the lowest level to use the tools and training they have received correctly in the heat of combat. The COIN manual highlights the need to "empower the lowest levels":

---

37 A device designed to help subordinates understand the larger context of their actions. U.S. Marine Corps, MCDP 1, *Warfighting* (Washington, DC: Department of the Navy, June 1997), 88.
38 U.S. Marine Corps, MCWP 6-11, *Leading Marines* (Washington, DC: Department of the Navy, November 2002), 77–78.

Mission command is the conduct of military operations through decentralized execution based upon mission orders for effective mission accomplishment. Successful mission command results from subordinate leaders at all echelons exercising disciplined initiative within the commander's intent to accomplish missions. . . . It is the Army's and Marine Corps' preferred method for commanding and controlling forces during all types of operations. Under mission command, commanders provide subordinates with a mission, their commander's intent, a concept of operations, and resources adequate to accomplish the mission. Higher commanders empower subordinates to make decisions within the commander's intent. They leave details of execution to their subordinates and expect them to use initiative and judgment to accomplish the mission.[39]

ROE are at the center of this system of decentralized command and individual initiative—the parameters and guidance that ROE provide to senior and junior commanders enable them to internalize the intent of the senior leadership and implement it on the ground. In counterinsurgency, these imperatives and challenges are magnified severalfold, both for operational and ethical reasons.

The dynamic and ambiguous environment of modern counterinsurgency places a premium on leadership at every level, from sergeant to general. Combat in counterinsurgency is frequently a small-unit leader's fight; however, commanders' actions at brigade and division levels can be more significant. Senior leaders set the conditions and the tone for all actions by subordinates.[40]

---

39 FM 3-24, *Counterinsurgency*, 1-145.
40 Ibid., 7-1. The COIN manual further emphasizes that "effective COIN operations are decentralized, and higher commanders owe it to their subordinates to push as many capabilities as possible down to their level." Ibid., 1-146.

As this statement from the *U.S. Army Counterinsurgency Manual* explains, the current conflict in Afghanistan, a complex counterinsurgency operation, offers an excellent opportunity to explore how ROE and leadership interact and reinforce each other, in effect. To do so, this section will therefore explore the current debate over the ROE in Afghanistan and the interaction between ROE and LOAC in these situations. Using this discussion as the foundation, this section will then highlight how understanding leadership and its role in military operations and counterinsurgency in particular is essential to addressing this debate and the role of ROE.

THE DEBATE OVER ROE IN AFGHANISTAN

Over the past few years, media coverage of the conflict in Afghanistan has exposed what appears to be a growing debate over the parameters of the ROE for U.S. and NATO forces in Afghanistan. U.S. operations against the Taliban, conducted in and among the civilian Afghan population, often led to unintended civilian casualties. As these incidents grew and the political and diplomatic ramifications increased in the face of anti-U.S. sentiment, the United States began to reconfigure the parameters on its use of force. In June 2009, then-ISAF Commanding General Stanley A. McChrystal issued a tactical directive setting out restrictions on the use of air power in order to hold civilian casualties to a minimum. The directive explained the new parameters as follows:

> We must fight the insurgents, and will use the tools at our disposal to both defeat the enemy and protect our forces. But we will not win based on the number of Taliban we kill, but instead on our ability to separate the insurgents from the center of gravity—the people. That means we must respect and protect the population from coercion and violence—and operate in a manner which will win their support.

This is different from conventional combat, and how we operate will determine the outcome more than traditional measures, like capture of terrain or attrition of enemy forces. We must avoid the trap of winning tactical victories—but suffering strategic defeats—by causing civilian casualties or excessive damage and thus alienating the people.

While this is a legal and a moral issue, it is an overarching operational issue—clear-eyed recognition that loss of popular support will be decisive to either side in this struggle. The Taliban cannot militarily defeat us—but we can defeat ourselves.[41]

The directive thus highlights the policy and operational imperatives driving the restrictions on the use of force and illustrates the components of ROE: legal, operational, and policy.

Over the next year, reports spread about "soldiers frustrated about the increasingly restrictive rules of engagement under which they have to operate [and] troops explain[ing] that they are hamstrung, unable to protect themselves and use their superior firepower to fight the enemy."[42] According to media reports in the United States and other allied countries, there was great discontent among enlisted men and junior officers about the supposed inability to "fight back" in response to Taliban attacks. For example, one British noncommissioned officer was quoted saying, "I agree with [the restrictions] to the extent that previous-

---

41 Memorandum from General Stanley A. McChrystal, *Tactical Directive*, Afghanistan, ISAF Headquarters, 6 July 2009, www.nato.int/isaf/docu/official_texts/Tactical_Directive_090706.pdf.
42 Celeste Ward Gventer, "Why U.S. Soldiers in Afghanistan Are So Frustrated; Restrictive Rules of Engagement Reflect a Deeper Problem: It's Not Altogether Clear Why U.S. Soldiers Are Trying to "Win Over" the Population," *Christian Science Monitor*, 30 June 2010. See also Wesley Morgan, "Weighing Threats and Rules of Engagement in Afghanistan," *New York Times*, 23 August 2010; and Diana West, "Afghan War Rules Endanger U.S. Troops," *Washington Examiner*, 12 September 2009.

ly too many civilians were killed but we have got people shooting us and we are not allowed to shoot back."[43] Soldiers complained about feeling prevented from responding to threats unless they were directly fired upon, and there were numerous allegations that the tight restrictions on air support meant that units had to fight back without this key weapon in the U.S. and allied arsenal against the Taliban. In other situations, some soldiers and Marines have said that they have heard—or been given—two or more conflicting interpretations of the ROE and the tactical directive. In situations where clarity of purpose and clear parameters are critical to efficient and effective military operations, such reports and complaints raise concern. Indeed, uncertainty and misinterpretations of the ROE and the basic parameters of the mission and the use of force seemed to grow amid claims that junior and senior commanders, unsure how to apply the rules, were tightening the restrictions further to avoid the risk of overstepping the bounds set in the tactical directive.

These developments offer a useful example of the dangers in viewing ROE as a restriction, leaving commanders looking for clearance from above or giving too much direction to their subordinates. Doing so can take away critical judgment from the trigger-puller and therefore create either unnecessary dangers (putting Marines or soldiers in harm's way) or automatons blindly following instructions. At times, the result can be cascading problems of getting permission or seeking cover for decisions, particularly when this perception is created at the top and magnified as it goes down the chain of command. Ultimately, this pattern runs directly counter to the fundamental concept of ROE.

Although the concerns within the military remained contained, the media coverage fostered a much greater dissatisfaction with

43 Thomas Harding, "Curbs on Firing at Taliban are Putting Us at Risk, Troops Warn," *Daily Telegraph*, 7 July 2010, 14.

the ROE among the general public—the second audience mentioned above. The American public, and even top civilian leaders and politicians, seemed to believe that U.S. troops in Afghanistan were unable to defend themselves and were forced to withstand attacks by the Taliban without the authority to shoot back.[44] The parents of one U.S. service member killed in Afghanistan charged that "our soldiers are forced to fight with one hand tied behind their backs. They're not allowed to take care of business."[45] In a modification of the actual ROE and the objectives of the tactical directive, such reports presented the parameters for the use of force as requiring U.S. troops to actually take enemy fire before engaging with kinetic force. Nonetheless, in the past few years there has been a growing perception that the United States is struggling in Afghanistan because of the rules we impose on ourselves, because we define the mission in a way that requires limits on the use of force and on the situations in which force is appropriate. This perception itself can be problematic, even if it is based on an inaccurate or partial understanding of the actual situation on the ground. In effect, "[a]s soldiers feel more restricted in using force and as friendly deaths mount, public support for a foreign deployment may fade quickly in a nation that abhors American casualties."[46] Even if tactical victories continue, this erosion of support can spell strategic victory for the opposing side, demonstrating the importance of the general public and the civilian leadership as an audience for the communication of the ROE.

At the same time that these concerns were spreading in the U.S. media and—at least as reported—among U.S. troops in Afghanistan, General David H. Petraeus took over command of ISAF

---

44 One congressman has declared that the ROE "have proved too often to be fatal" and "def[y] our fundamental belief in the right of self-defense." Dan Lamothe, "Rep: Hold Rules of Engagement Hearing Now," *Army Times*, 15 April 2010; see also Chandrasekaran, "This is Not How You Fight a War."
45 David Zucchino, "Military Families Fault Rules of War; Some Call Limits on Troops a Key Reason for the Growing U.S. Toll in Afghanistan," *Los Angeles Times*, 2 September 2010.
46 Martins, "Rules of Engagement for Land Forces," 14–15.

forces in Afghanistan. In his confirmation hearing, General Petraeus promised to engage in a comprehensive and serious review of the ROE in Afghanistan in direct response to such reports. As the media reported widely at the time in the summer of 2010,

> The controversy pits the desire of top military officers to limit civilian casualties, something they regard as an essential part of the overall counterinsurgency campaign, against a widespread feeling among rank-and-file troops that restrictions on air and mortar strikes are placing them at unnecessary risk and allowing Taliban fighters to operate with impunity.[47]

Statistics gathered and analyzed throughout the past few years have demonstrated that the U.S. focus on limiting air power, indirect fires, and minimizing civilian casualties has accomplished just that goal: civilian casualties caused by U.S. and other ISAF forces have decreased significantly.[48] The question thus was not whether the U.S. strategy and ROE effectively achieved the stated goal of reducing casualties among the Afghan civilian population, but whether the parameters set forth to achieve that goal were appropriate for enabling the United States to achieve its overall mission in Afghanistan.

After his review, General Petraeus issued a tactical directive in August 2010 that renewed the focus on limiting the use of force to protect civilians and reduce civilian casualties. Amid a broader discussion of strategic and operational goals in Afghanistan, the *COMISAF Counterinsurgency Guidance* maintains continuity in the area of use of force:

> **Fight hard <u>and</u> fight hard with discipline.** Hunt the enemy aggressively, but use only the firepower needed to win a fight. We can't win without fighting, but we also cannot kill or capture

---

47 Chandrasekaran, "This is Not How You Fight a War."
48 Condra, *The Effect of Civilian Casualties*; Ken Dilanian, "Study: Military Efforts to Prevent Afghan Casualties Help U.S. Troops Too," *Los Angeles Times,* 2 August 2010; Jason Motlagh, "Petraeus' Rules of Engagement: Tougher Than McChrystal's," Time.com, 6 August 2010.

our way to victory. Moreover, if we kill civilians or damage their property in the course of our operations, we will create more enemies than our operations eliminate. That's exactly what the Taliban want. Don't fall into their trap. We must continue our efforts to reduce civilian casualties to an absolute minimum.[49]

This framework maintains the dual strategic imperatives of defeating the Taliban and reducing civilian casualties, and reinforces that the United States and ISAF cannot win and accomplish the mission simply by killing every Taliban militant they find. General Petraeus also issued clear instructions that prohibited junior and mid-level commanders from making the guidance any stricter without his approval, in an effort to address directly concerns about inconsistent application of the ROE and the tactical directive.[50] This additional component to the August 2010 tactical directive highlights the key theme addressed here in this chapter: the role of leadership in the development, implementation, and communication of the ROE.

Understanding the Interplay Between LOAC and ROE

As the previous section explains, LOAC forms the basic parameters for the conduct of hostilities and the treatment of persons and objects during armed conflict. ROE operate within that framework to set the rules for the use of force in the circumstances of the particular military mission at hand, the operational imperatives, and national command policy. The inherent population-focused nature of counterinsurgency is an ideal venue to explore the interplay and distinction between LOAC and ROE.

At the most basic level, counterinsurgency is a strategy employed during armed conflict against one or more insurgent groups. Like other armed conflicts, counterinsurgency operates within the legal

---

49 *COMISAF's Counterinsurgency Guidance*, 1 Aug 2010.
50 Motlagh, "Petraeus' Rules of Engagement: Tougher Than McChrystal's."

framework of LOAC, which applies in all situations of armed conflict. One of the fundamental incidents of armed conflict is the authority to use force against enemy persons and property as a first resort. LOAC thus contemplates—and accepts—that parties will seek to destroy enemy personnel and capabilities and seek the complete submission of the enemy, in accordance with distinction, proportionality, and other key obligations under LOAC. In both international and noninternational armed conflict, enemy personnel can be—and are—targeted on the basis of their status as the enemy, as hostile forces.[51] Although LOAC certainly allows status-based targeting in counterinsurgency, just like any other conflict, as long as parties adhere to the rules of LOAC, strategic and policy imperatives often lead to a narrower approach to fighting insurgents. It is here that we see the distinction between LOAC and ROE in counterinsurgency in particular. The law allows for the killing of enemy personnel, and accepts the incidental civilian casualties that occur as a result of such lawful attacks, but those casualties may well undermine the military's ability to accomplish its broader mission. For this reason, "[s]ome military options, available under both international and national law, may not come within national policy intent, either generally or with respect to a specific operation."[52] Thus,

> When NATO soldiers are doing their best to avoid civilian casualties, tribal elders are more likely to work with them. Intelligence flows over cups of green tea. When the environment is soured by civilian death, however, these elders retreat behind mud walls—brooding and aloof. Likewise, villagers become more likely to respond to insurgent requests to bed down with

---

51 See Nils Melzer, "Interpretive Guidance on the Notion of Direct Participation in Hostilities under International Humanitarian Law," *International Review of the Red Cross* 90 (2008), 991 (adopted by ICRC Assembly, 26 February 2009), http://www.cicr.org/web/eng/siteeng0.nsf/html/review-872-p991; and Jimmy Gurulé and Geoffrey S. Corn, *Principles of Counter-Terrorism Law* (Eagan, MN: West Group, 2011), 70–76.
52 International Institute of Humanitarian Law, *Sanremo Handbook on Rules of Engagement*, 2.

new gun posts. A vicious cycle of killing ensues. This, in turn, creates a lockdown of development resources, which undermines the planned-for American exit.[53]

Or, as one U.S. soldier stated succinctly, "If we could exercise a great deal more violence of action, I think we would be more successful at killing. I don't know how much more successful we'd be at winning the war."[54] The specific limitation on the violence of action here is not LOAC but the ROE, is not the legal framework governing the conduct of hostilities but the policy and strategic objective of protecting the civilian population in order to facilitate the safe and secure environment for the legitimate government to grow and consolidate authority.

In today's conflicts, where one often hears complaints that LOAC-compliant militaries are at a grave disadvantage when fighting enemies who do not abide by the laws of war, and in fact exploit them for tactical and strategic gain, this distinction between LOAC and ROE is often blurred. There is no doubt that fighting an enemy that deliberately intermingles with the civilian population, uses innocent civilians as human shields, and launches attacks from civilian and protected buildings is an extraordinarily difficult and dangerous task. Claims that LOAC is obsolete and cannot work in such conflicts are unfounded, however, and manifest a misunderstanding of the purposes of LOAC and the operational capabilities of the U.S. and other advanced militaries.[55]

---

53 Philip Smucker, "Don't Take the Taliban's Bait; It Would Be Tempting—and a Mistake—to Alter the U.S. Rules of Engagement in Afghanistan. Here's Why," *USA Today*, 18 August 2010.

54 C.M. Sennott, "Petraeus to Review 'Rules of Engagement'; Video: In Afghanistan, US Soldiers Say Lives Are Put in Danger by Rules Intended to Save Civilians." *GlobalPost*, 8 July 2010, http://www.globalpost.com/dispatch/afghanistan/100707/petraeus-review-rules-engagement.

55 See e.g., Laurie R. Blank and Amos N. Guiora, "Teaching an Old Dog New Tricks: Operationalizing the Law of Armed Conflict in New Warfare," *Harvard National Security Journal* 1 (2010), 45–85; Laurie R. Blank, "New Wars, New Rules? Not So Fast," *JURIST*, 12 January 2010.

These claims also rest on a fundamentally incorrect interpretation of the LOAC principle of proportionality and strategic goals
to limit civilian casualties and the relationship between the two.
As explained above, proportionality requires that commanders refrain from launching attacks in which the expected civilian
casualties will be excessive in relation to the military advantage
gained. A disproportionate attack, therefore, is one in which civilian casualties are *excessive*, not one in which civilians are
killed. LOAC thus accepts that there will be civilian casualties,
even while requiring efforts to minimize—but not completely
eliminate—such casualties. Strategic policy and mission imperatives, in contrast, may well seek to eliminate civilian casualties
as much as possible, particularly in counterinsurgency. Although
such ROE have a critical mission purpose, they do not mean that
civilian deaths necessarily constitute violations of LOAC. And
yet the debate surrounding the ROE in Afghanistan, both within
the military and in the broader U.S. public, suggests that some
military personnel and a large proportion of the general public
believe exactly that. These beliefs, like recent claims that because
of the ROE in Afghanistan, U.S. soldiers and Marines are handcuffed in fighting an enemy that does not adhere to the law, show
a misunderstanding of the purpose of the ROE and how the ROE
relate to LOAC. COIN and ROE are ultimately inseparable, and
the relationship between ROE and LOAC is a major component of
education regarding COIN.

COMMUNICATING THE ROE AND THE MISSION

The above discussion points out the components of the ROE and
the misunderstandings and contradictions inherent in the current
debate about the ROE in Afghanistan. Above all else, however,
all of these factors point to leadership: to the fact that ROE are a
key leadership tool—and opportunity—that enables an effective

commander to communicate the relationship between mission accomplishment, the law of armed conflict, and the tactical needs on the ground. For example, no member of the military would hesitate or complain about a highly risky mission to assault a hill if it accomplished a larger mission, even knowing the likelihood of survival would be low. And yet, when the tactical directives and the ROE create risks and dangers for an outcome that is harder to measure ("hearts and minds"), the connection between the two becomes attenuated. Acceptance of risk for mission accomplishment is a core military function; since counterinsurgency requires the assumption of short-term risk for longer-term benefit, that benefit must be articulated and operationalized more effectively.

The senior civilian and military leadership understand the value of ROE in their everyday operations, and how these rules improve relationships with local military and government officials. However, the fact that many complain about the rules as unreasonable restrictions on the ability to use force is clear evidence that the message is not being communicated effectively down the chain of command or out to the general public. In the current conflicts, when strategic counterinsurgency goals of minimizing civilian casualties are mistaken for legal rules that do not allow for civilian deaths in wartime, the differences between law and policy are not understood by either a significant portion of the military or the U.S. civilian population at large.

In the face of the apparent confusion and misinterpretations detailed above, viewing ROE as a leadership tool, as a critical component of military leadership at all levels, is a fundamental aspect of today's military operations. It is established doctrine that "while ROE should never drive the mission, the political, military, and legal forces that may impact the mission and inhibit the use of force must be considered and planned for throughout the planning process."[56] This concept carries directly through to the communication, training, and implementation of ROE as well, especially in COIN, where the nature of individual Marines' or soldiers' interactions with the local population is significant. Although it is axiomatic that ROE must be clear and succinct to be effective, translating the above concepts from the planning process to the dissemination and implementation process becomes a key facet of effective leadership in COIN operations.

A Commander When the Commander's Not There

ROE are "commanders' rules."[57] Judge advocates and others play an important role in the planning, development, drafting, and training of ROE, but commanders must retain control over and authority for ROE throughout the process and, critically, during training and military operations. From the senior leadership to the soldier or Marine on the ground, ROE do more than give guidance for specific uses of force and other actions during military operations. They provide the link to communicate the commander's intent, a critical component of any effective mission and the key to effective operations in COIN. According to warfighting doctrine, commander's intent is "a device designed to help subordinates understand the larger context of their actions. The purpose of providing intent is to allow subordinates to exercise judgment and initiative—to depart from the original plan when the unforeseen occurs—in a way that is consistent with higher commanders' aims."[58] As the Marine *Warfighting Manual* explains,

There are two parts to any *mission*: the task to be accomplished and the reason or intent behind it. The intent is thus a part of

---

56 Center for Law and Military Operations, *Rules of Engagement (ROE) Handbook for Judge Advocates*, (Charlottesville, VA: Center for Law and Military Operations, 2000), 1-1.
57 Ibid., 1-2. See also Joint Pub 1-02, *Operational Law Handbook*, 73 ("ROE ultimately are the commander's rules that must be implemented by the Soldier, Sailor, Airman, or Marine who executes the mission.").
58 MCDP 1, *Warfighting*, 88.

every mission. The task describes the *action* to be taken while the intent describes the *purpose* of the action. The task denotes *what* is to be done, and sometimes *when* and *where*; the intent explains *why*. Of the two, the intent is predominant. While a situation may change, making the task obsolete, the intent is more lasting and continues to guide our actions. Understanding the intent of our commander allows us to exercise initiative in harmony with the commander's desires.[59]

In COIN, the intent—or the why—is not limited to the reason for taking a particular hill or capturing a particular strategic objective. Rather, the purpose of specific missions will be linked directly to the broader purpose of the U.S. mission. Put another way, the nature of COIN means that individual events on the ground can have a significant impact on the broader mission—in effect, the strategic corporal writ large. ROE are thus overall a strategic issue.

In all missions, context is an important piece of ROE development and, in particular, ROE training and dissemination. Context includes understanding who the enemy is, whether there even is an enemy, the nature of the civilian population, and its relationship with hostile forces—all considerations that enable soldiers to take a set of rules and apply them in a given situation. Past analysis of ROE in both training exercises and actual operations has shown that "often the source of confusion in an eighteen-year-old private lies not in the rules of engagement themselves, but in a lack of understanding of the situation."[60] Although this statement is directed at ROE and the soldier or Marine on the ground, it is equally relevant along the chain of command, especially in COIN. If a company commander is not getting the necessary context and message from higher-ups along the chain of command, then we cannot expect that he or she will be able to provide that context

---

59 Ibid., 88–89.
60 Center for Law and Military Operations, *ROE Handbook for Judge Advocates*, 3-22.

to subordinates. As one Marine general explained, understanding how to apply the ROE is not sufficient.

> More importantly, Marines must understand why the ROE exist. Marines need to be "educated" in the application of these rules and the consequences of their application, as well as the potentially disastrous consequences of improper application. . . . [Not] educat[ing] as to the reasoning and applicability of those rules may cause the force to inappropriately act.[61]

This is a two-part leadership issue: first, at the micro level, each level of leadership has an important role to play and contribution to make to ensure that the ROE are communicated effectively; and second, at the macro level, officers of every rank need to be trained in the culture of ROE and in the notion of ROE as critical component of every aspect of military operations, just like a weapons system.

On the micro, or mission-specific, level, the challenges in Afghanistan suggest a lack of sufficient internal communication and a failure to blend the broader mission imperatives with the specific ROE in disseminating the ROE. Like any other battle task, "ROE application is as important to a soldier's success in today's complex environment as the ability to fire and maintain a weapon."[62] All too often, unfortunately, commanders face multiple mission imperatives and delegate ROE training to the judge advocate, which undermines the ability to communicate the commander's intent and to reinforce the essential contextual aspects from the commander's perspective. Indeed, communicating ROE unrelated to self-defense has traditionally been difficult, both from the complexity of the issues and the lack of consistent approach from unit to unit and mission to mission.[63] So how does com-

---

61 Neller, "Lessons Learned," 14.
62 Center for Law and Military Operations, *ROE Handbook for Judge Advocates*, 2-2.
63 Mark Martins, "Deadly Force Is Authorized, but Also *Trained*," *Army Lawyer*, September/October 2001, 16.

mander's intent get communicated? The first step is to identify the intent at the highest level—that found in the tactical directive and other guidance. The natural tendency is to focus on the need for better ROE instruction and training at the unit level, which is important, but doing so fatally ignores the key prerequisite: the communication of the senior leadership's intent to the battalion commander, junior commander, and senior enlisted soldiers and Marines. Here lies the first component in the notion of the ROE as the commander when the commander is not there.[64] COIN demands seamless integration of mission and ROE, and depends on the ability of the junior commander to effectuate the intent of the top leaders and the overall strategic imperatives during daily interaction with the enemy and the local population.

> That's where it's going to be won—corporal, sergeants, lieutenants. That's where you have to focus on because that's who is going to be way out there on the edge of the empire, the pointy end of the spear, like we say. Those are the Marines that are going to make those tough calls and if they're not trained to deal with that type of decision making, if they don't have the requisite excellence and their weapons handling and their small unit tactics, they're not going to be able to do that job.[65]

Leadership in the ROE context cannot start at the junior commander level, therefore, but must come from the top. In order for subordinates to understand the intent of those at the top, i.e., those more than one level directly above them, the senior leadership needs to communicate intent clearly while still facilitating

---

64 See e.g., "The Art of Battalion Command," Counterinsurgency Leadership in Iraq, Afghanistan and Beyond, Marine Corps University Symposium, 23 July 2009, 5 ("And you try to operationalize it because you want them to understand it so that when they're in that point where they have to make a decision and no one's around and it's corporal so and so, he can do it. He knows what Furness would want him to do and that's probably the only thing—if that's the only thing he can remember, it's something he can fall back on and hopefully it gets him through that difficult decision.")
65 Ibid., 4.

the use of initiative at the individual level. "Combat in counterinsurgency is frequently a small-unit leader's fight[, but] commanders' actions at brigade and division levels can be more significant [because] senior leaders set the conditions and the tone for all actions by subordinates."[66] This includes not only communicating the mission and disseminating the ROE, but—and this is the critical step—clearly conveying and demonstrating how the mission, LOAC, and the ROE coalesce into a coherent whole rather than remain three separate imperatives with unclear coordination.

The senior military leadership has a second audience for communication about the ROE: the civilian leadership and the general public. Communicating consistently with both the military chain of command and the general public is a challenge, but in today's interconnected world, it is essential. Further, although the general public is not traditionally a relevant audience regarding ROE, the 24/7 interconnected Internet culture has made it so. Current efforts to convey the imperatives of COIN and how they are carried out through and preserved in the ROE and the tactical directive have fallen short, as evidenced by the debates and misunderstandings over the parameters of the U.S. and ISAF mission in Afghanistan. Diminished public support for the war effort—for whatever reason—among both the general public and the civilian leadership is highly problematic. The fact that the ROE themselves are directly tied to COIN and based on LOAC demonstrates that the source of the debate over the ROE is not the foundations but the message, not the components but the com-

---

66 FM 3-24, *Counterinsurgency*, 7-1. In some cases, the commander's rejection of or disdain for the strategy can create the opposite climate, one of disrespect for the ROE and the mission and even one that enables criminal behavior, such as in the case of Colonel Harry Tunnell, commander of the Army's 5th Stryker Combat Brigade in Afghanistan. . Five soldiers from the brigade have been charged with war crimes, accused of killing unarmed Afghan civilians for sport. . Many attribute the crimes in part to the culture of aggression and disregard for counterinsurgency strategy that Colonel Tunnell fostered. . Craig Whitlock, "Brigade's Strategy: 'Strike and Destroy'," *Washington Post*, 14 October 2010.

munication. That message and communication are the essence of leadership.

At the level of junior commanders and Marines and soldiers on the ground, ROE also play the role of the commander when the commander is not there. Because ROE provide the soldier or Marine with parameters for action, that individual will be operating in a vacuum without an effective understanding of the commander's intent and the way the ROE operationalize the mission and the broader strategy. Simply knowing the rule that indirect fires are prohibited without direct observation is not sufficient. Understanding why and how this rule advances the overall mission and the unit's local efforts creates the key link between the ROE and the soldier's actions to carry them out, between knowing the rule and internalizing it.[67] Eliminating unobserved indirect fires surely helps fulfill LOAC's goal of minimizing civilian casualties and obligation to refrain from attacks that will result in excessive civilian deaths, but is not required to adhere to the law. The rule thus marries LOAC obligations with the COIN goals of putting the local population first and, even more important, plays a role in protecting our own troops. Studies have shown that when U.S. and ISAF attacks cause civilian casualties, insurgent groups gain members and attacks against U.S. and allied forces increase.[68] All of these pieces come together and must be woven together into a complete message for troops on the ground. Those who question or complain about the ROE in Afghanistan argue that these parameters place the safety of Afghan civilians above the safety

---

67 See e.g., FM 3-24, *Counterinsurgency*, 2-14-15 ("2-72: Knowledge of the ROE itself is not sufficient to help Soldiers make informed decisions regarding the appropriate application of force. . . . Effective communication is equally essential. Leaders must ensure that every Soldier completely understands the mission and commander's intent, and has comprehensive situational understanding at all times. The appropriate level of situational understanding, realistic training, and disciplined adherence to basic troop leading procedures equips Soldiers with the tools necessary to make informed decisions regarding the decision to use or refrain from the use of force.").
68 Condra, et al, *The Effect of Civilian Casualties*, 3–4.

of U.S. troops. These claims show both a fundamental misunderstanding of the role of the ROE, COIN, and LOAC, as explained above, but also an absence of effective leadership in communicating the role of the ROE. Still more, these claims demonstrate how ROE themselves can be an effective leadership tool to communicate the broader mission; the safety of Afghan civilians and the safety of U.S. troops often go hand in hand, as studies on the linkage between civilian casualties and insurgent attacks show. ROE make this case and as such, should not only be the subject of leadership efforts but also the venue for leadership, communication, and dissemination.

Leadership and ROE at the Macro Level

Marrying ROE and leadership effectively at the micro level also requires comprehensive thinking and efforts at the macro level. In order to create a truly effective system in which ROE are seamlessly integrated with mission objectives and LOAC, ROE must take on a more central role in military education and training. Officers and NCOs must be trained in the culture of ROE itself, not just the planning, drafting, and dissemination process. In this way, teaching officers about the role of ROE in the broader system—particularly in today's conflicts where COIN and ROE are nearly inseparable in many ways—is leadership at the macro level. Just as officers and NCOs must completely master a weapons system or other component of military operations, so must they fully master ROE as a critical component of everything the military does. One step is to explore the curriculum at the war colleges to see how ROE are currently taught and how to integrate the concept of ROE, and the interplay with COIN and LOAC, more deeply into the curriculum. Military exercises offer another rich opportunity for broader ROE training, beyond the training and instruction in the specific ROE for the exercise. It is often too easy for commanders and planners to turn ROE over to the judge advocates, both for drafting and training, but doing so can take ROE right out of the leadership equation. Exercises, which can be a key tool for learning beyond the training specific to the exercise, thus provide a venue for inculcating ownership over the ROE process from the earliest planning through the constant and repetitive instruction, training, and conversations that make ROE an effective tool in the hands of a commander.

CONCLUSION

Training, dissemination, and implementation of ROE implicate leadership at every level, particularly in a counterinsurgency environment like the current conflict in Afghanistan. The nature of the mission in Afghanistan and the debate over the ROE offer an ideal venue to examine the role that ROE can play as a leadership tool, as well as the opportunity ROE provide to enhance leadership across the spectrum of command, from the top strategic levels to the small unit level. Indeed, the leadership challenges are greater still in multinational operations, where the command structure and organizational resources are more diffuse and dispersed. In such situations, the top commanders must exert significant powers of persuasion and the ability to convince others in the coalition to focus on the same mission objectives and implement that mission in the same manner.

Recognizing the contribution that ROE make to leadership efforts as the critical intersection between law, strategy, and policy during armed conflict—and that leadership makes to the effective implementation of ROE—is an essential step. In keeping with the focus on decentralized leadership and the need to communicate the commander's intent, ROE serve as a commander when the commander is not there, providing the link to the commander's intent and enabling soldiers and Marines to exercise judgment and initiative in accordance with strategic mission imperatives.

For these reasons, it is essential to highlight that ROE is a key aspect of leadership at both the micro and macro levels. First, ROE training, dissemination, and implementation must include clear and concrete explanations and demonstrations of how the commander's intent, the ROE, and LOAC are thoroughly integrated. Second, on the macro level, officers must be trained, educated, and proficient in the culture of ROE as a central component of all military operations. Although the debates over the ROE in Afghanistan may not get it right on the relationship between LOAC, ROE, and COIN, they do reinforce the essential connection and interrelationship between ROE and leadership, one that demands further emphasis.

Humanity in War

Leading by Example: The Role of the Commander in Modern Warfare

Jamie A. Williamson

Even war has its limits. In situations where the parties pitted against each other have resorted to armed force, there are restrictions to the extent and the type of violence that can and should be used, against whom, and what can be targeted. The exercise of humanity in armed conflicts is a thread which, as this chapter explores, needs to run through all aspects of planning and battlefield conduct, with military commanders having to make important and difficult decisions, and crucially, lead by example.[1]

At the outbreak of armed conflicts, international humanitarian law (IHL), also known as "the laws of war," becomes the regulating legal framework for the parties to the conflict.[2] This body of law, termed *jus in bello*, includes the four Geneva Conventions of 1949 and Additional Protocols of 1977 and 2005, the Law of The Hague, and customary international law.[3] It recognizes that in

---

1 The terms "commander" and "superior" are used interchangeably in this paper. The paper focuses on military commanders, not civilian commanders, of all ranks who have members of armed forces under their command.

2 The terms "international humanitarian law" and "laws of war" are used interchangeably in this paper.

3 The United States has not signed either of the Additional Protocols. However, a large portion of Additional Protocol I—including those provisions and principles highlighted in this chapter—are considered to be customary international law by both the U.S. and other countries around the world. See Gary D. Solis, *The Law of Armed Conflict* (Cambridge: Cambridge University Press, 2010), 138: "Having once accepted that 65 percent of the Protocol is customary international law, and necessarily forced to comply with the remaining portion . . . U.S. rejection of Additional Protocol I nears irrelevance."

armed conflicts there will be loss of life and damage to property—incidents that have no place in times of peace. Yet it attempts to limit the suffering caused by conflict and hostilities by inserting certain basics of humanity and minimum standards of behavior, while allowing the parties to the conflict to achieve their military objectives. IHL is about compromise between humanitarian principles and military necessity.[4]

In many regards, the military commander is left to decide how aggressively to engage the enemy. The law will, of course, set the outer framework and determine what actions are deemed lawful even if they result in widespread killing. Yet in exercising that right of action in hostilities, commanders and, by extension, their subordinates, are also to be guided by their own morals and personal standards to act with restraint. In the same way as the medieval Codes of Chivalry and the 19th-century Lieber Code, modern day IHL calls for combatants to behave in a civilized manner, not to cause unnecessary suffering and wanton destruction. In sum, one should act as an "officer and a gentleman," affording humane treatment to all.[5]

In conflict, if the enemy is dehumanized and demonized, it becomes easier for combatants to justify acts of untoward vio-

---

4 International Committee of the Red Cross (ICRC), *Commentary on the Additional Protocols of 8 June 1977 to the Geneva Conventions of 12 August 1949* (Geneva: Martinus Nijhoff, 1987), xxxiv. Put more bluntly by a U.S. commander in Iraq in 2006, "I really had to work to convince them, 'Dude, not everybody needs to get the crap kicked out of him. In fact, beating the crap out of people is wrong, you know? Geneva Conventions? Look it up. It's a concept.'" Jim Frederick, *Black Hearts* (New York: Broadway Paperbacks, 2010), 189.

5 Codes of Chivalry, at their peak in the 12th and 13th centuries, were in essence codes of conduct for knights, and indirectly provided some protection to civilians. See http://www.britannica.com/EBchecked/topic/721819/law-of-war/52915/Roots-of-the-international-law-of-war; Instructions for the Government of Armies of the United States in the Field (Lieber Code), 24 April 1863. The "Lieber Instructions" were prepared during the American Civil War by Francis Lieber, then a professor of Columbia College in New York, revised by a board of officers and promulgated by President Lincoln. They reflected in large part the then-prevailing laws and customs of war. See D. Schindler and J.Toman, *The Laws of Armed Conflicts* (Dordrecht: Martinus Nihjoff, 1988), 3–23.

lence against the adversary. With research showing that moral disengagement of combatants is a gradual process, the threshold of acceptable levels will be constantly pushed. The challenge for commanders is to lead by example and prevent their troops from committing acts of heedless destruction, unnecessary violence, and acts of barbarity.

This chapter recalls that military superiors at all levels have a clear legal obligation to respect and to ensure the respect of the tenets of IHL by their subordinates. Where they fail to do so, they leave themselves open to possible criminal liability. Today's commanders also have to assume a particular role in contemporary irregular-driven conflicts; not only must they ensure compliance with the laws of war, but in addition to their traditional military role, they often must also act as diplomats and politicians in complex, culturally diverse environments to increase chances of achieving the overall mission objective of winning hearts and minds. In so doing, they work toward sustainable peace, which is best served by conducting a "clean war," namely one within the boundaries of IHL.

DISCRETION OF THE COMMANDER IN APPLYING IHL

Conflict is a fickle beast, wreaking havoc, causing suffering, death, and destruction, yet allowing for acts of survival, courage, and humanity. International humanitarian law is its leash, seeking to place restraints on how this beast behaves, limiting the manner in which those engaged in the fighting conduct themselves vis-à-vis the enemy and seeking to ensure a minimum of protection will be afforded to specific categories of persons and property. Commanders are expected to hold the leash and to decide how tightly it needs to be gripped, all the time balancing military necessity and humanity.

The 1949 Geneva Conventions and their Additional Protocols of 1977 and 2005 provide the baselines that must be respected at all times. With over 600 provisions, the Conventions and their Protocols are detailed in many ways and specific to many issues. The Third Geneva Convention, which explains meticulously how prisoners of war are to be handled when they have fallen into the hands of the adversary, exemplifies how detailed IHL can be. Likewise, the Fourth Geneva Convention is extensive in its description of the treatment of civilians, with particular focus on more vulnerable groups, such as children and women. Read alongside the commentaries to the Conventions, military manuals, and doctrine, there seems little left to discretion. The parties to the conflict will struggle to argue against the common sense understanding of many of these provisions.

Yet, despite the best efforts of the drafters of the Conventions to provide clear and measurable rules, more significant areas of IHL, especially in the conduct of hostilities, continue to require a balancing act between, on the one hand, military necessity and operational requirements, and the other, the spirit of the law and humanitarian considerations. Some of the more important IHL concepts, seemingly obvious at face value, such as distinction, precaution, proportionality, and humane treatment, are also subject to this balancing act and the inevitable exercise of discretion by commanders.

The Principles

As noted in the Pictet Commentaries to the Additional Protocols of 1977, there were many differences of opinion during the drafting of these provisions "due to the heavy burden of responsibility imposed . . . on military commanders, particularly as the various provisions are relatively imprecise and open to a fairly broad margin of judgment."[6] The commanders, being at the forefront of

---

6 ICRC, *Commentary on the Additional Protocols of 8 June*, 679.

military operations, therefore translate the law into practice. They have to determine in the fog of war who, what, and how to target during military engagements, always evaluating myriad factors to ensure that their actions are compliant with IHL. A brief review of some of these concepts helps to illustrate this.

The principle of distinction is at the core of IHL, with the parties to a conflict having to direct their attacks only against military objectives and combatants, and not against civilians and civilian objects. Therefore, a party has to assess, before launching an attack, whether a potential target is a military objective, defined as "objects which, by their nature, location, purpose, or use make an effective contribution to military action and whose total or partial destruction, capture or neutralization, in the circumstances ruling at the time, offers a definite military advantage."[7] Any decision about whether an object is protected against attack depends on the information available to the commander and the circumstances at the time.[8]

Once it has been determined that a potential target is a military objective, the party considering launching the attack must make a proportionality assessment, as well as review the pattern of life, so as to limit any collateral damage. Thus, if an attack is expected to cause incidental loss of civilian life, injury to civilians, damage to civilian objects, or a combination thereof, which would be excessive in relation to the concrete and direct military advantage anticipated, it is to be cancelled or suspended.[9] Some have argued that proportionality is often "inexact in application" requiring

---

7 Additional Protocol I, to the Geneva Conventions, Art. 52(2).

8 Referred to as the *Rendulic* rule by the U.S. JAG, "the circumstances justifying destruction of objects are those of military necessity, based upon information reasonably available to the commander at the time of his decision." International and Operational Law Department, *U.S. Operational Law Handbook* (Charlottesville, VA: The Judge Advocate General's Legal Center and School, 2010), www.loc.gov/rr/frd/...Law/pdf/operational-law-handbook_2010.pdf

9 Additional Protocol I, to the Geneva Conventions, Art. 51(5)(b) and Art. 57(2)(b).

combatants to make "estimates" of likely harm.[10] Put another way, "it must be difficult for people who have not served in the military themselves to judge what is, or is not, 'reasonable' in confused and dangerous conditions."[11]

Throughout the planning and execution phase of operations, the parties are required to exercise the appropriate level of precaution. Care shall be taken in warfare to protect the natural environment against widespread, long-term, and severe damage;[12] to spare civilians and civilian objects;[13] and the parties are to exercise all "feasible precautions" in the choice of means or methods of attack in order to avoid and minimize collateral damage.[14] As with the proportionality assessment, such exercises of "care" and "precaution" are left to commanders to appreciate.

Civilians shall enjoy protection unless and for such time as they take a direct part in hostilities.[15] Determining at what point a civilian is said to be taking a direct part in hostilities is a difficult task, especially when one is in a "troops in contact" situation. As the International Committee of the Red Cross's (ICRC) guidance on direct participation and related articles published in response to it have shown, consensus was difficult to reach on a number of key issues. Practitioners, military operators, and academics, for instance, expressed differing views on which acts are "preparatory," when participation ends, if the person "regains" civilian status and the related protection after having participated in hostilities. Deciding on acceptable criteria as to who is a member of an organized group and of a declared hostile group was likewise a chal-

---

10 Bill Rhodes, *An Introduction to Military Ethics: A Reference Handbook* (Santa Barbara, CA: ABC-CLIO, 2009), 109.
11 Mike Jackson, *Soldier: The Autobiography* (London: Bantam Press, 2007), 421.
12 Additional Protocol I, to the Geneva Conventions, Art. 55(1).
13 Ibid., Art. 57(1).
14 Ibid., Art. 57(2)(b), and also Art. 58, generally.
15 Ibid., Art. 51(3).

lenging exercise for the experts.[16] Whereas experts might disagree on the legal aspects, the hard decisions on whether or not a particular individual has lost his protection are left to the commanders and their troops in theater. Most cases are clear-cut, yet there are times when a line cannot easily be drawn, and an inkling of doubt subsists. For the superior, the choice may be between jeopardizing the success of a particular mission and other goals, such as winning hearts and minds, to benefit the overall campaign.

All persons who do not or who no longer take part in hostilities are to be treated humanely in all circumstances.[17] Likewise, individuals who have been deprived of liberty are to be afforded humane treatment.[18] Although a commander responsible for detainee operations will be guided by his society's standards as to what constitutes "humane," these may nonetheless contradict cultural and religious dictates of the detainee, as was amply demonstrated over the last decade. Again, apart from obvious acts of ill treatment or torture, the commander will have to assess a multitude of factors, including behavior and attitude of guards vis-à-vis the detainees, appropriate disciplinary measures, and force protection constraints and will have to take into account religious, cultural, and dietary considerations of detainees in ensuring that any detention regimen as laid out in the relevant operating procedures is humane.

The Commander's Choice

The commander's exercise of discretion in applying these principles not only impacts how operations are conducted but can also have a bearing on how the war is viewed, on swaying public

16 Nils Melzer, *Interpretive Guidance on the Notion of Direct Participation in Hostilities under IHL* (Geneva: ICRC, 2009), http://www.icrc.org/eng/resources/documents/feature/direct-participation-ihl-feature-020609.htm.
17 Additional Protocol II to the Geneva Conventions, Art. 4(1) and Common Article 3 to the Geneva Conventions.
18 Common Article 3 of the Geneva Conventions.

opinion at home, on keeping troops' morale high, and on gaining the support of the local population. The U.S. Army is very much of the view that compliance with the laws of war not only enhances public support of military missions but can also end conflict more quickly.[19] Similarly, the Pictet Commentaries to the Additional Protocols have argued that the faithful application of IHL, by limiting the effects of hostilities, can contribute to reestablishing peace.

The amount of force and methods of warfare that are used on the adversary and the perception of the local population are central to counterinsurgency and can have major repercussions in winning hearts and minds. As the British army field manual underscores, "A society's view of acceptable levels of coercion and force are often closely linked to its cultural norms. How a society or groups within it regard the use of coercion and force are likely to have a bearing on how military operations are perceived. It is therefore an important factor to take into account when planning and conducting operations and when reacting to events."[20] In addition to its obvious negativity, perception of an excessive use of force can markedly increase the risk of individuals deciding to retaliate by joining or actively supporting the insurgency, whether by becoming directly engaged or through provision of safe houses, information gathering, and hiding places for weapons caches:

> Force may solve a tactical problem—a firing point neutralized, a fleeting target engaged, or a strongpoint destroyed—but if the use of force is perceived as excessive or ill targeted, the neutral segment of the population may be antagonized or alienated and it may leave a lasting feeling of resentment and bitterness. Worse still, active support for the insurgents by those suffering or observing the effects of force may be engendered. This is

19 *U.S. Army Operational Law Handbook*, 37.
20 *British Army Field Manual, Volume 1, Part 10, Countering Insurgency,* Army Code 71876, October 2009, 3-15, *news.bbc.co.uk/2/shared/bsp/hi/pdfs/16_11_09_army_manual.pdf.*

particularly so in those tribal cultures where codes of personal and family honor, justice, and vengeance are strong. Here the killing or perceived ill treatment of a family member (especially females) could result in other members of the family joining the insurgency.[21]

Thus, how a commander conducts operations and exercises his discretion under IHL can either stoke or diminish an insurgency. As Whetham warns, "Allowing the gloves to be taken off anywhere in the chain of command is not only bad strategy, but it can also undermine the reasons for which you went to war in the first place."[22] In counterinsurgency situations, commanders are expected to show greater restraint than ever in how much force is used.

Commanders' individual leadership styles and operational decisions may benefit the winning of hearts and minds in an operation, and for the most part troop morale and discipline may be positively enthused. Yet the reverse effect is also true, to the extent that soldiers may feel disempowered as to how force is used and lose confidence in their military might. The negative reactions of some military personnel to the concept of "courageous constraint" that NATO and U.S. commanders introduced in Afghanistan in 2010 reflects the tightrope that commanders must walk in finding the right level of force that needs to be used to achieve the objective of winning the war, especially in counterinsurgency situations. In revising the escalation of force standard operating procedures, the International Security Assistance Force (ISAF) hoped to reduce collateral damage and civilian casualties. However, military personnel were reported to have been very

---

21 Ibid., 3-28.
22 David Whetham, "Taking the Gloves Off and the Illusions of Victory: How Not to Conduct a Counter-Insurgency," in *Warrior's Dishonour: Barbarity, Morality and Torture in Modern Warfare*, ed. George Kassimeris (Farnham UK: Ashgate, 2006), 135.

critical of the new policy of courageous restraint, suggesting it was forcing them to fight with one hand tied behind their backs and that it was eroding troop confidence.[23] One article quoted a junior officer saying, "It's a major bugbear for the British army, it affects us massively. Thank God we have the ANA (Afghan National Army) here because they have different rules of engagement to us and can smash the enemy."[24]

This illustrates that the evolving nature of the hostilities in irregular warfare regularly tests the application of IHL standards and principles in theater. Also, with counterinsurgency blending military, political, economic, and social objectives, the mission's strategy and rules of engagement will be regularly modified to best meet these. As David Kennedy wrote, "In today's asymmetric postcolonial wars, the terrain beneath a soldier's interpretations of what is and is not appropriate is constantly shifting."[25] The British Army field manual explains, "Today's hybrid threats—any adversaries that simultaneously and adaptively employ a fused mix of conventional weapons, irregular tactics, terrorism, and criminal behavior in the same battle space to obtain their political objectives—are constantly seeking to exploit what they perceive to be the vulnerabilities of regular forces."[26] How to best tackle these adversaries and win hearts and minds is complex and requires the military to be able to continually adapt and adjust its "line of attack."

For commanders, the nature of threats is always evolving as non-state actor insurgents modify their tactics and the means they use

23 Thomas Harding, "'Courageous Restraint' Putting Troops Lives at Risk," *Telegraph* (London), 6 July 2010, http://www.telegraph.co.uk/news/worldnews/asia/afghanistan/7874950/Courageous-restraint-putting-troops-lives-at-risk.html; Sean Naylor, "McChrystal: Civilian Deaths Endanger Mission," *Army Times*, 30 May 2010, http://www.armytimes.com/news/2010/05/military_afghanistan_civilian_casualties_053010w/.
24 Ibid.
25 David Kennedy, *Of War and Law* (Princeton: Princeton University Press, 2006), 132.
26 *British Army Field Manual, Volume 1, Part 10,* 1-1.

to overcome the more powerful—in terms of capacity—regular armed forces. Left with a certain amount of discretion in applying IHL, the commander has an essential role to play in ensuring the norms of IHL are respected despite the shifting sands of counterinsurgency and irregular warfare.

TO PREVENT, PUNISH, OR BE PUNISHED

As witnessed over the centuries, participation in armed conflicts can bring out the best as well as the worst in individuals as they seek to defeat their respective adversary. Those who do not comply with IHL can and must be held accountable. Commanders can be held personally responsible for having ordered the commission of violations of the laws of war. They can also be held liable for having failed to either prevent or punish their subordinates (omission liability).

As discussed above, a large part of the commander's responsibility is assumed through preventive action, with leadership style and exercise of authority central to creating an environment conducive to the constant respect of IHL. However, there may be situations where, despite all the best endeavors on the part of a commander to instill the appropriate behavior and discipline among the troops, combatants under their authority and control may transgress the laws of war. In these cases, the laws of war require superiors to punish the perpetrators accordingly. As General Sir Mike Jackson explains, "We do our best to screen out such individuals, but inevitably, some slip through the net. So abuses will occasionally happen—but they can never be tolerated."[27]

---

27 Jackson, *Soldier: The Autobiography*, 421. General Sir Michael David "Mike" Jackson, GCB, CBE, DSO, DL served with the British military from 1963 to 2006. In 1997, he was commander of NATO's Allied Rapid Reaction Corps (ARRC) in the Balkans; from 2000 to 2003, he was commander in chief, Land Command of the British Army; and was the chief of the General Staff (CGS), the professional head of the British Army, from 2003 to 2006. He retired from the army after serving for over 40 years.

Punishment can occur through either disciplinary or penal action, though the latter is to be preferred where serious violations of IHL are at issue.[28] An argument could be made that it is unfair to hold commanders responsible by way of omission liability for the actions of their subordinates; a stronger counterposition is that to allow a culture of acquiescence or impunity for acts of unnecessary violence by subordinates potentially encourages more violations and can undermine the overall mission: "How can we expect to win the support of the Iraqi people if they believe that we are abusing their compatriots?"[29] The responsibility of commanders to control subordinates and to ensure that they respect IHL is now engrained in case law.

Even before the adoption of the 1949 Geneva Conventions, the U.S. Supreme Court in the *Yamashita* case strongly stated that commanders could be held accountable for failing to discharge their duties to control the operations of persons under their command who had violated the laws of war.[30] This 1946 case concerned General Tomuyuki Yamashita, the commander of the Japanese forces in the Philippines in 1944–45. The majority judgment, delivered by Chief Justice Stone, enounced the principle that the laws of war impose upon an army commander a duty to take

---

28 See *Prosecutor* v. *Hadzihasanovic'*, IT-01-47-A, Appeal Judgement, 22 April 2008, para. 33. "It cannot be excluded that, in the circumstances of a case, the use of disciplinary measures will be sufficient to discharge a superior of his duty to punish crimes under article 7(3) of the Statute. In other words, whether measures taken were solely of a disciplinary nature, criminal, or a combination of both, cannot in itself be determinative of whether a superior discharged his duty to prevent or punish under article 7(3) of the Statute."
29 Jackson, *Soldier: The Autobiography*, 421.
30 In *Re Yamashita* No. 61, Misc. Supreme Court of the United States 327 US 1; 66 S. Ct. 340; 90 L. Ed. 499; 1946 U.S. LEXIS 3090. The relevant charge held against General Yamashita was that "the law of war imposes on an army commander a duty to take such appropriate measures as are within his power to control the troops under his command for the prevention of acts which are violations of the law of war and which are likely to attend the occupation of hostile territory by an uncontrolled soldiery; and he may be charged with personal responsibility for his failure to take such measures when violations result."

such appropriate measures as are within his power to control the troops under his command and prevent them from committing violations of the laws of war. In the view of the court, the absence of such an affirmative duty for commanders to prevent violations of the laws of war would defeat the very purpose of those laws:

> It is evident that the conduct of military operations by troops whose excesses are unrestrained by the orders or efforts of their commander would almost certainly result in violations which it is the purpose of the law of war to prevent. Its purpose to protect civilian populations and prisoners of war from brutality would largely be defeated if the commander of an invading army could with impunity neglect to take reasonable measures for their protection. Hence the law of war presupposes that its violation is to be avoided through the control of the operations of war by commanders who are to some extent responsible for their subordinates.[31]

Similarly in the case of the *United States v. Wilhelm von Leeb et al.* (High Command Case), it was underscored that "under basic principles of command authority and responsibility, an officer who merely stands by while his subordinates execute a criminal order of his superiors, which he knows is criminal, violates a moral obligation under international law. By doing nothing he cannot wash his hands of international responsibility."[32] The concept of command or superior responsibility is now firmly ingrained in the prosecution of war crimes and has been extended to other grave international crimes, notably genocide and crimes against humanity.

International criminal tribunals prioritize the prosecution of commanders over that of subordinates, and a perusal of relevant judg-

---

31 Ibid.
32 *United States* v. *Wilhelm von Leeb et al.*, Trials of War Criminals before the Nuremberg Military Tribunals under Control Council Law No. 10, Vol. XI (Washington, DC: U.S. Government Printing Office, 1950), 1230, 1303.

ments highlights that these courts are inclined to sanction commanders more harshly than the foot soldier who wielded the fatal blow. The International Criminal Tribunal for the Former Yugoslavia (ICTY) has underscored that with rank comes responsibility, which can aggravate the sentence to be handed down. The tribunal emphasized that commanders, by virtue of their positions, set the example for their troops and that failure to act appropriately could have a negative effect on the how their subordinates behave.

As the ICTY explained in *Obrenovic*, "When commanders, through their own actions or inactions, fail in the duty, which stems from their position, training, and leadership skills, to set an example for their troops that would promote the principles underlying the laws and customs of war and thereby—either tacitly or implicitly—promote or encourage the commission of crimes, this may be seen as an aggravating circumstance."[33] This reasoning has been followed by other international and hybrid tribunals in sentencing superiors, civilian or military.

The International Criminal Tribunal for Rwanda stipulated that "as a general principle, this Appeals Chamber agrees with the jurisprudence of ICTY that the most senior members of a command structure, that is, the leaders and planners of a particular conflict, should bear heavier criminal responsibility than those lower down the scale, such as the foot soldiers carrying out the orders."[34] The Special Court for Sierra Leone has followed suit "where the accused has actively abused his position of command or participated in the crimes of his subordinates, however, such conduct can be considered to be aggravating."[35]

---

33 ICTY Sentencing Judgement, *Prosecutor* v. *Dragan Obrenovi*, Case No.: IT-02-60/2-S, 10 December 2003, 32.
34 ICTR Appeals Judgement, *Prosecutor* v. *Alfred Musema*, Case No. ICTR-96-13-A, 16 November 200, 125.
35 SCSL Sentencing Judgement, *Prosecutor* v. *Issa Hassan Sesay, Morris Kallon, and Augustine Gbao*, Case No. SCSL-04-15-T, 8 April 2008, 17.

In terms of responsibility, the principle has been set in stone that commanders, because of their hierarchical positions, are to be held to higher standards than those of lower rank. This trend is very much influenced by the belief—in part based on common sense—that leaders set the example for the actions of their subordinates and are specifically tasked under the laws of war to assume this responsibility. It is not strict liability as such, but as the courts argue, there is a strong presumption that commanders at all times should know or have known of the actions of their subordinates.

Of course, much of the case law has evolved in the context of conventional warfare, where clear lines of responsibility could be determined. With the development of such warfighting concepts such as "distributed operations," the challenge will arguably be even greater on commanders at all levels to exercise and to be seen to exercise the required level of supervision over their subordinates.

A Better Peace by Waging a Cleaner War

In today's complex asymmetrical situations involving one or a number of nonstate actor groups engaged against regular armed groups, insurgents may aim to overthrow their governments through violent means. Such groups can use a mélange of political, religious, and criminal methods and tools, seeking to weaken a government's power as well as increase their own control over territory and the local population. Commanders have had to adapt to these new realities and the expectations of counterinsurgency warfare.

Fighting these belligerents and overcoming their adversity is no longer just a question of inflicting the greatest amount of death upon the enemy in the least possible time, as General Patton would argue, but is also significantly about winning hearts and minds, thus depriving the groups of potential support of the local population.[36] Conflict is more than the use of lethal force on the adversary, and more a blending of military, political, and economic means to achieve long-lasting peace. As General Sir Rupert Smith explained,

> In our new paradigm, which I call "war amongst the people," you seek to change the intentions or capture the will of your opponent and the people amongst which you operate, to win the clash of wills and thereby win the trial of strength. The essential difference is that military force is no longer used to decide the political dispute but rather to create a condition in which a strategic result is achieved. . . . In large measure, the strategic objective is to win the hearts and minds of the people. In other words, this isn't a supporting activity of your tactical battle. It is the purpose of what you are doing. So arriving afterwards to paint a school or deliver toothpaste isn't helping if you've blown the school away in the first place.[37]

Modern-day thinking on counterinsurgency reflects this shift in attitude and strategy in dealing with new threats in multifaceted noninternational armed conflicts.[38] Defeating the enemy is also gaining the support of the local population, and this cannot be achieved by the use of force alone. "Counterinsurgency is warfare; it is distinctly political, not primarily military; and it involves the people, the government, and the military. The strength of the relationship between these three groups generally determines the outcome of the campaign."[39] Commanders in the field are expected to juggle among "security," "development," and "sustainment" and to "always execute good judgment, tactical patience, and innovation to defeat an insurgency."[40] They must navigate

---

36 See for instance General Patton's speech to the Third Army, given in 1944.
37 General Sir Rupert Smith, interview, *International Review of the Red Cross* 88 (2006): 719, 724.
38 *British Army Field Manual, Volume 1 Part 10*, 1.1.
39 Ibid.
40 Department of the Army, FMI 3-24.2 (FM 90-8, FM 7-98), *Tactics in Counterinsurgency* (Washington, DC,: Headquarters Department of the Army, 2009), ix, www.fas.org/irp/doddir/army/fmi3-24-2.pdf.

through various domestic political structures, sometimes based on complex tribal customs and allegiances. In such contexts, commanders often have to interact and cooperate with a plethora of different actors, including international, intergovernmental, and nongovernmental organizations; community representatives; coalition partners; traditional leaders; religious leaders; and the media.

Besides the capacity building and structural support that parties to a conflict can bring to affected communities, the success or failure of a counterinsurgency depends substantially on the ability to win the confidence of the local population in order to undermine the support for the insurgency. In counterinsurgency, ultimately, whoever gains the support of the local population has the better chance of defeating the enemy.

To achieve such an objective, it is important that all appropriate steps are taken to ensure that the laws of war are fully respected. With the 24/7 media coverage of hostilities, the reputation of armed forces can be made or broken by a single image of an act of humiliation, of ill treatment, or of wanton disregard for life. These images may sway public opinion, can be used to undermine public support, and more dangerously can generate greater animosity toward a country and its armed forces. Locally, military commanders are not only expected to understand the local communication system, as it "influences local, regional, national, and international audiences,"[41] but must also be aware that the insurgents can use media and propaganda to gain credibility and undermine the adversary. According to the U.S. COIN field manual, insurgents "will use all available means, including the media, nongovernmental organizations, and religious and civic leaders, to get their information out to all audiences. Successful insurgents strive to seize the moral high ground on any counterin-

---

41 Ibid., 1-38.

surgent mistakes, both real and perceived. This includes political, military, economic, social, religious, cultural, or legal errors."[42] For any commander, finding the right mechanisms and tools, besides the use of force, to counter these means is crucial to ensure success of the mission.

Even though force used in an operation may be lawful and meet proportionality requirements, it may still be perceived as unacceptably violent in a given community, leading to a negative perception by the affected population.[43] In counterinsurgencies, using restraint and culturally appropriate and accepted forms of violence can help more effectively combat and overcome the insurgency and achieve sustainable peace. At the same time though, not using the right amount of force can undermine the credibility of the campaign in the eyes of the local population and also encourage the insurgency to continue to act unchecked.

Counterinsurgencies evolve and change with peaks and troughs of violence. Certain situations, such as riots, may call for law enforcement action in support of local police, whereas others may require a more traditional resort to force in the military sense, where there is obvious hostile insurgent action. Ultimately, as recognized by the GB-COIN, the decision as to the level of force required is at the discretion of commanders, who have to weigh the prevailing circumstances at the time and associated risks within the overall objectives of the counterinsurgency operation.[44] Deciding on the course of action is only part of the equation; ensur-

---

42 Ibid., 1-26.
43 *British Army Field Manual, Volume 1, Part 10*, 3-15, 3-28 "Force may solve a tactical problem—a firing point neutralised, a fleeting target engaged, or a strongpoint destroyed—but if the use of force is perceived as excessive or ill targeted, the neutral segment of the population may be antagonised or alienated and it may leave a lasting feeling of resentment and bitterness. Worse still, active support for the insurgents by those suffering or observing the effects of force may be engendered. This is particularly so in those tribal cultures where codes of personal and family honour, justice and vengeance are strong."
44 Ibid., 3-29.

ing that the troops under their command act within the limits is another, especially when they are called to exercise substantial restraint in the use of force. The *Aitken Report* highlights this, noting that commanders, through Mission Command, will communicate their overall intent—what and why—but that it will be left to the subordinates to determine how to accomplish this:

> Those subordinates are not, of course, given completely free rein—the commander's orders are to be obeyed, but he will stipulate the constraints under which his subordinates are to be bound, and he should personally supervise the execution of those tasks in an appropriate fashion, in order to satisfy himself that they are being carried out correctly. And whilst tasks can be delegated, responsibility for them can never be delegated; that responsibility remains with the commander.[45]

In very much the same way as the commanders, their subordinates should have a good understanding of the contexts in which they operate, possess situational awareness, and be cognizant of the "four corners" of their rules of engagement and of the consequences—including possibly criminal liability—of operating outside these. How much force each commander decides to use very much depends on the nature of the threat faced, and threats vary and evolve. Yet any engagement, even if spontaneous, needs to respect IHL, which again calls upon the superiors to exercise a certain amount of discretion in the conduct of hostilities, effectively control their subordinates, and call on them to use restraint where appropriate.

The Commander's Influence

History is littered with examples of atrocities committed by armed

---

45 British Army, *The Aitken Report: An Investigation into Cases of Deliberate Abuse and Unlawful Killing in Iraq in 2003 and 2004*, 25 January 2008, 8, *mod.uk/NR/ rdonlyres/7AC894D3-1430-4AD1...*/aitken_rep.pdf.

individuals in conflict. In many contexts, the example set by commanders makes the difference between maintaining discipline and committing crimes. It has been documented that without the buy-in of the higher command, junior officers struggle to enforce discipline among their troops.

In the context of the Soviet military advance into the East in 1945, mass rape was committed against German women. The failure by the hierarchy in Moscow to denounce these atrocities allowed for the offenders to continue in their offences and go unpunished. The combination of alcohol and lack of discipline made it virtually impossible for lower-ranked officers to put an end to the rapes. Once the leash has been removed, it can be very difficult, and at times virtually impossible, to bring offenders back in line:

> The more intelligent junior officers were deeply disturbed, and they knew that they were virtually powerless to stop it. . . . Marshal Rokossovsky issued order No. 006 in an attempt to direct "the feelings of hatred at fighting the enemy on the battlefield." It appears to have had little effect. There were also a few arbitrary attempts to exert authority. The commander of one rifle division is said to have "personally shot a lieutenant who was lining up a group of his men before a German woman spread-eagled on the ground." But either officers were involved themselves, or the lack of discipline made it too dangerous to restore order over drunken soldiers armed with submachine guns.[46]

Reacting to many of the horrors seen during the Second World War, more efforts were undertaken to have states better assume their responsibilities to respect IHL and ensure its respect.[47] States

---

46 Antony Beevor, *War and Rape: Germany 1945* (Lees-Knowles Lectures, Cambridge, England, 2002–03, http://www.culturahistorica.es/beevor/war_and_rape.germany.pdf .
47 Article 1 Common to the Four Geneva Conventions and Article 1(1) of Additional Protocol I.

are to disseminate the text of the Geneva Conventions in times of peace and war as widely as possible, particularly in military programs,[48] and to adopt necessary legislation to effectively suppress violations of IHL.[49] Additional Protocol I also requires states to ensure that legal advisers are available when necessary to advise military commanders at the appropriate level on the application of the Geneva Conventions and Protocol and on the appropriate instruction to give to the armed forces.[50]

The adoption of these articles was aimed at finding the appropriate mechanism to enable the states' undertakings to be effectively enforced at the field level before and during actual military operations. As commanders were deemed to have the means at their disposal to ensure the respect of IHL, Article 87 of Additional Protocol I requires states and parties to the armed conflict to require military commanders to prevent as well as suppress breaches of IHL by members of the armed forces under their command and other persons under their control. It was recognized that in military environments, subordinates will be influenced by the actions or inactions of their commanders and that for a superior to allow a culture of acquiescence or impunity for acts of unnecessary violence has damaging consequences.

Placing this responsibility on superiors reflects the fact that "everything depends on commanders, and without their conscientious supervision, general legal requirements are unlikely to be effective."[51] Commanders are on "the spot and able to exercise control over the troops and the weapons which they use."[52] The commander is not expected in complex battlefield environments to be able to exercise control over his troops all of the time. Disci-

---

48 Articles 47 GCI, 48 GCII, 127 GCIII , 144 GCIV, and 83 API.
49 Articles 49 GCI, 50 GCII, 129 GCIII, 146 GCIV and 85 API.
50 Article 82 API.
51 ICRC, *Commentary on the Additional Protocols of 8 June 1977 to the Geneva Conventions of 12 August 1949*, 1018.
52 Ibid.

pline, though, must be imposed to a sufficient degree, to enable the enforcement of the Conventions and their Protocols "even when he may momentarily lose sight of his troops."[53] Commanders by virtue of their hierarchical responsibility over their subordinates "have the authority, and more than anyone else they can prevent breaches by creating the appropriate frame of mind, ensuring the rational use of the means of combat and by maintaining discipline."[54] The crucial role of commanders in creating the right environment has been underscored by the UK chief of General Staff, who wrote,

> When I took up my previous appointment as commander in chief of Land Command in April 2005, I reminded and required all commanders to set the example to their subordinates and, within the context of Mission Command, provide the leadership and supervision that will ensure the delivery of required outcomes as well as professional behavior. I underline again the responsibility of all leaders from chief of the General Staff down to the most junior lance corporal to both delegate, as necessary, but also to supervise, where appropriate, the execution of tasks. That is the responsibility of command.[55]

Conclusions in the *Aitken Report* make for similar reading, underscoring the importance of the army's core values, which should not be seen as mere empty symbolism:

> The Army's core values—selfless commitment, courage, discipline, integrity, loyalty, and respect for others—articulate the code of conduct within which the Army conducts its unique business. They reflect the moral virtues and ethical principles which underpin any decent society but which are particularly important for members of an institution with the responsibility

---

53 Ibid., 1018.
54 Ibid., 1022.
55 UK chief of the General Staff, writing to the army in April 2007, as cited in the *Aitken Report*, 27.

of conducting military operations—including the use of lethal force—on behalf of the nation. The Army requires that all its people understand these values and live up to their associated standards. It does this in part by mandating annual training for all ranks but also by requiring its leaders to set a personal example to their subordinates.[56]

Coupled with creating an environment conducive to respecting IHL, Article 87(2) of Additional Protocol I requires states to ensure that commanders, commensurate with their level of responsibility, ensure that members of their armed forces under their command are aware of their obligations under the Geneva Conventions and Protocols.[57] Article 87(3) goes on to stipulate that commanders have the duty to prevent and suppress violations by their subordinates as well as to initiate disciplinary or penal action as may be necessary.[58]

In a study entitled "The Roots of Behaviour in War: Understanding and Preventing IHL Violations," the ICRC sought to identify those factors that condition the behavior of combatants in armed conflicts.[59] In the same vein as Article 87 of Additional Protocol I, the conclusions of the study serve to reinforce the relevance and importance of the role of the commander as outlined in ensuring the respect of IHL.

As a premise, the study suggested that in war there will be excesses, blunders, and acts of violence. Despite the privilege that combatants have to use lethal force in armed conflicts, the study noted that in general they are reluctant to kill another human being. To overcome this normal human response, soldiers have

---

56 British Army, *The Aitken Report*, 24.

57 Article 87 (2) API.

58 Article 87 (1) and (3) API.

59 Daniel Muñoz-Rojas and Jean-Jacques Frésard, "The Roots of Behaviour in War: Understanding and Preventing IHL Violations," International Review of the Red Cross 86 (2004), www.icrc.org/eng/assets/files/other/irrc_853_fd_fresard_eng.pdf.

to be trained, prepared, and conditioned before in-contact situations when engaging the adversary. It is this training and preparation that can make the difference between acting within the law or outside it.

The challenge then is to be able to engrain the "right way" to act among the troops, limiting behavior that deviates from acceptable norms, while at the same time allowing combatants sufficient flexibility and aggression to fulfill their missions. In irregular warfare, finding this balance is an intricate challenge.

As the study explains, there may be individuals who will always pose a behavioral risk. Yet, in general, the conduct of most troops will be molded by the group ethos, the actions of their peers, and the example set by their superiors. Calling for the destruction of the adversary through his dehumanization must be balanced with the need to place limits on the means and methods used, and not causing unnecessary suffering. Failure to instill a sense of professional and personal responsibility and morality early on in the basic training and predeployment cycles could have serious negative implications for the conduct of hostilities in theater. The challenge of creating the appropriate mind-set among the combatants has taken on greater amplitude in the context of contemporary irregular warfare, which can blend peace operations, counterterrorism, and law enforcement.

One of the main conclusions of the ICRC study concerns the impact of authority and the obedience to orders in ensuring the respect of IHL. Basing itself on some of the results of Stanley Milgram's experiments reported in *Obedience to Authority*, the study submitted that in a military setting, individuals will be less inclined to disobey orders than an ordinary citizen and will not allow external influences to come in the way of the relationship he has with his superiors.[60]

---

60 Muñoz-Rojas and Frésard, "The Roots of Behaviour in War," 83.

Interestingly, the study concluded that combatants have a tendency to shift responsibility for their actions to their superior chains of command. They would adapt their conduct to that which would be expected of them, even when contrary to their moral obligations. Through training and collective training, the sense of the individual is subordinated to the concern of showing one's self as being worthy of expectations of the command. Thus, the foot soldier accepts the commander's view of the situations and will conform willingly to what is expected of him.

The study found that violations of IHL were not for the most part committed as a result of direct orders but rather because of a lack of specific orders *not* to violate the law, or because of an implicit authorization to behave reprehensibly. The study added tentatively that prevention of IHL violations requires an explicit order not to violate the laws of war.[61] The *U.S. Army Operational Law Handbook* makes a similar statement about clarity of orders: "Clear, unambiguous orders are the responsibility of good leadership. Soldiers who receive ambiguous orders or who receive orders that clearly violate [laws of war] must understand how to react to such orders. . . . Troops who receive unclear orders must insist on clarification."[62]

Through dissemination, training, the presence of legal advisers, and underscoring responsibility on commanders, the Geneva Conventions and their Protocols have thus placed understanding of the law at the center of IHL. In so doing, the drafters of the Geneva Conventions aimed to generate the reflex among properly trained soldiers to fully respect IHL at all times in the conduct of hostilities. Training and good leadership are essential as junior soldiers are the ones on the frontline who will have to find the courage to uphold the law under extreme pressure.[63] Creating the

61 Ibid., 85.
62 *U.S. Army Operational Law Handbook*, 36.
63 Jessica Wolfendale, "What is the Point of Teaching Ethics in the Military," in *Ethics Education in the Military*, ed. Paul Robinson, Nigel De Lee, and Don Carrick (Farnham UK: Ashgate, 2008),168–69.

appropriate mind-set is as much a matter of prevention before any hostile action as it is in combat where the decisions and actions taken by the commanders will have a bearing on the comportment of their subordinates.

Not only must the commander be the standard-bearer, but he will also need to be in a position to effectively influence his subordinates, even if they are out of sight. Finding the right mechanism, whether through training or the taking of disciplinary measures, for encouraging his troops to remain well-ordered even in the more testing and hostile of environments can be a challenge, but it is one that must be met. As has been shown in many contexts, losing moral influence and control of one's subordinates can have unwanted and arguably, at times, serious repercussions.[64]

CONCLUSION

With the universal ratification of the Geneva Conventions, states have sent out the strong message that they intend to abide by their IHL obligations as they engage in hostilities. With today's irregular warfare blending military power with political and economic development in complex counterinsurgency operations, winning hearts and minds is essential not only to emerge victorious but also to bring stability over the longer term. When winning the war and hearts and minds intertwine, respecting IHL is not only a legal obligation and a moral imperative, but it also becomes an operational requirement. Disobedience of the laws of war brings dishonor, and rather than weakening the enemy's will to fight, it can often strengthen it.

---

64 "If every time you go down to see your soldiers, you tell them that they're fucked up, then guess what. They don't want to see you anymore. And they will do just enough to not get your attention. But they aren't going to trust in you as a commander, and as a leader you have no influence. And when the formal chain of command breaks down, the informal command steps up, and then you are entering dangerous territory, because nobody has any idea where the informal leaders will take the group." Jim Frederick, *Black Hearts*, 199.

As British Army Lieutenant General John P. Kiszely noted, it has been recognized that "in the eyes of the warrior, counterinsurgency calls for some undecidedly unwarrior-like qualities, such as emotional intelligence, empathy, subtlety, sophistication, nuance, and political adroitness."[65] These are all traits that today's commander must bring to the table and to the battlefield. Compliance with the law is one important element to ensure success and much will be left to commanders to lead their troops to success by example.

---

[65] John Kiszely, "Learning about Counter-Insurgency," *Royal United Services Institute Journal* 151 (December 2006).

AGENCY OF RISK

THE BALANCE BETWEEN PROTECTING MILITARY FORCES AND THE CIVILIAN POPULATION

CHRIS JENKS

Leaders prepare to indirectly inflict suffering on their soldiers and Marines by sending them into harm's way to accomplish the mission. At the same time, leaders attempt to avoid, at great length, injury and death to innocents. This requirement gets to the very essence of what some describe as "the burden of command.". . . Ultimate success in [Counterinsurgency Operations] is gained by protecting the populace, not the [military] force . . . [yet] combatants are not required to take so much risk that they fail in their mission or forfeit their lives.

- Field Manual 3-24, *Counterinsurgency*[1]

In 2010, nine years into Operation Enduring Freedom in Afghanistan, casualties among both Afghan civilians and members of the U.S. military were at their highest levels to date. Taliban insurgents employing suicide and improvised explosive devices directly caused the vast majority of those casualties.[2] Yet attitudes

---

1 Department of the Army, FM 3-24 *Counterinsurgency* (Washington, DC: Department of the Army, 2006) 7-14, 1-149, 7-23.
2 By the middle of 2010, Leon Panetta, then-head of the Central Intelligence Agency, estimated that there were only 50 to 100 al-Qaeda militants operating in Afghanistan while the Taliban were "engaged in greater violence," including improvised explosive devices and "going after our troops. There's no question about it." "Fewer Than 100 Al Qaeda in Afghanistan: CIA Chief," ABCNews.net, 28 June 2010, http://www.abc.net.au/news/stories/2010/06/28/2938358.htm.

and perceptions among both the Afghan population and the U.S. military reflect differing conceptions of blame and responsibility. Afghans blame the mere presence of the United States as the underlying cause for Taliban attacks and resulting civilian casualties, while members of the U.S. military question whether self-imposed limitations on employing force has led to increasing numbers of American service members wounded and killed.

One of the origins of this angst is the degree and manner by which risk is borne by the two groups, Afghan civilian and U.S. military, as a result of the ongoing counterinsurgency operations. At the strategic level, military doctrine, including the language above from the U.S. military's 2006 counterinsurgency manual, provides fundamental principles that govern the conduct of military operations and, in so doing, shapes the parameters of this risk. Moving from the strategic level through the chain of command to the tactical level—the soldier and Marine on the ground—the contours of risk are operationalized through the war-fighting command promulgating rules of engagement and, in Afghanistan, the tactical directive.

The tactical directive "provides guidance and intent for the use of force by" U.S. and Coalition military forces operating in Afghanistan.[3] Beginning in 2008, three successive commanders of the International Security Assistance Force (ISAF) in Afghanistan have made portions of the directive—previously classified to protect the force and still classified in parts—public.

The 2008 tactical directive stressed minimizing death or injury of innocent civilians and reinforced the idea of proportionality, "requisite restraint, and the utmost discrimination in our application of firepower."[4] This iteration, however, didn't place any specific

3 International Security Assistance Force, Tactical Directive, 1 August 2010, http://www.isaf.nato.int/article/isaf-releases/updated-tactical-directive-emphasizes-disciplined-use-of-force.html.
4 International Security Assistance Force, Tactical Directive, 2 December 2008, http://www.nato.int/isaf/docu/official_texts/Tactical_Directive_090114.pdf.

limitations on certain types of force, relying instead on "[g]ood tactical judgment" to minimize civilian casualties.[5] In contrast, the 2009 directive dictated that "use of air-ground munitions and indirect fires against residential compounds is only authorized under very limited and prescribed conditions."[6]

The 2009 tactical directive "de-emphasized airstrikes, artillery, and mortars. This transferred some of the risk in skirmishes from Afghan civilians to Western combatants. In the past, American patrols in contact often quickly called for and received fire support. Not anymore. Many firefights . . . are strictly rifle and machine gun fights."[7] In accordance with this perspective, in not providing the fire support that otherwise may be available, the tactical directive increased engagement times with the enemy, which in turn heightened the risk to troops on the ground. The resulting concern of some U.S. service members wasn't so much that the tactical directive transferred risks away from the civilians to the U.S. military, but that it transferred risk away from the enemy.[8] Yet a 2010 review of the tactical directive "found no evidence that the rules restricted the use of lifesaving firepower" or even "a single situation where a soldier has lost his life because he was not allowed to protect himself."[9]

At first glance, the current iteration of the tactical directive, which General David H. Petraeus issued in 2010, differs only slightly from the 2009 version. And those differences seem little more than an alternatively worded means to the same conceptual end—the

5 Ibid.
6 International Security Assistance Force, Tactical Directive, 2 July 2009, www. nato.int/isaf/docu/official.../Tactical_Directive_090706.pdf.
7 C.J. Chivers, "What Marja Tells Us of Battles Yet to Come," *New York Times*, 10 June 2010, http://www.nytimes.com/2010/06/11/world/middleeast/11marja. html?_r=1.
8 Ibid.
9 Rajiv Chandrasekaran, "Frustration Over War Rules of Engagement Grows: Petraeus Will Review Policy to Limit Civilian Deaths That Some Troops Say Hinders Military Activity," *Washington Post*, 9 July 2010, http://www.washingtonpost.com/ wp-dyn/content/article/2010/07/08/AR2010070806219.html.

importance of protecting the Afghan population. But the 2010 tactical directive seemingly alters the risk relationship and balance between Afghan civilians and the U.S. military. The 2009 directive expressly acknowledged that "the carefully controlled and disciplined use of force entails risk to our troops," recognizing that protecting the force must, on some level and at some times, be subordinate to protecting the civilian population. In partial yet profound contrast, the 2010 tactical directive lists protecting Afghan civilians and the men and women in uniform as coequal moral imperatives.

Utilizing both the tactical directive and doctrinal concepts from the counterinsurgency manual, this chapter will explore the allocation of risk between the military force and Afghan civilian population. The chapter first reviews civilian and military casualty figures and then uses those numbers as a touchstone against which to consider each group's perception of the risk it faces.

To set the conditions for that comparison, the chapter discusses the allocation of risk outlined in recent counterinsurgency doctrine and how that allocation translates from the conceptual or strategic level to the operational reality of soldiers and Marines in harm's way at the tactical level. This chapter examines whether that translation is conceptually consistent and tactically viable.

While the concept of the U.S. military accepting increased risk in order to protect the civilian population is codified as doctrine, how well is the military translating, and training, that doctrine? As one commentator stated, "[n]o one wants to advocate loosening rules that might see more civilians killed. But no one wants to explain whether the restrictions are increasing the number of coffins arriving at Dover Air Force Base and seeding disillusionment among those sent to fight."[10] This chapter seeks not to provide that explanation but to prompt a discussion on whether

---

10 C.J. Chivers, "General Faces Unease Among His Own Troops, Too," *New York Times*, 22 June 2010, http://www.nytimes.com/2010/06/23/world/asia/23troops. html.

there is consistency in risk tolerance between U.S. military counterinsurgency doctrine and the execution of that doctrine at the
tactical level in Afghanistan.

CIVILIAN AND MILITARY CASUALTIES AND THEIR CAUSES

Civilian Casualties

According the United Nations Assistance Mission in Afghanistan
(UNAMA), 2,777 civilians were killed in 2010 as a result of the
ongoing conflict in Afghanistan, a 15 percent increase from 2009.
[11,12] This follows a four-year trend during which each year more
civilians were killed in Afghanistan than the year prior.[13] In terms
of wounded civilians, UNAMA documented 4,343 conflict-related
injuries in 2010, a 22 percent increase from 2009.[14]

While the total number of civilian casualties has been increasing, in both 2009 and 2010 the percentage of civilian casualties
caused by progovernment forces (including the Afghan, U.S., and
Coalition militaries) *decreased*.[15] In 2010, UNAMA claimed that
progovernment forces were responsible for 440 civilian deaths or
16 percent of the total, a 26 percent decrease from 2009.[16] Progovernment forces were also purportedly responsible for 400 civilian

---

11 This chapter relies on data from the United Nations Assistance Mission in
Afghanistan (UNAMA). While the United States compiles its own statistics,
certainly, for U.S. casualties, for Afghan casualties the U.S. refers to UNAMA for
Afghan civilian casualties and to the Special Inspector General for Afghanistan
Reconstruction for Afghan National Army and police casualties. See Susan G.
Chesser, "Afghanistan Casualties: Military Forces and Civilians," *Congressional
Research Service* R41084, 14 January 2011.
12 UNAMA Human Rights Unit, *Afghanistan Annual Report 2010 on Protection
of Civilians in Armed Conflict* (Kabul, Afghanistan: UNAMA, 2010), i, http://
unama.unmissions.org/Portals/UNAMA./human%20rights/March%20PoC%20
Annual%20Report%20Final.pdf.
13 Ibid, according to UNAMA, between 2007–10, 8,832 civilians have been
killed in Afghanistan.
14 Ibid., ii.
15 Ibid., i.
16 Ibid.

injuries or 9 percent of the total, a 13 percent decrease from 2009.[17] In terms of the manner by which Afghans were wounded or killed by Coalition forces, UNAMA stated that "[a]erial attacks claimed the largest percentage of civilian deaths caused by progovernment forces in 2010, causing 171 deaths (39 percent of the total number of civilian deaths attributed to progovernment forces)."[18] However, that figure represents a 52 percent decrease in Afghan civilian fatalities stemming from Coalition air strikes from 2009.[19]

Conversely, antigovernment elements, such as the Taliban, were responsible for 2,080 deaths in 2010 or 75 percent of the total civilian deaths, a 28 percent increase from 2009.[20] This continues, and widens, the trend recognized by UNAMA in 2009, that "more civilians are being killed by AGEs [antigovernment elements] than by PGF [progovernment forces]."[21] Antigovernment elements injured some 3,366 civilians or 78 percent of the total, a 21 percent increase from 2009. Antigovernment element suicide and improvised explosive device attacks caused the greatest overall number of killed and wounded Afghan civilians.[22]

U.S. Military Casualties

Not until 2008 did U.S. troop levels in Afghanistan exceed

---

17 Ibid., ii.
18 Ibid., i.
19 Ibid.
20 Ibid.
21 UNAMA, Human Rights Unit, Afghanistan Mid Year Bulletin on Protection of Civilians in Armed Conflict, 4 July 2009, http://unama.unmissions.org/Portals/ UNAMA/human%20rights/09july31-UNAMA-HUMAN-RIGHTS-CIVILIAN-CASUALTIES-Mid-Year-2009-Bulletin.pdf. According to UNAMA, "[i]n the first six months of 2009, 59 percent of civilians were killed by AGEs and 30.5 percent by PGF. The UN characterized this as "a significant shift from 2007 when PGF were responsible for 41 percent and AGEs for 46 percent of civilian deaths." (AGE: antigovernment elements; PGF: progovernment forces.)
22 UNAMA, *Afghanistan Annual Report 2010*, i. UNAMA claimed that the most alarming trend in 2010 was "the huge number of civilians assassinated" by AGEs. The 462 civilians AGE forces assassinated represent "an increase of more than 105 percent from 2009." The majority of the assassinations occurred in southern Afghanistan, with Helmand Province experiencing a 588 percent increase and Kandahar Province, a 248 percent increase compared to 2009.

30,000.[23] But by November 2009, there were 68,000 troops in Afghanistan and roughly 100,000 by mid-2010. Such variance either skews quantitative comparison, or at a minimum, renders statistical analysis of U.S. casualty rates over time beyond the scope of this chapter. Accordingly, this section will refer to U.S. casualty data only from 2010 to provide a frame of reference and comparison and not as the basis for empirical analysis.

In 2010, 499 U.S. troops were killed in Afghanistan.[24] Improvised explosive devices were responsible for 268 of those fatalities, or roughly 54 percent.[25] In terms of injuries, 5,173 U.S. service members were wounded in Afghanistan in 2010.[26]

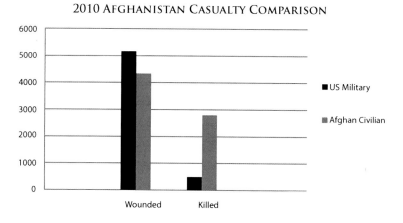

FIGURE 13. **Casualty Comparisons and Causes of Death and Wounds.**

23 Brian Montopoli, "Chart: Troop Levels in Afghanistan Over the Years," CBS News.com, 1 December 2009, http://www.cbsnews.com/8301-503544_162-5855314-503544.html.
24 Coalition Military Fatalities by Year, iCasualties.org, http://icasualties.org/OEF/Index.aspx.
25 In 2010, Afghan insurgents emplaced 14,661 IEDs, "a 62 percent increase over 2009 and more than three times as many as the year before." Craig Whitlock, "IED Casualties in Afghanistan Spike: Big Increase From 2009 to 2010 Is Result of U.S. Troop Surge," *Washington Post*, 25 January 2011.
26 Operation Enduring Freedom: U.S. Wounded Totals, iCasualties.org, http://icasualties.org/OEF/USCasualtiesByState.aspx.

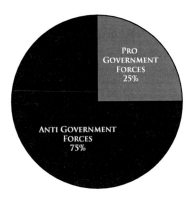

FIGURE 14. **2010 Causes of Death for Afghan Civilians.**

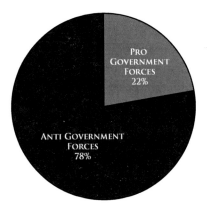

FIGURE 15. **2010 Causes of Wounds to Afghan Civilians.**

Using 2010 for comparison purposes, and speaking only in total numbers and ratios, more U.S. forces were wounded (5,173) than the total number of Afghan civilians wounded (4,343). Within that total number, the UN claims that U.S. and Coalition forces were responsible for wounding 400 Afghan civilians. Thus, U.S. forces were wounded almost eleven times for every one instance when they wounded an Afghan civilian.

Compared to the numbers of wounded, the fatalities discussion is almost flipped. In 2010, 2,412 Afghan civilians were killed compared to 499 U.S. service members.[27] Within that total number, the UN claims that U.S. and Coalition forces were responsible for the deaths of 440 Afghan civilians. Thus, U.S. forces were killed at roughly the same ratio by which they killed Afghan civilians.[28]

That comparison provides a reference point for a normative discussion on how the two sides, Afghan civilian and U.S. military, perceive the risk that the casualty rates depict, the extent to which those perceptions are consistent with the numerical indicia of risk, and ultimately how that impacts the overall counterinsurgency effort. From the perspective of the U.S. military, examining U.S. service member perceptions of risk requires first briefly reviewing the modern doctrine by which the U.S. purports to conduct counterinsurgency campaigns, filtered through the ISAF and actualized through the tactical directive.

U.S. Military Counterinsurgency Doctrine

In 2006, the U.S. Army and Marine Corps filled a doctrinal gap that spanned two decades or more by issuing a publication exclusively devoted to counterinsurgency operations, the type of armed conflict soldiers and Marines were—and are—fighting in

27 Indeed, the number of Afghan civilians killed in 2010 considerably exceeds the sum total of all U.S. fatalities in Afghanistan to date:1476. iCasualty.org.
28 These figures seem incongruent with historical trends. In a story on the 60th anniversary of the 1949 Geneva Conventions, the British Broadcasting Corporation claimed that "in World War I, the ratio of soldiers to civilians killed was ten to one. In World War II it became 50–50, and today the figures are almost reversed— up to ten civilians for every one soldier." Imogen Foulkes, "Geneva Conventions' Struggle For Respect," *BBC News*, 8 August 2009, http://news.bbc.co.uk/2/hi/europe/8196166.stm. In Vietnam, the last major counterinsurgency operation the U.S. military conducted, the ratio of monthly combat fatalities per 100,000 U.S. service members was 667, or more than 15 times higher than the casualty rate in Afghanistan. Marcus Baram, "Overall, Afghanistan More Lethal for U.S. Soldiers Than Iraq (Chart)," Huffington Post.com, 18 March 2010, http://www.huffingtonpost.com/2009/10/15/overall-afghanistan-more_n_319194.html.

Iraq and Afghanistan.[29] The publication provides principles and guidelines for counterinsurgency operations, including lessons learned thus far in Iraq and Afghanistan and those learned and to some extent forgotten, or at least neglected, from Vietnam.

The doctrine details a variety of guidance on the risk relationship between military forces and the civilian population, including the following:

- "The military forces' primary function in COIN is protecting the populace."[30]

- "The importance of protecting the populace, gaining people's support by assisting them, and using measured force when fighting insurgents should be reinforced and understood."[31]

- "In conventional conflicts, balancing competing responsibilities of mission accomplishment with protection of noncombatants is difficult enough. Complex COIN operations place the toughest of ethical demands on soldiers, Marines, and their leaders."[32]

- "Limiting the misery caused by war requires combatants to consider certain rules, principles, and consequences that restrain the amount of force they may apply. At the same time, combatants are not required to take so much risk that they fail in their mission or forfeit their lives. As long as their use of force is proportional to the gain to be achieved

---

29 FM 3-24, *Counterinsurgency*, para 7-14, 1-149. The U.S. military defines an insurgency as "the organized use of subversion and violence by a group or movement that seeks to overthrow or force change of a governing authority" and counterinsurgency as "comprehensive civilian and military efforts taken to defeat an insurgency and to address any core grievances." U.S. Department of Defense, *Joint Publication 1-02, Department of Defense Dictionary of Military and Associated Terms*, 15 February 2012, 161, 77, http://www.dtic.mil/doctrine/dod_dictionary/.
30 FM 3-24, *Counterinsurgency*, para 2-4.
31 Ibid., para 5-65.
32 Ibid., para 7-21.

and discriminates in distinguishing between combatants and noncombatants, soldiers and Marines may take actions where they knowingly risk, but do not intend, harm to noncombatants."[33]

The doctrine also identifies paradoxes of counterinsurgency operations, including:

- "Sometimes, the more you protect your force, the less secure you may be."[34]

- "Sometimes, the more force is used, the less effective it is."[35]

- "The more successful the counterinsurgency is, the less force can be used and the more risk must be accepted."[36]

- "Sometimes doing nothing is the best reaction."[37]

Suffice to say that doctrinal guidance on the use of force in a counterinsurgency is markedly different than in conventional armed conflicts where the focus is to "concentrate the effects of combat power at the decisive place and time."[38]

The counterinsurgency doctrinal guidance is then filtered through the International Security Assistance Force mission:

[i]n support of the government of the Islamic Republic of Afghanistan, ISAF conducts operations in Afghanistan to reduce the capability and will of the insurgency, support the growth in capacity of the Afghan National Security Forces, and facilitate improvements in governance and socioeconomic development

---

33 Ibid., para 7-23.
34 Ibid., para 1-149.
35 Ibid., para 1-150.
36 Ibid., para 1-151.
37 Ibid., para 1-152.
38 Department of the Army, FM 3-0 *Operations, w/change 1* (Washington, DC: Department of the Army, 2011), para A-6. This manual describes the principle of war of "mass" and how "[c]ommanders mass the effects of combat power in time and space to achieve both destructive and constructive results . . . " in order to overwhelm an opponent or dominate a situation.

in order to provide a secure environment for sustainable stability that is observable to the population.[39]

The question then becomes how the United States should reduce the capability and will of the insurgency. The answer is both through offensive or kinetic operations as well as through denying the insurgency the ability to operate by improving governance and socioeconomic development. Protecting the civilian population is inextricably linked to both. The tactical directive provides the means, or parameters on force as a means, to accomplishing the mission, shaped by the counterinsurgency doctrine.

TACTICAL DIRECTIVE

After revising the tactical directive in the summer of 2009, General Stanley A. McChrystal made portions public "to ensure a broader awareness of the intent and scope" of his guidance to the force.[40] While there had always been limitations on the use of force by U.S. military forces in Afghanistan and at least one prior version of the tactical directive released to the media, General McChrystal's modifications sparked discussion and controversy, much of which has continued and remained not just unresolved but unaddressed by the U.S. military.[41]

---

39 Afghanistan International Security Assistance Force, *About ISAF*, http://www.isaf.nato.int/mission.html.

40 Tactical Directive, 2 July 2009, 7.

41 For example, in September 2011, President Obama awarded former Marine Sergeant Dakota Meyer the Medal of Honor for his heroic actions in Afghanistan in September 2009, in saving the lives of fellow Marines and Afghan army soldiers during a Taliban ambush. The award ceremony retriggered questions of whether the tactical directive and/or risk-averse commanders contributed to some of the deaths of U.S. Marines in the ambush. See Larua Rosen, "Medal of Honor Recipient Highlights Marine's Valor as Well as Risks U.S. Troops Faced under Controversial Rules of Engagement," *The Envoy blog*, 14 September 2011, http://news.yahoo.com/blogs/envoy/medal-honor-recipient-highlights-marine-valor-well-risks-220928608.html.

As discussed in the introduction, contrary to his predecessor, General McChrystal allowed the use of air-to-ground munitions and indirect fires under "very limited and prescribed conditions." In his iteration of the directive, General McChrystal acknowledged the implicit trade-off inherent in limiting the use of force at the tactical level in support of the broader strategic goal of Afghan civilian support.

> We must fight the insurgents and will use the tools at our disposal to both defeat the enemy and protect our forces. But we will not win based on the number of Taliban we kill, but instead on our ability to separate insurgents from the center of gravity—the people. That means we must respect and protect the population from coercion and violence—and operate in a manner which will win their support.
>
> This is different from conventional combat, and how we operate will determine the outcome more than traditional measures, like capture of terrain or attrition of enemy forces. We must avoid the trap of winning tactical victories—but suffering strategic defeats—by causing civilian casualties or excessive damage and thus alienating the people.
>
> While this is also a legal and a moral issue, it is an overarching operational issue—clear-eyed recognition that loss of popular support will be decisive to either side in this struggle. The Taliban cannot military defeat us—but we can defeat ourselves.
>
> *I recognize that the carefully controlled and disciplined employment of force entails risk to our troops* [emphasis added]—and we must work to mitigate that risk wherever possible. But excessive use of force resulting in an alienated population will produce far greater risks. We must understand this reality at every level in our force.[42]

---

42 Tactical Directive, 2 July 2009.

Some U.S. service members were critical of the 2009 tactical directive. As a junior U.S. Army officer in Afghanistan queried,

> [m]inimizing civilian casualties is a fine goal, but should it be the be-all and end-all of the policy? If we allow soldiers to die in Afghanistan at the hands of a leader who says, "We're going to protect civilians rather than soldiers," what's going to happen on the ground? The soldiers are not going to execute the mission to the best of their ability. They won't put their hearts into the mission. That's the kind of atmosphere we're building.[43]

One noncommissioned officer sent an e-mail to a member of the U.S. Congress complaining that the rules of engagement (ROE) that flowed from the tactical directive were "too prohibitive for Coalition forces to achieve sustained tactical success."[44] Another noncommissioned officer commented to a reporter that he "wish[ed] we had generals who remembered what it was like when they were down in a platoon. . . . Either they have never been in real fighting, or they forgot what it's like."[45] One news story quoted U.S. service members in Afghanistan as complaining that the tactical directive "handcuffed" them.

The directive's limitations on the employment of indirect fire and close air support received much of the front-line criticism. One Marine infantry officer said he had stopped requesting air support during ground engagements as the approval process was too time consuming and tethered him to a radio. Moreover, the officer claimed that air support didn't arrive, was late when it did arrive, or that pilots were hesitant to conduct the requested air strike of ground targets. Alternatively, some units describe "decisions by patrol leaders to have fellow soldiers move briefly out into the open to draw fire once aircraft arrive, so the pilots might

---

43 Chandrasekaran, "Frustration Over War Rules of Engagement."
44 George Will, "An NCO Recognizes a Flawed Afghanistan Strategy," *Washington Post*, 20 June 2010.
45 Chivers, "What Marja Tells Us of Battles Yet to Come."

be cleared to participate in the fight."[46] While those are perhaps anecdotal and isolated examples, they occurred within a time frame in which both U.S. military and Afghan civilian casualty rates increased. And while at the same time the percentage of Afghan casualties caused by the U.S. military significantly decreased, there doesn't appear to have been a corresponding increase in Afghan civilian perceptions of safety, security, and the legitimacy and utility of the U.S. military presence in their country. From the Afghan civilian perspective, that the casualty-producing entity is predominantly the Taliban is little comfort and doesn't constitute protection. From the U.S. service member perspective, if the Afghan civilians refuse to place blame for civilian casualties on the entity actually causing them, and the only real beneficiary of U.S. restraint is the enemy, then why should those service members accept more risk?

The criticism of the tactical directive seemed to reach its zenith in the spring and summer of 2010. In the spring came word of an ISAF proposal to award a medal to U.S. service members for "courageous restraint for holding fire to save civilian lives."[47] According to an ISAF statement: "[w]e routinely and systematically recognize valor, courage, and effectiveness during kinetic combat operations. . . . In a COIN campaign, however, it is critical to also recognize that sometimes the most effective bullet is the bullet not fired."[48] As one story noted, "[a] combat medal to recognize a conscious effort to avoid a combat action would be unique."[49] The courageous restraint effort, which would seemingly recognize the tactical application of some of the counterinsurgency doctrinal points listed above, was short lived.

---

46 Ibid.
47 William H. McMichael, "Hold Fire, Earn a Medal," *Navy Times*, 11 May 2010, http://www.navytimes.com/news/2010/05/military_restraint_medal_051110mar/.
48 *Honoring Courageous Restraint*, ISAF COIN Analysis News, ISAF website, http://www.isaf.nato.int/article/caat-anaysis-news/honoring-courageous-restraint.html.
49 McMichael, "Hold Fire, Earn a Medal."

By the summer, one U.S. Army colonel claimed the troops "hated" the tactical directive and that "right now we're losing the tactical-level fight in the chase for a strategic victory."[50] At this point and unrelated to the tactical directive, General McChrystal resigned as the ISAF commander.[51] But the tactical directive loomed large in the interim while General Petraeus was awaiting confirmation to succeed General McChrystal. At his confirmation hearing, General Petraeus submitted an opening statement acknowledging that some U.S. service members were concerned over the tactical directive:

> Our efforts in Afghanistan have appropriately focused on protecting the population. This is, needless to say, of considerable importance, for in counterinsurgency operations, the human terrain is the decisive terrain. The results in recent months have been notable. Indeed, over the last 12 weeks, the number of innocent civilians killed in the course of military operations has been substantially lower than it was during the same period last year. And I will continue the emphasis on reducing the loss of innocent civilian life to an absolute minimum in the course of military operations.
>
> Focusing on securing the people does not, however, mean that we don't go after the enemy; in fact, protecting the population

---

50 Chivers, "What Marja Tells Us of Battles Yet to Come."
51 General McChrystal resigned following the release of a *Rolling Stone* story. Seemingly lost in the controversy over remarks made by members of his staff in the story which prompted the resignation, much of the article dealt with the tactical directive. See Michael Hastings, "The Runaway General," *Rolling Stone*, 22 June 2010, http://www.rollingstone.com/politics/news/the-runaway-general-20100622. The article quoted one U.S. Special Forces operator as claiming, "I would love to kick McChrystal in the nuts. His rules of engagement put soldiers' lives in even greater danger. Every real soldier will tell you the same thing." While the article quotes General McChrystal as acknowledging that "[w]inning hearts and minds in COIN is a coldblooded thing" the general also explains how ISAF can't kill its way out of Afghanistan, that "[t]he Russians killed one million Afghans, and that didn't work."

inevitably requires killing, capturing, or turning the insurgents. Our forces have been doing that, and we will continue to do that. In fact, our troopers and our Afghan partners have been very much taking the fight to the enemy. Since the beginning of April alone, more than 130 middle- and upper-level Taliban and other extremist element leaders have been killed or captured, and thousands of their rank and file have been taken off the battlefield. Together with our Afghan counterparts, we will continue to pursue relentlessly the enemies of the new Afghanistan in the months and years ahead.

On a related note, I want to assure the mothers and fathers of those fighting in Afghanistan that I see it as a moral imperative to bring all assets to bear to protect our men and women in uniform and the Afghan security forces with whom ISAF troopers are fighting shoulder-to-shoulder. Those on the ground must have all the support they need when they are in a tough situation. This is so important that I have discussed it with President Karzai, Afghan Defense Minister Wardak, and Afghan Interior Minister Bismullah Kahn since my nomination to be COMISAF, and they are in full agreement with me on it. I mention this because I am keenly aware of concerns by some of our troopers on the ground about the application of our rules of engagement and the tactical directive. They should know that I will look very hard at this issue.[52]

In early August 2010, General Petraeus issued an updated tactical directive. In the first paragraph of the revised directive, General Petraeus cautioned that "[s]ubordinate commanders are not authorized to further restrict this guidance without my approval."[53] This requirement sought to address concerns that

---

52 U.S. Central Command, General Petraeus ISAF confirmation, 29 June 2010, http://www.centcom.mil/from-the-commander/gen-petraeus-isaf-confirmation-hearing.
53 Tactical Directive, 1 August 2010.

the issue with the tactical directive under General McChrystal was not so much the limitations he imposed, but that those limitations were a floor, not a ceiling, and that several layers of command between ISAF and a U.S. Army or Marine Corps unit in contact were adding additional restrictions or requirements.

But contrary to General McChrstyal's express acknowledgement of the limitations on the use of force equating to increased risk for U.S. service members, General Petraeus' version placed protecting Afghan civilians and the force on the same level.

> We must balance our pursuit of the enemy with our efforts to minimize loss of innocent civilian life, and with our obligation to protect our troops. Our forces have been striving to do that, and we will continue to do so.
>
> *In so doing, however, we must remember that it is a moral imperative both to protect Afghan civilians and to bring all assets to bear to protect our men and women in uniform and the Afghan security forces* [emphasis added] with whom we are fighting shoulder-to-shoulder when they are in a tough spot.[54]

This language seems if not inconsistent with counterinsurgency doctrine, an avoidance or obfuscation of the subordinate relationship between protecting the force and the civilian population which the doctrine emphasizes and the harsh consequences that flow from that subordination during use-of-force situations.

Yet the current tactical directive is not receiving as much open criticism as its predecessor, at least from U.S. service members. Its impact on the "hearts and minds" of the population is unclear though. Despite the report that in the third quarter of calendar year 2010, antigovernment elements caused 90 percent of Afghan civilian deaths and injuries, in the minds of many Afghans, the true cause of the casualties was ISAF's presence in Afghanistan.[55]

---

54 Tactical Directive, 1 August 2010.
55 David Nakamura, "Afghans Reject Good Guy-Bad Guy Narrative," *Washington Post*, 14 August 2010.

So to the extent, notwithstanding the changed wording, that the tactical directive is still implemented in a way to emphasize protecting civilians, those civilians don't seem to agree. This seems due in part to continuing challenges ISAF faces in strategic communications—a 2010 poll revealed that 40 percent of those interviewed believed ISAF was in Afghanistan "to destroy Islam or to occupy or destroy the country." Additionally, "only 8 percent of interviewees in the south knew the story of the 9/11 attacks and as a result had no understanding of the justification for the conflict with the Taliban and al-Qaeda."[56] Further supporting an argument that Afghan perceptions of their safety and future in Afghanistan are not positive, beginning in 2008 and continuing through 2010, more Afghans are seeking asylum in foreign countries than at any point since the 2001 U.S. invasion.

As one commentator aptly noted, "[a]n American counterinsurgency campaign seeks support from at least two publics—the Afghan and the American. Efforts to satisfy one can undermine support in the other."[57] Overt, or at least publicized, U.S. military criticism of the current directive seems to have abated. But is that indicative of greater acceptance of the implicit risk trade-offs the directive represents? A lesser dislike of the current directives requirements than those in prior versions? Or is the absence of overt comment simply masking continued divergence between doctrinal counterinsurgency guidance and tactical realities on

---

56 Norine MacDonald, "The Good, The Bad, and the Ugly in Afghanistan," *Foreign Policy*, 16 December 2010, http://afpak.foreignpolicy.com/posts/2010/12/16/the_good_the_bad_and_the_ugly_in_afghanistan. Further supporting an argument that Afghan perceptions of their future in Afghanistan are not positive, beginning in 2008 and continuing through 2010, more Afghans are seeking asylum in foreign countries than at any point since the 2001 U.S. invasion. David Nakamura, "More Afghans Pursue Asylum Disillusioned by War and Instability, Many Have Become Economic Migrants," *Washington Post*, 28 November 2011. From 2004 through 2007, fewer than 10,000 Afghans a year sought asylum, while from 2008 through 2010, the numbers are closer to, if not in excess of, 20,000. Nonetheless, even this recent spike of asylum applicants is significantly less than the 30,000 plus from 2000, the last full year of Taliban rule.
57 Chivers, "What Marja Tells Us of Battles Yet to Come."

the ground? General McCrystal subordinated the military to the civilian population in terms of risk allocation. General Petraeus seemed to consider them coequal. What then of the U.S. Army platoon leader who, prior to leading his soldiers on patrol in Afghanistan, told them "[w]e are going to go up there and take care of each other. That is going to be our number one priority."[58] Thus at the same point in time, the operational commander considered protecting the military and civilian population equally important, a tactical leader on the ground considered protecting the military more important, and the doctrine stating that protecting the civilian population was paramount. The first needed step is to recognize this cognitive dissonance. From there, the task becomes reassessing how the U.S. military operationalizes counterinsurgency doctrine.

TRANSLATING AND TRAINING DOCTRINE

The counterinsurgency doctrine lays out competing interests: protection of the force and the civilian population. On the one hand, the doctrine claims that "ultimate success in [counterinsurgency operations] is gained by protecting the populace, not the [military] force"[59] while "[a]t the same time, combatants are not required to take so much risk that they fail in their mission or forfeit their lives."[60]

The U.S. military acknowledges that "[e]thically speaking, COIN environments can be much more complex than conventional ones" and that "[t]he fortitude to see soldiers and Marines closing with the enemy and sustaining casualties day in and day out requires resolve and mental toughness in commanders and units.

---

58 Greg Jaffe, "Combat Generation Elusive Victory: Fighting to Get Out of the Way: U.S. Troops Battle to Hand Off a Valley Strongly Resistant to Afghan Governance," *Washington Post*, 27 December 2010, .
59 FM 3-24, *Counterinsurgency*, para 1-149.
60 Ibid., para 7-23.

Leaders must develop these characteristics in peacetime through study and hard training. They must maintain them in combat."[61]

There have now been three iterations of how doctrinal concepts of risk allocation in counterinsurgency are implemented through the tactical directive and a wide range of service member responses. There are a few lingering questions regarding the evolution of this doctrine, primarily: Are the implications of this tactical directive conceptually consistent with doctrine and has the military discussed the ethics of risk allocation necessary to conduct counterinsurgency operations?

The service member complaints about the tactical directive and the short-lived nature of the courageous restraint medal, suggests a disconnect between doctrine and practice. What kind of discussions is the military fostering through its professional military education and other training? The U.S. Army's Center for the Army Profession and Ethic (CAPE) provides one example.[62] One of the CAPE training vignettes involves a new platoon leader who overhears one of his noncommissioned officers telling the soldiers in the platoon that protecting each other, that ensuring every member of the platoon returns home uninjured, is their mission. While the vignette sets the conditions for the very discussions this chapter suggests, CAPE training is relatively new, is not widespread and, more importantly, not required.

That U.S. service members are in harm's way in Afghanistan and are trying to protect the civilian population is implicitly understood. But is there an explicit training component in place that will ensure all service members in the U.S. military have a shared understanding when doctrine espouses the dichotomy that "[u]ltimate success in [counterinsurgency operations] is gained by

---

61 Ibid., para 7-24, 7-149.
62 The U.S. Army chief of staff established the CAPE at the U.S. Military Academy at West Point, New York, to reinforce the Army profession and its ethic. See http://cape.army.mil/.

protecting the populace, not the [military] force . . . [yet] combatants are not required to take so much risk that they fail in their mission or forfeit their lives."[63]

The U.S. military inculcates its service members with the concepts of selfless service and duty, among other values.[64] The military ethos at its core focuses on service members looking out for the safety and welfare of their fellow service member—the proverbial, but in combat, literal—soldier or Marine on your left and right. While that focus is understandable and even commendable, without frank and candid discussions on where and how risk is apportioned and accepted in counterinsurgency operations, is the military unintentionally sowing the seeds of that dissonance?

---

63 *Counterinsurgency,* para 7-14, 1-149, 7-23.
64 See *Army Values,* http://www.army.mil/values/.

ACCOUNTABILITY OR IMPUNITY
RULES AND LIMITS OF COMMAND RESPONSIBILITY
KENNETH HOBBS

Discipline is the soul of an Army. It makes small numbers formidable, procures success to the weak, and esteem to all.

-George Washington[1]

The subject of command responsibility strikes familiar tones for commanders and legal advisors in their national military frameworks. However, for better or worse, the armed conflicts and other military operations that the United States currently engages in seldom happen solely within the context of a single nation's force.[2] Far more common today is the joint and multinational coalition force. For example, this may take the form of a NATO-led coalition force like the International Security Assistance Force (ISAF) in Afghanistan.[3]

---

1 George Washington, Letter of Instructions to the Captains of the Virginia Regiments, 29 July 1759, http://www.revolutionary-war-and-beyond.com/george-washington-quotes-1.html.

2 U.S. commitments in Afghanistan (both ISAF and Operation Enduring Freedom), Kosovo, Bosnia-Herzegovina, Iraq, off the coast of Somalia, and elsewhere are either under a multinational joint command as with the NATO operations, part of some other coalition or collaborative effort, or sometimes a mixture of both national and multinational commands.

3 ISAF currently includes the forces of 49 Troop-Contributing Nations (TCNs), website of the International Security Assistance Force, Troop Numbers and Contributions, http://www.isaf.nato.int/troop-numbers-and-contributions/index.php.

This chapter will look at the general principles that apply in cases of national command, followed by some of the challenges inherent in multinational environments. Next, these paradigms and current developments in international law will be compared.[4] The international community is justifiably interested in minimizing the effects of war, especially in the case of civilians.[5] The use of criminal prosecutions serves both to hold violators accountable and to deter future misconduct.[6] However, there are indications that this desire to protect may lead to an expansion of the scope of liability for commanders that has only rarely been seen before.

This nature and scope of command responsibility has significant effects not just on the individuals involved, but also on the organization of military operations and the degree and nature of communications and information flow up and down the chain of command. Expanding the scope of command responsibility has the potential to reduce criminal misconduct, but it could also substantially alter the nature of commands and the idea of leadership within the military.

In the context of multinational operations, expanding the criminal liability of commanders brings into sharp focus the differences between national and multinational command structures and the limitations of the control and authority of the multinational commanders over the members of their force. If the multinational

---

4 The focus is primarily concerned with the law of armed conflict, also known as the law of war or international humanitarian law; however, these international law developments cannot be fully considered without at least some discussion of international law outside of armed conflict and international human rights law (IHRL).
5 The international community includes nations, acting individually and collectively through organizations like the United Nations, organizations including the International Committee of the Red Cross (ICRC) and nongovernmental organizations (NGOs), as well as educational and religious institutions, commentators, etc. The expressions of this concern include the development of international and domestic law, the investigations and reports of various organizations and the efforts of the international community as described above in areas of conflict to minimize the effects of armed conflict.
6 Guenael Mettraux, *The Law of Command Responsibility* (Oxford and New York: Oxford University Press, 2009), 15.

commander is to be criminally liable for the actions of subordinates from partner nations in the same manner as for those from the commander's own national forces, a fundamental change should also take place in the way coalitions are established to give the commander the ability to enforce compliance and demand accountability of subordinates to a degree that can be fairly described as exercising "effective control" over those subordinates.[7]

As significant as the possibility of individual criminal liability is, however, other developing trends, from new technologies like live video feeds, Blue Force Tracker, and YouTube, in the development of international human rights law may also result in dramatic changes in the nature of command and the way leaders command in future military operations.

COMMAND RESPONSIBILITY, DEFINED

Before going further, one needs to understand the idea of command responsibility and what it means. In the legal context, command responsibility is the potential criminal liability commanders have for serious crimes committed by their subordinates resulting from their failure to prevent the crimes or to deal with the subordinates after the crimes have been committed. While commanders are responsible for other actions as well, including actions that are not crimes, the focus will be on this unique form of criminal liability for most of the discussion.

Criminal Responsibility, Generally

When we speak about commission of crimes, in most cases the crime involves two concepts: the *actus reus* ("guilty act") and *mens*

---

7 "Effective control" has been required by courts in order to hold commanders accountable for crimes committed by their subordinates. See, e.g., *Celebici Appeal Decision*, Case No.: IT-96-21-A , 20 February 2001, Paragraph 196. This term "effective authority and control" is used in Article 28 of the Rome Statute, the treaty creating the International Criminal Court (ICC). See *Rome Statute of the International Criminal Court*, 17 July 1998, Art. 28, U.N.T.S., vol. 2187, No. 38544.

rea ("guilty mind").[8] *Actus reus* is the criminal act itself. In the case of a murder, the *actus reus* is the killing of a person by the accused. *Mens rea* is the mental state of the accused during the relevant time period, which is often at the time the *actus reus* takes place. For most criminal offenses, the completion of an act in and of itself does not constitute the commission of a crime.

The crime of murder is an appropriate example to use when looking at criminal acts in the context of the military and armed conflict. Killing the enemy is one of the things that service members are specifically trained to do from the earliest days of basic training. If all that was required to commit a crime was the act of killing a person, any life taken during an armed conflict or other military operation would be a crime. However, this is not the case since troops who kill the enemy in an armed conflict— provided they comply with the law of armed conflict—are protected by combatant immunity.[9] So, we know that in certain cases the killing of a person is not a crime. The other component to the act of killing a person that makes killing a crime is the mental state of the actor, the actor's *mens rea*.

For most crimes, there must be either a specific intent to commit the crime or knowledge that an action will lead to the commission of the crime. In some cases, a person's recklessness or extreme disregard for the consequences of their actions will establish the necessary mental state. Also, the potential punishment for crimes is often more or less severe depending on the level of *mens rea* on the part of the actor committing the crime.

Murder is only one of several crimes that can be committed when someone kills another person.[10] Murder generally requires a specific intention to kill. If the intent to kill was formed in advance

---

8 Henry Black and Joseph Nolan, *Black's Law Dictionary*, 6th Ed. (St Paul, MN: West Publishing, 1990).
9 *10 U.S.C*, Section 914 (2011); see also, *Manual For Courts-Martial* (Washington, DC: Government Printing Office, 2008), iv–63
10 Ibid., Sections 918, 919.

of the killing, the murder might be considered premeditated, making the crime even more serious. If the actor did not specifically intend to kill, but knew that death was the likely result of his action, the killing would still be a crime, but not as egregious as a premeditated murder. Killing another person through recklessness can also be a crime.[11] This is sometimes called negligent homicide, but requires recklessness, also called criminal negligence, rather than mere negligence. However, in each of these cases, the actor has acted in some way with a "guilty mind."

Criminal Responsibility Applied to Commanders

Commanders intuitively understand that they are responsible for the consequences of the orders they give, just as they would be for the actions they take themselves. For example, a commander who orders a subordinate to shoot a group of unarmed civilians is as responsible for the multiple murders that result as the troop that pulled the trigger.[12] This is the most basic level of responsibility for commanders, although this is not "command responsibility" as it is defined in international law. There is no real question about the commander's intention; the evidence of *mens rea* is at its clearest. When a commander gives an order to attack an objective during an armed conflict, the commander can potentially be held accountable for the resulting damage. To avoid potential legal liability, the commander must have taken appropriate steps to ensure that the target met the requirements imposed by the law of armed conflict.

"Command responsibility" as described in international law generally does not stem from actions taken by a commander but rather by *omissions* on the part of the commander.[13] These omis-

---

11 Ibid., Section 919.

12 See generally *United States* v. *Calley*, 46 C.M.R. 1131 (C.M.A. 1973); Michael L. Smidt, "Yamashita, Medina, and Beyond: Command Responsibility in Contemporary Military Operations," *Military Law Review* 164 (2000): 155, 232, note 289.

13 Website of the International Criminal Tribunal of the Former Yugoslavia, http://www.icty.org/x/file/Legal%20Library/Statute/statute_sept09_en.pdf. , Art. 7, *Rome Statute*, Art. 28

sions can involve failures by a commander to prevent crimes committed by subordinates or failures to properly investigate and punish crimes committed by subordinates. The commander either knows that the criminal acts of the subordinates are about to be committed or that they have been committed. We return to the case of a soldier or Marine murdering a group of unarmed civilians using his machine gun. In this case, however, there was no order from the commander to take this action. The commander has not taken any action that would have prompted the killing, which would indicate that the commander does not have the *mens rea* to kill. Now, suppose that the commander is informed that this is about to take place; for instance, by hearing the troop explaining his intentions on the radio. The commander at that moment has a responsibility to take action to stop the act and prevent the murder of the civilians. Failure to act not only violates the commander's responsibility to exercise control over subordinates, it may also create the impression of tacit approval. The commander may be liable for the killings.[14]

Suppose the commander did not hear about the killings before they took place over the radio. Instead, the commander received an after action report and learned about the killings after the fact. The commander did not order the action and was unaware that it took place. At this point there is no *mens rea* to kill the civilians on the part of the commander. However, the commander still has a responsibility to exercise control and supervision over subordinates, so the commander is faced with a choice: to take action or to take no action. In this example, the action to be taken may be initiating an investigation potentially leading to a court-martial, along with the possibility of other additional remedial measures (which may include conducting training, issuing orders, or taking other actions to prevent further violations). Taking no action,

---

14 Mettraux, *The Law of Command Responsibility.*

or taking action that would attempt to hide the misconduct or fail to address it adequately, would put the commander in legal jeopardy.[15] It should not come as a surprise to a commander that there would be legal consequences for failure to investigate, prosecute, or otherwise address violations of the law of armed conflict because the commander has taken an affirmative step, by commission or by omission, to thwart the administration of justice. There is a "guilty mind," a *mens rea*, in this case just as when the commander orders the illegal act directly.

Finally, imagine if the commander neither ordered the killings nor was even aware that the civilians were killed. Can the commander be held liable in this situation? If so, where is the *mens rea* on the part of the commander here? Although it may sound strange, the answer may be that commanders can be prosecuted for atrocities committed by subordinates even though they had no knowledge of the wrongdoing. To explore this idea, we will begin by going back to World War II.[16]

DEVELOPMENT OF COMMAND RESPONSIBILITY DOCTRINE SINCE WORLD WAR II

During the Second World War, atrocities were perpetrated upon civilians and soldiers on a large scale in both Europe and Asia.[17] In the aftermath of the war, the victorious Allied powers prosecuted many alleged war criminals in both the European and Pacific theatres. We begin with one group of atrocities, which was committed during the final months of the war in the Pacific by Japanese

---

15 Ibid.

16 There are other examples of the discipline of commanders for acts of subordinates prior to World War II, by the United States and others. W. Hays Parks, "A Few Tools in the Prosecution of War Crimes," *Military Law Review* (1995): 73–74; Smidt, "Yamashita, Medina, and Beyond," 155.

17 Estimates of casualties in WWII range from 20 to 60 million, an extremely large percentage of whom were civilians (37,215,153), WWII Multimedia Database, http://www.worldwar2database.com/html/frame5.html.

soldiers, largely against Filipino civilians. The soldiers were from units under the command of General Tomoyuki Yamashita.

The Yamashita Precedent

In September 1944, General Yamashita was transferred from Manchuria to command the Japanese 14th Army, which was responsible for the Japanese defense of the Philippines, including the island of Luzon where the Philippine capital was located.[18] Yamashita arrived at his new command at a particularly inopportune time: the Allied forces were 11 days away from beginning the campaign to retake the Philippines.[19] Yamashita hastily developed plans for repelling the Allied forces even before he could assemble his general staff.

The Allied advance was extremely successful, beginning with the retaking of the island of Leyte, which moved the battle to the island of Luzon. Yamashita's troops were under the command of subordinate generals in the southern, central, and northern sections of the island. In the center of Luzon sat Manila, the capital. Approximately 20,000 Japanese military personnel, mostly sailors under the command of Admiral Soemu Toyoda, were located in Manila. In spite of Yamashita's order to withdraw and abandon Manila, Admiral Toyoda chose to remain. The capital and the Japanese military located there were cut off from the remaining Japanese units, including General Yamashita. During the days and weeks that followed, the Japanese forces in Manila committed atrocities on a large scale, including thousands of rapes, indiscriminate property destruction, and the murder of an estimated 25,000 civilians. Atrocities also took place in other parts of Luzon in smaller numbers. The Allies succeeded in their campaign to retake the Philippines and General Yamashita surrendered to the Allied forces on 2 September 1945.

---

18 Richard L. Lael, *The Yamashita Precedent: War Crimes and Command Responsibility* (Wilmington, DE: Scholarly Resources, 1982), 6; Smidt, "Yamashita, Medina, and Beyond," 155.
19 Ibid.

A U.S. military commission was presented with the allegation that General Yamashita "unlawfully disregarded and failed to discharge his duty as commander to control the operations of the members of his command, permitting them to commit brutal atrocities and other high crimes against people of the United States and of its allies and dependencies, particularly the Philippines."[20] Although Yamashita faced only a single charge, his prosecutors provided a list of 123 alleged war crimes perpetrated by Yamashita's subordinates. The trial took place before the military commission and concluded on 5 December 1945. On 7 December, Yamashita was found guilty and sentenced to death by hanging.[21] Yamashita filed a petition following his conviction for a writ of habeas corpus,[22] requesting his release and contesting the propriety of the military commission proceeding.[23] The U.S. Supreme Court ultimately decided the matter in one of many controversial decisions of the period involving the war and national security.[24]

The Supreme Court found that the commanders were required under the law of war to exercise control over their subordinates to prevent the commission of war crimes.[25] Although the Court deferred almost completely to the military commission in its procedures and the determinations made concerning whether the evidence actually established Yamashita's guilt, the decision has been widely cited for the proposition that commanders can be criminally liable for the crimes of subordinates even without actual knowledge of their wrongdoing. Some commentators call the Yamashita decision "absolute liability" for commanders, but

---

20 *In Re Yamashita*, 327 US 1 (1946), 13–14.
21 Ibid.
22 Black and Nolan, *Black's Law Dictionary*, 6th ed., s.v. habeas corpus (literally, "you have the body"). Also known as the Great Writ, the ability to have one's confinement questioned is one of the cornerstones ensuring our liberty. The right to habeas corpus is the only individual right found in the original Constitution before adoption of the Bill of Rights.
23 *In Re Yamashita*, 327 U.S. 1 (1946).
24 W. Hays Parks, "A Few Tools in the Prosecution of War Crimes," 73–74.
25 *In Re Yamashita*, 327 US 1 (1946).

at the very least, it established one precedent for the imposition of liability when the commander "should have known" of the subordinate's actions and taken steps to prevent them.[26]

Nuremberg

The trial of General Yamashita was merely one of the scores of prosecutions that took place following World War II. Many were conducted, as Yamashita's was, by U.S. military tribunal; many also took place in international tribunals composed of judges, prosecutors, and defense counsel from multiple nations. The most famous war crimes trials of this period took place in Nuremburg, Germany. International tribunals conducted some of the Nuremburg trials, while others were conducted by the United States as military commissions under the provisions of the Control Council Law No. 10. While the Nuremberg trials are well-known for rejecting the defense of "just following orders," command responsibility was also featured in the Nuremberg trials. Especially significant for this discussion is a case often referred to as the *High Command* trial.[27]

High Command involved the trial of 14 senior general officers who held positions at the highest levels of the German armed forces during the Second World War. These senior officers included commanders of the German army, air force, and navy who were accused of a range of crimes, including war crimes, crimes against the peace, and crimes against humanity.[28]

The tribunal in *High Command* described the concept of command responsibility, what it was and was not, and provided an alternative to the *Yamashita* decision that offered a much less controversial view of the doctrine, one that more closely reflected traditional notions of criminal justice:

---

26 Jennifer S. Martinez, "Understanding Mens Rea in Command Responsibility: From Yamashita to Blaaki and Beyond," *Journal of International Criminal Justice* 5 (2007): 638–64; see also Smidt, "Yamashita, Medina, and Beyond,"184–85.
27 United Nations War Crimes Commission, *Law Reports of Trials of War Criminals*, vol xii (London: His Majesty's Stationery Office,1948), 1, 76
28 Ibid. (All accused were acquitted of crimes against peace.)

Criminality does not attach to every individual in this chain of command from that fact alone. There must be a personal dereliction. That can occur only where the act is directly traceable to him or where his failure to properly supervise his subordinates constitutes criminal negligence on his part. In the latter case it must be a personal neglect amounting to a wanton, immoral disregard of the action of his subordinates amounting to acquiescence. Any other interpretation of international law would go far beyond the basic principles of criminal law as known to civilized nations.[29]

This is a significantly more restrictive view on command responsibility than *Yamashita*, especially as viewed by the decision's detractors.[30] There is a clear focus on the connection between the criminal acts of the subordinate and the omission on the part of the commander. Further, the level of culpability is much higher than mere negligence, rising instead to a wanton disregard more consistent with recklessness.

The *High Command* tribunal also made other significant findings concerning command responsibility. The tribunal determined that staff officers generally were not subject to conviction for the actions of subordinates since it was generally the commander that actually had the authority to take action and the commander was the individual who made decisions and issued orders.

In the case of areas under occupation by the German forces, the tribunal convicted commanders with responsibility for occupied territory for not only the actions of the crimes committed by troops in their own units, but also for the crimes committed by troops under other commands but operating within their areas of responsibility because of the enhanced obligations of forces in occupation.[31]

---

29 Ibid.

30 Ibid. The *High Command* decision made specific reference to *Yamashita*, stating that while the decision is entitled to deference, the latter situation was factually distinct from the cases against the German accused.

31 Ibid., 77.

The "Ad Hoc" Tribunals: Tribunals Created by United Nations Security Council Resolutions

In the decades following World War II, the number and prominence of war crimes tribunals diminished until the 1990s. More recently, over the last 20 years, the United Nations Security Council created a series of special tribunals to investigate criminal allegations in the former Yugoslavia, Sierra Leone, Rwanda, Congo, and elsewhere.[32] The first and longest running of these special tribunals is the International Criminal Tribunal for the Former Yugoslavia, generally referred to as the ICTY. Yugoslavia disintegrated with the fall of the communist regimes in Eastern Europe, torn apart by war and ethnic divisions. The conflicts were bitter and many atrocities were committed, but there was significant doubt within the international community that anyone would be prosecuted for war crimes and other crimes against humanity, including genocide. The United Nations Security Council, acting pursuant to its responsibility to maintain or restore international peace and security under Article 24 of the UN Charter,[33] established the ICTY to investigate and prosecute these types of crimes in the former Yugoslavia.

The governing document for the ICTY, called the ICTY Statute, directly addressed the question of command responsibility.[34] The ICTY Statute provides that commanders may be responsible for acts committed by subordinates if the commander "knew or had reason to know that the subordinate was about to commit such

---

32 The tribunals created for the former Yugoslavia and for Rwanda are the most important for this article, as most of these courts' decisions involved armed conflicts.

33 *Charter of the United Nations* and *Statute of the International Court of Justice*, 26 June 1945, Art. 24, 59 Stat. 1031, TS No. 993, 3 Bevans 1153.

34 (*ICTY Statute*, Article 7, paragraph 3: "The fact that any of the acts referred to in articles 2 to 5 of the present Statute was committed by a subordinate does not relieve his superior of criminal responsibility if he knew or had reason to know that the subordinate was about to commit such acts or had done so and the superior failed to take the necessary and reasonable measures to prevent such acts or to punish the perpetrators thereof."

acts or had done so and the superior failed to take the necessary and reasonable measures to prevent such acts or to punish the perpetrators thereof."[35]

There are three elements to this formulation. First, a superior-subordinate relationship must be present. This is not simply a question of whether there is a relationship on paper, but rather requires "effective control" of the subordinate by the superior.[36] Second, the commander must have known, or had reason to know, that the subordinate was about to commit or had committed the crimes in question. The ICTY and other ad hoc tribunals have interpreted this to mean that the commander was in possession of information that would put a reasonable commander on notice that crimes were about to be committed or had been committed. This is a more rigorous requirement for the prosecution than in the *Yamashita* case. The third and final requirement is the failure by the commander to take reasonable and necessary actions to either prevent the crimes from being committed or the failure to punish the perpetrators.

These failures are of a nature of a dereliction of duty, as described in the *High Command* case; however, command responsibility is not the same as dereliction of duty. While the omissions on the part of the commander are essentially derelictions of duty, the commander is actually held responsible for the acts committed by the subordinate, e.g., murder. Which actions would be reason-

---

35 The other authorities that should be mentioned at this point are the 1977 Additional Protocols to the 1949 Geneva Conventions. These treaties also addressed the idea of command responsibility. The United States has signed but not ratified these treaties; however, commentators contend that the provisions dealing with command responsibility are customary international law. See Article 86, paragraph 2, of Additional Protocol I:

"[t]he fact that a breach of the Conventions or of this Protocol was committed by a subordinate does not absolve his superiors from penal disciplinary responsibility, as the case may be, if they knew, or had information which should have enabled them to conclude in the circumstances at the time, that he was committing or was going to commit such a breach and if they did not take all feasible measures within their power to prevent or repress the breach."

36 Mettraux, *The Law of Command Responsibility*, 156.

able and necessary are, of course, situation dependent. Consider, for instance, what might be reasonable steps to punish subordinates for their crimes; in some cases the commander will have the authority to initiate criminal prosecutions, but in others the referral of the alleged crime to appropriate authorities could be adequate if that is the action reasonable and necessary under the circumstances.

Another significant aspect of the ICTY Statute is the way the ICTY related to national courts. Concerned that justice might not be served if national courts were responsible for prosecutions, the Security Council determined that the ICTY would have primacy over national courts.[37]

In the case of the conflict in Rwanda, a tribunal (known as the ICTR) was likewise formed to deal with the crimes involved in that situation. In Rwanda, an armed conflict erupted between the two major ethnic groups in the country, the Hutus and the Tutsis, for economic and political control over one of the poorest and most overcrowded nations within the African continent. Millions died in Rwanda during the armed conflict and a number of individuals in positions of leadership were prosecuted in the ICTR. One such person was Jean Paul Akayesu, who held some of the greatest power over the people in his prefecture. Akayesu was accused of a range of crimes, including superior responsibility for rape, murder, mutilation, and other atrocities. This was in significant part because of his status as the *bourgmestre* (similar to a mayor) of the villagers, as well as the unofficial authority he was able to exercise. Akayesu was determined to have been in control over his village and the persons who lived there, including the persons who killed, raped, tortured, and assaulted members of the village during Akayesu's tenure. It did not matter that Akayesu was not the military commander of these persons under his authority.

37 http://www.icty.org/x/file/Legal%20Library/Statute/statute_sept09_en.pdf.

The statute for this criminal court did not require that the person accused of crimes be the commander of the persons conducting the activities. The statute for the special criminal tribunal in Rwanda determined that civilian as well as military leaders could be held accountable under the doctrine of command responsibility.[38] This determination may have some significance as we look at the degree of authority exercised by multinational commanders. The idea of both military commanders and civilian leaders having liability under the rubric of command responsibility—also called superior responsibility to include those not acting as military commanders—was adopted by the treaty that created the International Criminal Court.

International Criminal Court

Unlike the ICTY and subsequent special tribunals, a UN Security Council resolution did not create the International Criminal Court, also known as the ICC. A treaty, called the Rome Statute, created this entity.[39] The ICC was created to establish a standing tribunal for investigation and prosecution of war crimes and crimes against humanity.[40] The jurisprudence of these international tribunals have added to the body of interpretive guidance in the area of international criminal law and the law of armed conflict and suggest potential trends in these areas that merit consideration. The section of the Rome Statute dealing with command responsibility is found in Article 28.[41] For the first time, there are two

---

38 *Prosecutor* v. *Akayesu*, case no. ICTR-96-4-T, Judgment of 2 September 1998.
39 The United States invested considerable effort in the negotiation of the Rome Statute. While the United States was unable to obtain all of the modifications it desired, it did sign the treaty. The treaty was not ratified by the Senate, and was subsequently "unsigned" by President George W. Bush. Following this denunciation, the United States entered into bilateral agreements with many of its allies and partner nations that provided that American personnel within their territories would not be turned over to the ICC. These agreements are known as Article 98 agreements. To view the document itself, see http://untreaty.un.org/cod/icc/statute/romefra.htm.
40 The Rome Statute undertakes to create offenses through the use of convention (treaty) law instead of customary international law.
41 Article 28, Responsibility of Commanders and Other Superiors.

different standards of liability. One standard, set out in subsection (a) of Article 28, applies to military commanders and those acting like a military commander. The second standard, set out in subsection (b), applies to those superiors that still exercise "effective control" over the subordinate but do not fit within the category of commanders or those acting like commanders. In both cases, the requirement of "effective control" is made explicit. The two standards differ in the degree of awareness on the part of the commander of the actions of subordinates and on the nature of the authority exercised by the superior. In the case of commanders or effective commanders, the Rome Statute returns to the objective "should have known" standard, whereas for the other category of superior, the much less rigorous and more difficult to prove, standard of actual knowledge or conscious disregard, also described as "willful blindness," is put into place.

The Rome Statute formulation also departs from the ad hoc tribunals in that it modifies the idea of "punishing perpetrators" to specifically also include "submitting the matter to competent authorities for investigation and prosecution." These changes create a significant possibility that the International Criminal Court will interpret cases involving command responsibility differently than the ICTY and other ad hoc tribunals.

At this point, there has only been one decision by any of the chambers of the ICC dealing with command responsibility. In the case of Jean-Pierre Bemba Gombo, the Pre-Trial Chamber of the ICC issued a detailed opinion approving his case to go to trial, which included a significant discussion of command responsibility as defined in Article 28 of the Rome Statute.[42] In its assessment of

---

42 *Prosecutor v. Jean-Pierre Bemba Gombo*, ICC-01/05 -01/08, Pre-Trial Chamber, 15 Jun 2009, http://www.icc-cpi.int/iccdocs/doc/doc699541.pdf. (Mr. Bemba is allegedly responsible, as military commander, of two counts of crimes against humanity: murder (article 7(1)(a) of the Statute) and rape (article 7(1)(g) of the Statute); and three counts of war crime: murder (article 8(2)(c)(i) of the Statute); rape (article 8(2)(e)(vi) of the Statute); and pillaging (article 8(2)(e)(v) of the Statute).

Article 28 of the Rome Statute, the Pre-Trial Chamber found that five elements were required to establish the responsibility of commanders for acts of their subordinates:

(1) The suspect must be either a military commander or a person effectively acting as such;

(2) The suspect must have effective command and control or effective authority and control over the forces (subordinates) who committed one or more of the crimes set out in Articles 6 to 8 of the Statute;

(3) The crimes committed by the forces (subordinates) resulted from the suspect's failure to exercise control properly over them;

(4) The suspect either knew or, owing to the circumstances at the time, should have known that the forces (subordinates) were committing or about to commit one or more of the crimes set out in Articles 6 to 8 of the Statute; and

(5) The suspect failed to take the necessary and reasonable measures within his or her power to prevent or repress the commission of such crime(s) or failed to submit the matter to the competent authorities for investigation and prosecution.[43]

This decision, assuming that the trial and appeals chambers of the ICC adopt its conclusions, would provide a great deal of clarity in a number of areas that were unclear in the decisions of the ad hoc tribunals.[44] Regarding the issue of who can be held responsible, the court determined that superiors must both be commanders (or acting as a commander, for example, in the case of irregular or militia groups that lack a legal command relationship) and the superior must exercise "effective control" over the subordinates.

---

43 Ibid 141–42.
44 Ibid., 148–51. These areas of clarity include the requirement that "effective control" existed prior to the commission of the crimes of the subordinates and the unequivocal establishment of a causal connection between the failure on the part of the commander and the crimes committed by subordinates.

In considering what establishes "effective control," the court described a fact-specific, case-by-case analysis that considered whether the following indicia of control were present:

- Official position

- Power to issue or give orders

- Capacity to ensure compliance with orders issued

- Position within the military structure and actual tasks carried out

- Capacity to order forces to engage in hostilities

- Capacity to resubordinate units or make changes to command structure

- Power to promote, replace, remove, or discipline any member of the forces

- Authority to send forces where hostilities take place and withdraw them at any given moment

As one will see, the command responsibility framework established by the Rome Statute may have significant impacts in the multinational environment.

Application of Command Responsibility to Multinational Commands

Case Study: NATO International Security Assistance Force, Afghanistan

As complex as the legal issues may be for national commands and their military commanders, for the multinational commander, the complexities are much greater. This is an important issue to consider since many recent military operations were and are organized as multinational commands. This is especially true in NATO operations. NATO's military command structure is by nature multinational. Operational commands in NATO are routinely multinational as well. NATO's largest current operational commitment is the International Security Assistance Force (ISAF) in Afghanistan. We will take a look at this command to illustrate some of these additional complexities.

The ISAF commander is an American four star general, and the vast majority of the military personnel supporting ISAF are from the United States.[45] The day-to-day conduct of military operations is the responsibility of the ISAF Joint Command, commanded by an American three star general. However, as mentioned above, 49 nations provide troops to ISAF.[46] The area of operations includes the entirety of Afghanistan, divided into six Regional Commands, three of which are commanded by non-U.S. commanders. Many of the contributing nations have assigned senior commanders as part of their contributions. Some nations' contingents are organized so that they fall under the command of one of their own commanders, but many nations have their personnel interspersed within multinational units.

INTERNATIONAL SECURITY ASSISTANCE FORCE, AFGHANISTAN: KEY FACTS AND FIGURES[47]

A variety of other teams also operate within Afghanistan as part of ISAF. Provincial Reconstruction Teams (PRTs) are arrayed across Afghanistan to provide assistance with reconstruction, security, governance, and rule of law activities, operating with a significant level of autonomy.[48] Operational Mentor and Liaison

---

45 Troop Numbers and Concentration, website of the International Security Assistance Force (ISAF), http://www.isaf.nato.int/leadership.html; http://www.isaf. nato.int/troop-numbers-and-contributions/index.php.
46 Ibid.
47 Key Facts and Figures, website of the International Security Assistance Force (ISAF), http://www.isaf.nato.int/images/stories/File/Placemats/9%20September%202011%20ISAF%20Placemat%281%29.pdf.
48 About ISAF, website of the International Security Assistance Force (ISAF), http://www.isaf.nato.int/mission.html.

Teams (OMLTs) are located with Afghan National Security Forces (ANSF) units to provide assistance and training. Police Operational Mentor and Liaison Teams (POMLTs) fulfill a similar role in developing the Afghan National Police (ANP).

Regardless of the lower-level organizational structures, all personnel from all 48 nations fall under the command of the ISAF commander.[49] The ISAF commander has all the apparent attributes of a national commander. He has the title and the associated command structure. He develops policies and procedures and issues orders and other guidance. He also receives direction from and reports to higher-level commanders in the NATO military command structure. But does he have "effective control" over his subordinates? In our national context, the commander issues orders and if the orders are violated, then the commander has the clear authority to take immediate action to deal with the violation. This authority may include prosecution and confinement or other disciplinary and administrative sanctions. This authority exists because the nation has enacted laws to create the authority and has given it to commanders.

NATO does not have the ability to enact laws. Like most international organizations, NATO is created by a treaty, in this case the North Atlantic Treaty of 1949 (also referred to as the Washington Treaty).[50] The nations that are members of NATO each have a representative in Brussels at NATO headquarters and make up the North Atlantic Council (NAC). Although it serves as the decision-

---

49 Operation Enduring Freedom (OEF), the effort that involved the initial Coalition campaign against the Taliban and al-Qaeda following the attacks of 11 September 2001, also continues in Afghanistan, but is not a NATO operation. The same U.S. general in command of ISAF is also in command of the personnel supporting OEF. Many of the rules and procedures are the same for both ISAF and OEF. However, there are differences, some of them significant.
50 North Atlantic Treaty between the United States of America and other governments, T.I.A.S. No. 1964, (24 August 1949), http://www.nato.int/cps/en/natolive/official_texts_17120.htm.

making body for NATO, the NAC does not operate as a congress or parliament. The NAC does not have the authority to make laws as nations do. NATO also lacks a judicial apparatus—there are no judges, no prosecutors, and no ability to punish misconduct in the same way that a nation can.

Therefore, the NATO commander in many ways lacks the ability to arrest, initiate prosecutions, and enforce discipline that a U.S. commander has. What tools does the NATO commander have? The NATO commander has the ability to prevent a soldier from participating in operations and has the ability to notify the soldier's national authorities. The NATO commander cannot require that a soldier be disciplined or prosecuted—that is left to the discretion of the sending nation.

One might suggest that regardless of the particular authorities the multinational commander has, he or she still has the ability to issue orders that all must follow. But is that accurate? It is undeniably true that the members of the U.S. military have an obligation to follow lawful orders.[51] Failure to follow a lawful order is a crime. Even without an express order, it is still a crime for a U.S. service member to fail to fulfill his or her military duties.[52] American commanders have the force of law and the threat of potential prosecution to back up their orders. However, even with this authority, members of the military still violate orders and fail to comply with their duties. Only a small percentage of the force falls into this category; nevertheless, it shows that even where commanders have the force of law backing up their orders, it is not always enough to ensure compliance.

How are orders perceived when issued by a NATO commander? The answer may vary. For the personnel from that commander's nation, for example, the order will likely be ethically and legally

---

51 *10 U.S.C.*, Sections 890, 892.
52 Ibid., Section 892.

binding. But this is not always the case. Unlike in the national context, multinational commanders must cope with the influences and direction from the nations that provide their forces to the operation. As mentioned above, only national authorities have the ability to discipline and punish, even though authority over the forces is transferred to the multinational commander. This is not unique to NATO, but rather is part of the nature of multinational military operations.

Nations provide forces voluntarily and may attach conditions or restrictions to the employment of the forces they contribute. In some cases, these conditions, commonly referred to as "caveats," may result from international or domestic legal requirements. For example, many NATO members and partners are parties to the treaties prohibiting the use of cluster munitions and antipersonnel land mines. They may stipulate that their forces will not be required to use these munitions, since to do so would violate their international legal obligations. Sometimes these caveats are based on policy or political concerns. Regardless, nations continue to exercise control over their forces in multinational operations. Considering this, and considering that the essentially all-national authorities have coercive tools at their disposal, the multinational commander appears to be in a position of authority. However, it is certainly arguable that the commander lacks the means to exercise "effective control" over the forces ostensibly subject to the commander's orders.

One sees this in the way that investigations are carried out as well. In the U.S. military, commanders generally have the ability as part of their command authority to order investigations of any matter that takes place within their command. In a multinational command, the commander can still conduct investigations, but the nature of the multinational command requires additional considerations. The multinational commander does not have the ability to compel forces to cooperate with investigators because of the lack of disciplinary authority discussed above. In a similar vein, because national authorities must handle any potential prosecution for a violation of law, any investigation must be handled very carefully.

Aspects of an investigation like the gathering of evidence and the questioning of potential suspects can be ruinous to a prosecution if handled incorrectly. Because the rules and procedures will vary depending on the nationality of the individual concerned, the multinational commander risks the possibility that an investigation undertaken by the command will taint evidence and may result in a crime going unpunished. In the case of ISAF, these concerns have resulted in a command policy that acknowledges the dangers of tainting evidence and jeopardizing prosecutions. Investigations conducted by ISAF are designed to identify problems in the organization, training, and direction of ISAF personnel and units but are not intended to serve as criminal investigations. In fact, ISAF policy states that if evidence suggests a crime may have been committed, the investigation is to cease and the national authorities of the individuals involved are to be notified. The intention is that national authorities will then conduct any criminal investigations in accordance with their own legal requirements.

Assessing the NATO Approach

Let us take a look at how the scenarios discussed above might play out in a multinational command like ISAF. If a U.S. commander orders the summary execution of civilians, regardless of the nationality of the subordinate that actually carried out the order, the commander would be subject to discipline, since the commander is still subject to U.S. jurisdiction under the Uniform Code of Military Justice (UCMJ).[53] If a U.S. commander engaged in any misconduct, American authorities would be able to investigate

53 Ibid., Section 802.

and take any necessary actions. If a potential crime was uncovered concerning a commander of any other nation, the national authorities for that commander would be notified and that nation would take any actions it deemed appropriate for the situation.

Suppose that an ISAF commander heard over the radio that a subordinate from another nation was about to kill a group of unarmed civilians in his custody. If that commander does nothing, that inaction is a tacit approval of the killings, subjecting the commander to possible jeopardy even though the subordinate is from a different nation. But what if the commander objects immediately and orders the subordinate to stop? The commander lacks the means to arrest, prosecute, or punish the "subordinate" of another nationality. Does the commander exercise "effective control"? Based on the decisions of the ad hoc tribunals and the *Bemba* Pre-Trial Chamber decision, which makes the ability to punish only one of a series of factors to consider, the answer may be yes.[54] If the commander does exercise "effective control," has the commander taken "all necessary and reasonable measures within his or her power" to prevent the misconduct? Since the subordinate is from another nation, is the commander required to contact a superior of the subordinate's nationality to ensure that the subordinate received the order to stop from an officer that has legal authority to do so?

In the event that the commander did not know the killings were taking place until afterward (and did not have reason to know under the circumstances at the time), what actions are required? Is notifying the subordinate's national authorities sufficient to meet the commander's legal responsibilities? Perhaps.[55]

---

54 As part of the legal courses conducted at the NATO School, the author has had several opportunities to present this type of scenario before groups of legal advisors from more than 20 nations. This unscientific anecdotal evidence reveals a general level of agreement with the proposition that in some cases, the relationship may be sufficient to rise to the level of "effective control" even in a multinational command.
55 Mettraux, *The Law of Command Responsibility*, 250, note 97.

Now consider a scenario where the multinational commander is completely unaware of the killings. The person who killed the civilians was from a subordinate command and of a different nationality than the commander. At what point does the commander "have reason to know" of the subordinate's actions? This is one of the key points of contention in *Yamashita* and in subsequent cases. In the Philippines, more than 20,000 innocent lives were taken and the military commission concluded that the atrocities were so widespread that Yamashita must have known that they were occurring. The defense argued that Yamashita was unaware of the actions of his subordinates due chiefly to the success of the Allied forces in cutting off Yamashita's command, control, and communications capabilities.

The multinational commander, like the ISAF commander, has the traditional impediments to complete information: distance, limited communications, and the time pressure of operational commitments. In addition, the multinational commander has significantly diminished legal authority. Reliance on persuasion and cooperation is required, while in a traditional military hierarchy, the commander's authority alone is generally sufficient to ensure compliance. The ISAF commander has issued policies and procedures for the reporting and investigation of potential violations of the law of armed conflict. In fact, ISAF procedures are in place requiring investigations of civilian casualties generally, regardless of how they are caused. Even though an investigation is initiated, the ISAF commander or commanders at lower echelons cannot conduct an investigation in the same fashion that a U.S. commander would for an incident involving subordinate U.S. persons. As described above, the multinational commander lacks authority to place witnesses under oath or to punish criminally witnesses that make false statements or attempt to conceal evidence.

As important to this discussion as the question of knowledge is the notion of what actions can reasonably be taken to prevent the commission of a criminal act. It is unrealistic to expect that a commander with significantly less authority than a commander in a national command will have the same menu of options as that of a national commander. In the context of ISAF, policies and procedures have been developed to ensure compliance with the law and rules governing the use of force. Some commentators have expressed discontent with these policies on the basis that they unnecessarily restrict the ability of subordinates to use force.[56]

The international precedents dealing with the idea of criminal liability for commanders range from the harshest interpretation of *Yamashita* as an imposition of "strict liability" for commanders, to simple negligence, to criminal negligence or recklessness, to something just short of actual knowledge or "willful blindness." The jurisprudence of the ICC, which most of the NATO members are subject to, will likely continue to develop this concept. However, there are significant differences between a national command and a multinational command that deserve careful consideration should this concept, developed with a national command structure in mind, be applied to a multinational command structure, especially one as complex as ISAF.

OTHER TRENDS IMPACTING COMMANDERS AND HOW THEY LEAD

Technological Developments

One of the considerations in modern warfare, and modern military operations generally, for the United States concerns the dra-

---

56 Dan Murphy, "Afghanistan War: Will the New Petraeus Rules of Engagement Make Troops Safer?," *Christian Science Monitor*, 5 August 2010, http://www.csmonitor.com/World/Asia-South-Central/2010/0805/Afghanistan-war-Will-the-new-Petraeus-rules-of-engagement-make-troops-safer.

matic technological advances that have taken place in recent years. During WWII, commanders used short-range radios and runners to communicate. The nature of land warfare was such that maneuver units could be out of contact for hours or days at a time. This required subordinate commanders to exercise significant authority and initiative. In the case of air warfare, senior commanders would often have to wait for the return of their fighters and bombers to be able to conduct a debriefing and determine what took place during the day's missions. Today, communications technologies have ushered in the age of the satellite phone, e-mail, internet chats, and other tools to ensure that commanders have constant communication with their subordinate commanders on a real-time or near real-time basis.

In the case of the modern Combined Air Operations Center (CAOC), the commander can, in many cases, obtain live video feeds from piloted or remotely piloted aircraft to inform decisions concerning whether or not to attack a potential target. This can allow the commander to make personal decisions and take responsibility directly for targeting. In a situation like modern-day Afghanistan, where the number of combat sorties is small, this level of engagement by a senior commander may be feasible. However, we must remember that sooner or later another major theater war or potentially another world war may take place. In that environment, where the number of daily combat sorties will climb to the hundreds or thousands, execution of daily missions will be the responsibility of lower-level commanders.

This applies to the land and maritime components, as well. Advances in communications and other technologies, like the Blue Force Tracker system that allows real-time awareness of the location of friendly forces, make it possible for commanders to have greater visibility over the area of operations than ever before. In a situation like ISAF, where only a limited number of engagements with the enemy take place on a daily basis, there may be a tendency for higher-level commanders to have a heightened degree of knowledge. Just as with the air component, the outbreak of a new large-scale armed conflict would require a much greater delegation of the operational details of combat activities.

The law of armed conflict must be flexible enough to deal with the range of potential conflicts. It is tempting to view armed conflict narrowly through the lens of our recent experience, but that temptation should be resisted.

Legal Developments in International Human Rights Law

Another developing area of the law is international human rights law. Soon after the negotiation of the UN Charter, the UN General Assembly approved a statement of principles called the Universal Declaration of Human Rights.[57] Treaties of various types followed; some were centered on a particular aspect of human rights, and some were more all-encompassing. Regional human rights treaties were negotiated that contained mechanisms for enforcement. For example, the European Convention on Human Rights (ECHR) created a court that has the ability to hear complaints directly from individuals throughout Europe and can require the state parties to the Convention to pay damages if the ECHR determines that the human rights enshrined in the Charter were violated.[58]

In the past, these human rights treaties were considered to have little impact on military operations because of the nature of the law of armed conflict (LOAC) as a special law, a *lex specialis*, which operated as a self-contained legal system that came into effect only during armed conflict and that displaced human rights law during this period.

---

57 UN General Assembly, *Universal Declaration of Human Rights*, 10 December 1948, 217 A (III), http://www.unhcr.org/refworld/docid/3ae6b3712c.html.
58 Council of Europe, *European Convention for the Protection of Human Rights and Fundamental Freedoms*, 4 November 1950, ETS 5, http://www.unhcr.org/refworld/docid/3ae6b3b04.html.

This understanding has been subjected to criticism over the decades, and there are now other conceptions of the relationship between LOAC and international human rights law.[59] In some regions, the current favored formulation is that international human rights law applies at all times, including during armed conflict. If this perspective is taken to its logical conclusion, for instance, concerning the right to life under the ECHR, the ability of armies to use deadly force would essentially be eliminated and replaced with a law enforcement model. The other perspective is to view LOAC and international human rights law as complementary. Under this view, where there is a specific provision in LOAC, it will apply during armed conflict. On the other hand, where there is no specific rule under LOAC, human rights law will be applied to fill in this gap. This may seem like an academic discussion; however, this has become a significant issue in Afghanistan in the context of detainee operations. Some nations, based on decisions by the European Court of Human Rights, have concluded that the ECHR applies to detainee operations in Afghanistan and that the possibility of ill-treatment of detainees by Afghan authorities prohibits the release of detainees to Afghan police. In addition to the practical concerns to ongoing operations, the expanding influence of international human rights law during armed conflict could also impact the doctrine of command responsibility.[60]

CONCLUSION

As illustrated, the principle of command responsibility is well established in the law, but the contours of this principle are still a

59 "Legality of the Threat or Use of Nuclear Weapons, Advisory Opinion," *I.C.J. Reports* (New York: United Nations, July 1996), 226, 240.
60 UN Human Rights Council, *Report of the Special Rapporteur on Extrajudicial, Summary or Arbitrary Executions, Philip Alston: Addendum: Mission to the United States of America*, 28 May 2009, A/HRC/11/2/Add.5, http://www.unhcr.org/refworld/docid/4a3f54cd2.html.

work in progress. The development of international criminal law from the ad hoc tribunals and the International Criminal Court as well as developments in human rights law have the potential to dramatically expand the exposure that commanders face. In the multinational context, commanders could be obligated to exercise ever-greater oversight, with ever-increasing accountability, but without an increase in authority over their subordinates. Based on current legal rules, it is unlikely that the multinational commander in an organization like ISAF is able to exercise "effective control" over his or her subordinates, which makes the imposition of command responsibility impossible. Other than the course correction following the controversial *Yamashita* decision, the trend since has been in the direction of expanding the scope of command responsibility. Considering the development of the doctrine of command responsibility and other current trends like technological advances and the development of international human rights law, the potential exists for criminal liability for commanders to continue to expand. Examining these potential developments raises grave questions about the concept of military leadership, especially for multinational commands.

If commanders will be held accountable for the misconduct of subordinates that they neither ordered nor were aware of, but rather should have been aware of, the potential for criminal liability will serve as a powerful incentive for commanders to insist upon a constant flow of information from subordinates regarding all aspects of activities. Similarly, this potential for criminal prosecution for the unknown actions of subordinates will encourage, if not require, the personal intervention of commanders at all echelons into the smallest level of detail concerning operations.

Combine this incentive structure with modern communication technologies and you have the very real possibility that general officers will be monitoring and directing patrols at the company and perhaps the platoon level to avoid the potential of liability because of the actions of subordinates. While a level of oversight of this sort would almost certainly limit the use of force by subordinates, the impact on military organizations and on the concept of leadership should be apparent. When the superior at the operational or potentially strategic level exercises tactical control over a subordinate leader's unit, the superior will dictate the actions taken by the subordinate. This invariably affects the subordinate's ability to develop plans, employ initiative, and make decisions. This would unquestionably impact the development of future leaders.

These are issues that American military personnel must bear in mind. It may seem that these international law developments are something of little consequence—the United States is not a party to the Rome Statute and the ad hoc tribunals take place half a world away. But this could not be further from the truth. One needs look no further than the most recent U.S. National Security Strategy to see that international law has a prominent place in current U.S. policy:

> From Nuremberg to Yugoslavia to Liberia, the United States has seen that the end of impunity and the promotion of justice are not just moral imperatives; they are stabilizing forces in international affairs. The United States is thus working to strengthen national justice systems and is maintaining our support for ad hoc international tribunals and hybrid courts. Those who intentionally target innocent civilians must be held accountable, and we will continue to support institutions and prosecutions that advance this important interest. Although the United States is not at present a party to the Rome Statute of the International Criminal Court (ICC) and will always protect U.S. personnel, we are engaging with State Parties to the Rome Statute on issues of concern and are supporting the ICC's prosecution of those

cases that advance U.S. interests and values, consistent with the requirements of U.S. law. [61]

The impacts of developing norms of international law will be felt in the United States, regardless of whether the nation is a party to particular treaties.[62] Moreover, U.S. forces, including commanders who must work together with superiors and subordinates from nations that are parties to the ICC and recognize these international legal rules, will find themselves significantly handicapped without a basic awareness of these principles.

---

61 Barack H. Obama, *National Security Strategy* (Washington, DC: White House, 2010), 48.

62 For example, the ICTY stated that the Rome Statute, even in 1999 before the treaty came into force, had real legal significance because it reflected the *opinio iuris* of the large number of states that initially signed the treaty. *Opinio iuris* reflects a sense of legal obligation and is part of the requirement for a rule to acquire the status of customary international law, which is generally binding on all states. *Prosecutor* v. *Dusko Tadic (Appeal Judgment)*, IT-94-1-A, International Criminal Tribunal for the former Yugoslavia, ICTY, 15 July 1999, para 233.

Part III
— Spirituality —

Spiritual Injuries

Wounds of the American Warrior on the Battlefield of the Soul

David Gibson and Judy Malana

War is the realm of physical exertion and suffering. These will destroy us unless we can make ourselves indifferent to them, and for this, birth or training must provide us with a certain strength of body and soul.

-Carl von Clausewitz, *On War*[1]

The United States is a nation at war. Since 11 September 2001, hundreds of thousands of American service members have deployed to Iraq or Afghanistan in support of U.S. Overseas Contingency Operations (formerly known as the Global War on Terrorism). At the time of this writing, the number of American deaths associated with these operations stands at 6,325, with the number of those wounded in action at 47,266.[2] These casualty numbers only reflect those whose wounds are visible. Many more suffer serious, invisible, psychological injuries resulting from combat and operational stress.

To what extent are American troops affected by these invisible wounds of combat? A recent study by the U.S. Army determined

---

1 Carl von Clausewitz, *On War*, ed. and trans. Michael Howard and Peter Paret (Princeton, NJ: Princeton University Press, 1976), 101.
2 Statistics provided by the Department of Defense as of 10 December 2010, http://www.defense.gov/news/casualty.pdf.

that "8 to 14 percent of infantry soldiers who served in Iraq or Afghanistan returned seriously disabled by mental health problems, between 23 and 31 percent returned with some impairment, and about half the soldiers with either posttraumatic stress disorder (PTSD) or depression also misused alcohol or had problems with aggressive behavior."[3] While much has been documented on the mental health aspect of troops, the nature of warfare itself is vastly complex, touching on not only the body and mind of a warrior, but also his/her spirit.

In describing the nature of warfare, Sir John Keegan wrote,

> What battles have in common is human: the behavior of men struggling to reconcile their instinct for self-preservation, their sense of honor and the achievement of some aim over which other men are ready to kill them. The study of battle is therefore always a study of fear and usually of courage; always of leadership, usually of obedience; always of compulsion, sometimes of insubordination; always of anxiety, sometimes of elation or catharsis; always of uncertainty and doubt, misinformation and misapprehension, usually also of faith and sometimes of vision; always of violence, sometimes also of cruelty, self-sacrifice, compassion; above all, it is always a study of solidarity and usually of disintegration—for it is towards the disintegration of human groups that battle is directed.[4]

The trauma and stress experienced on the battlefield can without a doubt affect the very core of a person's being. Warfighters are confronted with the issues of life and death in ways that the average citizen may never understand. The experience of combat can breed an intensity of emotions, such as rage, fear, guilt, and anxiety. Traumatic events "frequently call into question existen-

---

3 Scott Hensley, "PTSD and Depression Common in Returning Combat Soldiers," NPR blog, 7 June 2007, http://www.npr.org/blogs/health/2010/06/07/127541187/ptsd-depression-iraq-afghanistan-soldiers.
4 John Keegan, *The Face of Battle* (New York: Viking Press, 1978), 303.

tial and spiritual issues related to the meaning of life, self-worth, and the safety of life."[5] For those who hold religious beliefs, the question of redemption, forgiveness, good vs. evil, and the nature of God can also be called into question. If left unaddressed, the trauma and stress of warfare can ultimately shatter a person's soul. As one Vietnam veteran and former military chaplain so aptly observed, "A foxhole can make an atheist out of a believer."[6]

The purpose of this chapter is to examine the spiritual dimension of invisible injuries sustained on the battlefield. Spiritual injury is a concept that is relatively unexplored in clinical research but one that is crucial to the dialogue of restoration and healing. It results from a person's inner conflict between his/her worldview and what was experienced. The dynamics of this type of conflict are spiritual, moral, and/or ethical. The intent of the chapter is not to pathologize spiritual injury as a new medical condition, but to call attention to its effects on the American warfighter. This chapter will also address what leaders can do to mitigate the impact of spiritual injury on service members.

First, discussion will be given to the stressful environment that service members face. Identifying the uniqueness of battlefield/war zone stressors, as well as the stressors faced in an operational setting are a basis for understanding the context in which spiritual injury can occur. The next section will explore what is known and understood in the clinical realm regarding the psychological wounds of war. While there is a lack of clinical research about the connection between spiritual injury and combat-related trauma, what is known about combat trauma, combat and operational stress, reactions, injuries, and illness can provide a foun-

5 Mark. W. Smith and David W. Foy, "Spirituality and Readjustment Following War Zone Experiences," in *Combat Stress Injury: Theory, Research and Management*, ed. Charles R. Figley and William P. Nash (New York: Routledge, 2007), 295.
6 William P. Mahedy, *Out of the Night: The Spiritual Journey of Vietnam Vets* (New York: Ballatine Books, 1986), 142.

dation for, and inform the understanding of, spiritual injury. In the next section, discussion will focus on the processes and rites from various religious traditions that can help warfighters find restoration and healing from their experiences in combat. While the military's purpose is to win wars, the individual warfighter should not lose his or her soul in the process. The final section will examine strategies through which effective leadership can help lessen the effects of spiritual injury and foster spiritual healing. As the Marine Corps Reference Publication (MCRP) 6-11C on Combat Stress states, "Leaders at all levels are responsible for preserving the psychological health . . . (wellness of body, mind, and spirit)" of their personnel.[7]

WAR

Characteristics of the Battlefield

The current wars in Iraq and Afghanistan are characteristic of asymmetric warfare. While there is not a standard definition of asymmetric warfare among the branches of the U.S. Armed Forces or Coalition nations, it is generally held that it is "an armed conflict, in which the conventional armed forces of one party, which uses regular means, is opposed by an unconventional army using irregular means."[8] Because the United States "possesses overwhelming conventional military superiority," its enemies have resorted to "mixing modern technology with ancient techniques of insurgency and terrorism."[9]

---

7 Department of the Navy, MCRP 6-11C, NTTP 1-15M, *Combat and Operational Stress Control* (Washington, DC: Headquarters, Department of the Navy, 2010), 1, http://www.marines.mil/news/publications/Pages/MCRP6-11C%28PRELIM%29.aspx#.Tyaj9Eoxoio.
8 Ted van Baarda and Désirée Elisabeth Maria Verweij, *The Moral Dimension of Asymmetrical Warfare: Counter-Terrorism, Democratic Values and Military Ethics* (Leiden, The Netherlands: Martinus Nijhoff, 2009), 15.
9 Department of the Army, FM 3-24, *Counterinsurgency* (Washington, DC: Headquarters, Department of the Army, 2006), ix, http://www.fas.org/irp/doddir/army/fm3-24.pdf.

For those in the direct line of fire, the asymmetric nature of combat in Iraq and Afghanistan can pose great challenges. Adding to the complexity of the battlefield, the enemy is "invisible"; they do not wear uniforms, drive military vehicles, or muster in garrisons. They look like everyday citizens in ordinary towns. They can be men, women, and even children. The distinction between enemy combatants and noncombatants is blurred, escalating the complexity of the battlefield.[10] Because they are nondescript and blend seamlessly into a community, insurgents potentially cloud the American warfighters' ability to adhere to theater rules of engagement (ROE). The enemy is essentially anonymous, using the culture, dress, and everyday life as a way to conceal themselves. Much to the frustration of the American warfighter, an unknown enemy is not easily defeated on the battlefield.

Improvised explosive devices (IEDs) have been and will continue to be the weapons of choice for insurgents in Iraq and Afghanistan. The North Atlantic Treaty Organization (NATO) defines IEDs as devices "placed or fabricated in an improvised manner incorporating destructive, lethal, noxious, pyrotechnic, or incendiary chemicals and designed to destroy, incapacitate, harass, or distract. It may incorporate military stores, but is normally devised from nonmilitary components."[11] IEDs can be homegrown or factory made. They are a low-cost, technologically simple, and effective means to level the battlefield. With its variety of form and mechanisms, IEDs are difficult to detect and defend against. They can be hidden anywhere, and make anything or anyone suspect. Additionally, given their capability to detonate in close proximity to intended targets at a predetermined angle, IEDs can have the same effect as precision-guided weapons.[12] Even when IEDs

10 Alexandra Zavis and Garrett Therolf, "Militants Use Children to Do Battle in Iraq: As More Youths Are Recruited, Boys Outnumber Foreign Fighters at U.S. Detention Camps," *Los Angeles Times*, 27 August 2007, http://articles.latimes.com/2007/aug/27/world/fg-childfighters27.

11 James Bevan, ed., *Conventional Ammunition in Surplus: A Reference* (Geneva: Small Arms Survey, 2008), 171.

12 John Moulton, "Rethinking IED Strategies: From Iraq to Afghanistan." *Military Review* 89 (July 2009): 26–33.

do not wound or kill troops, the mere suspicion of an IED's presence restricts and complicates movements on the battlefield. This keeps service members on a heightened level of alert constantly, which in and of itself can cause stress injuries.

In his recent book, *Lethal Warriors,* journalist David Phillips describes the intensity of IEDs:

> Troops had few effective means of detecting or defeating these concealed weapons and no clear enemy to counterattack, leaving them only a tense, helpless fear they experienced day in and day out. Hopelessness, helplessness, and uncertainty are some of the most toxic emotions that lead to damaging doses of combat stress, which means the modern type of warfare may not be as loud or bloody as the invasions of D-Day or the Tet Offensive in Vietnam, but it is no less vicious, and because there are no battle lines in Iraq or Afghanistan, troops almost never get a break. Troops like to say it is 360/365-all around, every day.[13]

The Pentagon does not release the actual numbers of American service members killed or wounded by IEDs. However, it is estimated that since 2003, IED attacks in Iraq are responsible for roughly 70 percent of American casualties.[14] In Afghanistan, they are responsible for approximately 30 percent of Americans killed or wounded.[15] It is estimated that between August 2008 and August 2009, IED attacks doubled and continue to remain at their highest levels to date.[16] As insurgent use of IEDs increases, it can

---

13 David Philipps, *Lethal Warriors: When the New Band of Brothers Came Home* (New York: Palgrave Macmillan, 2010), 11.

14 "IED's the Insurgents Deadliest Weapon," *The Times* (London), 8 December 2008, http://www.thefirstpost.co.uk/46075,news-comment,news-politics,I.E.D.'s-the-insurgents-deadliest-weapon, 2.

15 Clay Wilson, *Improvised Explosive Devices (IEDs) in Iraq and Afghanistan: Effects and Countermeasures* (Washington, DC: Congressional Research Service, Library of Congress, 2006), 1.

16 Committee on Armed Services, Oversight and Investigations Subcommittee, *Update to HASC O&I Report: "The Joint Improvised Explosive Device Defeat Organization: DOD's Fight Against I.E.D.s Today and Tomorrow"* (Washington, DC, U.S. House of Representatives: 2008), 39.

be expected that the number of those who suffer stress and spiritual injury will also increase.

The trend toward asymmetric warfare will most likely not decrease in the near term. The 2010 Joint Operating Environment report, which highlights the current and future battle space for American warfighters, predicts that "future integrated close combat will place increased demands on the physical, psychological, and spiritual domains."[17] These demands will accumulate and continue to take a toll on service members.

War Zone Stressors

What are some of the other war zone conditions that service members must endure on the battlefield that create stress? According to the National Center for PTSD, war zone stressors can include "combat exposures (firing a weapon or being fired upon by enemy combatants or friendly fire), perceived threat, low-magnitude stressors (uncomfortable living and working conditions), exposure to suffering, witnessing civilians suffering, and exposure to death and destruction."[18] The cumulative effect of combat and operational stress takes a toll on service members, in which personnel are

> . . . taxed physically and emotionally in ways that are unprecedented for them. Although soldiers are trained and prepared through physical conditioning, practice, and various methods of building crucial unit cohesion and buddy-based support, inevitably, war zone experiences create demands and tax soldiers and unit morale in shocking ways. In addition, the pure physical demands of war zone activities should not be underestimated, especially the behavioral and emotional effects of circulating norepinephrine, epinephrine, and cortisol (stress

17 David J. Hufford, Matthew J. Fritts, and Jeffrey E. Rhodes, "Spiritual Fitness," *Military Medicine* (2010): 73–87.
18 Department of Veterans Affairs, *Iraq War Clinician Guide* 2nd Edition, June 2004, http://www.ptsd.va.gov/professional/manuals/iraq-war-clinician-guide.asp.

hormones), which sustain the body's alarm reaction (jitteriness, hypervigilance, sleep disruption, appetite suppression, etc.).[19]

War zone stressors contribute to the risk for chronic PTSD. With multiple deployments, the stress is compounded and the severity of stress injuries increases exponentially.

With wars on two fronts and numerous military operations other than war, U.S. military troops are stretched thin. The repeated combat deployment tours and the lack of decompression time between tours have taken a serious toll. In commenting on this high operational tempo, General George W. Casey, former U.S. Army chief of staff said, "The human mind and body wasn't made to do repeated combat deployments without substantial time to recover."[20]

In addition to combat exposure, concern on the home front has compounded stress in the lives of service members. Family life and disruptions are significant issues. Constant deployments with the addition of fear and uncertainty for loved ones are unsettling and disruptive to family routines and stability. Thus, American warfighters are no longer the only casualty of war; their families are now at risk of being worn down to the point where they too become susceptible to stress injuries and spiritual injuries.

At the time of this writing, it is estimated that since 9/11, over $1.283 trillion has been spent on the war effort thus far.[21] While the cost of such things as equipment and materiel can be measured in dollars, the cost of the war in terms of physical, psychological,

19 Brett Litz and Susan M. Orsillo, "The Returning Veteran of the Iraq War: Background Issues and Assessment Guidelines" in *Iraq War Clinician Guide*, 22.
20 Doug Stanling, "Army Wants to Reduce Combat Zone Deployments to 9 Months," *USA Today*, 21 June 2010, http://content.usatoday.com/communities/ondeadline/post/2010/06/army--wants-to-reduce-combat-zone-deployments-to-9-months/1.
21 Amy Belasco, *The Cost of Iraq, Afghanistan, and other Global War on Terror Operations Since 9/11*, 7-5700, RL33110 (Washington, DC: Congressional Research Service, 2011), 1, http://www.fas.org/sgp/crs/natsec/RL33110.pdf.

and spiritual damage is not as easily documented. One thing is known: America will be expending funds for the treatment of its warriors for many years to come. For instance, it was reported that in 2009 the Veterans Administration spent approximately $5.6 billion dollars in compensation to veterans with mental disorders as a result of their service in the Vietnam War[22]—a conflict that ended thirty-five years ago. In terms of veterans of the Global War on Terrorism (GWOT) era, the cost could be even more staggering. In their book, *The Three Trillion Dollar War*, authors Joseph Stiglitz and Linda Bilmes estimate that the cost of caring for Iraq and Afghanistan veterans over their lifetimes could topple $717 billion with a significant portion covering mental disorders (such as PTSD, which is considered a "signature wound" of the current war).[23] Of course, any dollar amount can never adequately describe the value of the loss of life or limb, or even characterize the potential devastation that combat and combat-related operations can have on each individual touched by war.

WOUNDS OF WAR

In order to better understand the concept of spiritual injury as it relates to combat, it is helpful to understand what is known about combat trauma in general. Most research and documentation about combat trauma primarily addresses the physiological and psychological dimensions. Likewise, military doctrine regarding combat trauma mainly focuses on these areas. Combat trauma is indeed complex, and no one body of research can address all the

---

22 Tim Jones and Jason Grotto, "Costs Soar for Compensating Veterans with Mental Disorders: PTSD and Other Psychological Disorders are Becoming a Costly Consequence of Wartime Service," *Chicago Tribune*, 12 April 2010, http://articles.chicagotribune.com/2010-04-12/health/ct-met-veterans-mental-illness-20100412_1_disorders-service-related-disability.
23 Joseph E. Stiglitz and Linda Bilmes, *The Three Trillion Dollar War: The True Cost of the Iraq Conflict* (New York: W.W. Norton, 2008), 61–113.

facets of the phenomenon. The next section of this chapter focuses on the current understanding of the invisible wounds of war.

Combat Trauma

What makes trauma related to combat unique? The word "trauma" is derived from the Greek word for "wound": "Trauma is a wound which can create distress in many 'systems': physiological, neurological, cognitive, behavioral, emotional, social, psychological, and spiritual."[24] It has a broad definition and is generally applied to events that are perceived to be life threatening.

The perception of threat is subjective to each individual experiencing the event.[25] Trauma can "shatter the assumptions" of one's core beliefs about the world concerning "benevolence, meaning, and self-worth."[26] Trauma then becomes a deeply personal phenomenon that splinters personal sustaining beliefs, floods an individual with a deluge of intense emotions, endangers a person's sense of control, and can alienate a person from him/herself and the world.[27] It can also threaten a person's capacity to sustain him/herself and to recover.

Combat trauma involves "multiple events (even daily) over an extended period of time (7–12 months) with multiple deployments."[28] Actions such as killing, which may be morally questionable outside the context of the battlefield, are sanctioned

---

24 Brian Hughes and Georgia Handzo, *Spiritual Care Handbook On PTSD/TBI* (Washington, DC: Department of the Navy, 2010), 6.

25 Charles W. Hoge, *Once a Warrior, Always a Warrior: Navigating the Transition from Combat to Home—Including Combat Stress, PTSD, and MTBI* (Guilford, CT: GPP Life, 2010), 18–20.

26 Ronnie Janoff-Bulman, *Shattered Assumptions: Towards a New Psychology of Trauma* (New York: The Free Press, 1992), 6.

27 Naval Chaplain's School, *Understanding and Addressing Combat Stress for Caregivers.* U.S. Navy Professional Development Training Course, 2008, 281.

28 Kent Drescher, "Suggestions for Including Spirituality in Coping with Stress and Trauma," PowerPoint PDF, 6, http://uwf.edu/cap/HCWMS/materials/Drescher%20-%20Suggestions%20for%20Including%20Spirituality%20in%20Coping%20with%20Stress%20and%20Trauma.pdf.

by one's nation and are incorporated into the rules of engagement. The multiple roles in combat add to the potential for a trauma injury. Service members may be pressed into roles such as observer, direct victim, and actor/perpetrator. Consequently, the range and intensity of symptoms may be expanded and increased.[29]

Combat and Operational Stress

The physical and psychological effects of the battlefield, unique war zone stressors, and constant deployments are a springboard from which a person can experience combat and operational stress. Combat and operational stress is the term currently used among the different branches of service. The U.S. Army identifies combat and operational stress as "the physiological and emotional stresses encountered as a direct result of the dangers and mission demands of combat."[30] The U.S. Marine Corps defines it as "the mental, emotional, or physical tension, strain, or distress resulting from exposure to combat and combat-related conditions."[31]

Not all stressors in the war zone are from the direct exposure to combat. Operational stress can be just as damaging as being in the direct line of fire. The Navy and Marine Corps define operational stress as "changes in physical or mental functioning or behavior due to the experience or consequences of military operations other than combat—during peacetime or war, and on land, at sea, in the air, or in the home."[32] Operational stress can occur with individuals in support positions, such as those handling human remains or body parts. Military members serving in peacekeeping and peace-enforcing operations are also vulnerable to highly traumat-

---

29 Ibid.

30 Department of the Army, FM 4-02.51, *Combat and Operational Stress Control* (Washington, DC: Headquarters, Department of the Army, 2006), 1, www.fas.org/irp/doddir/army/fm4-02-51.pdf. .

31 Department of the Navy, FM 90-44/6-22.5/MCRP 6-11C, *Combat Stress* (Washington, DC: Headquarters, Department of the Navy, 2000), iii, www.au.af.mil/au/awc/awcgate/usmc/mcrp611c.pdf.

32 FM 4-02.51, *Combat and Operational Stress Control*, 3.

ic and stressful situations. Depending on the mission, they may "witness death and dying, be charged with the clearing of civilian corpses, be confronted with unexploded land mines, be fired upon as a result of misunderstandings, and witness destruction of property, or atrocities committed against fellow peacekeepers and civilians."[33]

Historically, the terminology used to describe physiological and psychological injuries has varied. During the Civil War, these conditions were known as "soldier's heart" and "nostalgia." A soldier was said to be suffering from a "soldier's heart" if he experienced increased cardiovascular activity or blood pressure with no present medical condition. "Nostalgia" was a term to describe emotional conditions such as homesickness, explosive aggression, and disciplinary problems.[34] Mental health casualties in the First World War were labeled as "shell shocked" from the effects of artillery shelling on their nervous system. Soldiers during the Second World War were considered to have "battle fatigue" or "battle exhaustion."[35] During the Korean War, with better diagnostic methods and treatments available, the term "combat stress reaction" was coined. While the labeling of these physiological and psychological conditions has changed over time, their impact on the warfighter has not.[36]

Combat and Operational Stress Reactions, Injuries, and Illnesses

Reactions from combat and operational stress include fatigue; muscular tension; shaking and tremors; problems with digestive,

---

33 Elisa E. Bolton, *Traumatic Stress and Peacekeepers.* Department of Veterans Affairs website, 5 July 2007, http://www.ptsd.va.gov/professional/pages/traumatic-stress-peacekeepers.asp.
34 Todd C. Helmus and Russell W. Glenn, *Steeling the Mind: Combat Stress Reactions and Their Implications for Urban Warfare* (Arlington, VA: Rand Corporation, 2004), 10.
35 Hans Binneveld, *From Shellshock to Combat Stress: A Comparative History of Military Psychiatry* (Amsterdam: Amsterdam University Press, 1997), 2–5. .
36 Charles R. Figley and William P. Nash. *Combat Stress Injury Theory, Research, and Management,* Routledge Psychosocial Stress Series (New York: Routledge, 2007), 33.

circulatory, and respiratory systems; anxiety; irritability; and depression.[37] These symptoms can be mild and self-limiting. Combat and operational stress can be masked by negative behaviors and can lead to more severe mental health conditions. Disciplinary problems, alcohol abuse, illicit drug use, aggression, and the threat of self-harm are common reactions to stress. Other manifestations include criminal acts such as mutilating dead enemy bodies, brutality, and torture.[38] Jonathan Shay described these negative reactions to combat stress as "the painful paradox," in which "fighting for one's country can render one unfit to be its citizen."[39]

Combat and operational stress injuries can lead to mental illness, such as posttraumatic stress disorder. In 1980, the American Psychiatric Association described PTSD as a mental disorder with debilitating, chronic, and long-lasting symptoms. Much of the initial research on PTSD focused on the psychological trauma of Vietnam veterans.[40] The current research on PTSD indicates that stress and injury in the moral domain of the human psyche can lead one down the path of PTSD. In his book, *War and the Soul*, Edward Tick, a leading clinical psychotherapist working with combat veterans suggests, "Moral pain with its incumbent harm to the soul is a root cause of PTSD. If we do not address the moral issues, we cannot alleviate it, no matter how much therapy or how many medications we apply."[41]

THE NATURE OF SPIRITUALITY

The nature of spirituality is a broad topic that seeks out and incorporates transcendent relationships beyond the self. Spirituality

---

37 FM 90-44, *Combat Stress*, v.

38 U.S. Army Center for Health Promotion and Preventive Medicine, *USA-CHPPM Technical Guide 240, Combat Stress Behaviors*, 1999, http://www.docstoc.com/docs/7273919/USACHPPM-Technical-Guide-240-Combat-Stress-Behaviors, 2.

39 Jonathan Shay, *Achilles in Vietnam*. (New York: Scribner, 1994), xiii.

40 George Fink, ed., *Encyclopedia of Stress*, vol.2 (New York: Academic Press, 2000), 2.

41 Edward Tick, *War and the Soul: Healing Our Nation's Veterans from Post-Traumatic Stress Disorder* (Wheaton, IL: Quest Books, 2005), 117.

has been defined as "a person's pursuit to connect to something or someone beyond him or herself as a means of making meaning or significance."[42] This is of particular importance, for in finding meaning and significance one finds purpose and coping skills not only for a traumatic event but for life itself. "This meaning or significance can be found in relationship with self, others, ideas, nature, higher power, art, or music. These relationships are prioritized by the person seeking meaning (Navy Medicine)."[43]

According to military medicine and the doctrine for *Total Force Fitness for the 21st Century,* the concept of spirituality "includes an array of domains including values, feelings, aspirations, and so forth, typically reflecting common theological assumptions about the human spirit."[44] Spirituality is not the same as religion. Religion implies institutions based on a person's spirituality. The military's concept of total fitness for each war fighter emphasizes "spiritual fitness," which includes healthy spiritual beliefs (positive personal worldviews), values (which guide moral decision making), practices (both inward and outward expressions of faith), and core beliefs (purpose and meaning of life).[45]

SPIRITUALITY AND MENTAL HEALTH RESEARCH

The scientific study of spirituality and mental health is a relatively new field. Twenty years ago, of the health research journals that included the topic of religion/spirituality, less than 3 percent were psychology journals, less than 2 percent were psychiatry journals, less than 1 percent were medical journals, and approximately 12 percent were nursing journals.[46] However, since 1990, there has been a growing trend with a five-fold increase of research in this

---

42 Kent Drescher, Mark Smith, and David Foy, "Spirituality and Readjustment Following War-Zone Experiences" in *Combat Stress Injury,* 295–320.
43 Hughes, *Spiritual Care Handbook,* 17.
44 Hufford, "Spiritual Fitness," 74.
45 Ibid. 74–75.
46 *Understanding and Addressing Combat Stress for Caregivers,* 283.

area.[47] There is a mounting body of research that indicates spirituality and trauma interact.

In regard to the broader topic of the moral domain where spirituality is often clinically relegated, leading voices in the field of PTSD research and clinical psychology recognize that like spirituality, the moral aspects of trauma have been, until recently, unexplored. The general reasons for this absence of exploration have been that clinicians have not asked the right questions and have felt that patient shame might prevent disclosure; clinicians may also feel helpless or unprepared in this area, may be too frightened of their own reactions, or may be perceived as being judgmental.[48] Regardless of the reasons, this type of trauma "calls for particular attention, since it is so severe in veterans, so neglected by the therapeutic community, and under modern political and technological conditions, more endemic to the practice of warfare than ever before."[49]

SPIRITUAL INJURY

Inner Conflict

The current body of literature by scholars and clinical researchers working with combat veterans suggests that spiritual injury is very much a part of an inner conflict. The assumption is that an individual's moral, ethical, and spiritual dimensions interact to form meaning, purpose, values, and perceptions of oneself and the outside world. An inner conflict arises when these perceptions are in stark contrast to what is experienced.

Inner conflict injury can be defined as "damage to individuals' conceptions of themselves, other people, important institutions,

47 Hufford, "Spiritual Fitness," 73.
48 Brett Litz, "Moral Injury and Moral Repair." Presentation, Navy and Marine Corps Combat Operational Stress Control Conference, San Diego, CA, 18 May 2010.
49 Tick, *War and the Soul,* 110.

or the Divine resulting from betrayal of deeply held values and beliefs, either by the individuals, themselves, or by others they trust."[50] For example, warfighters may be involved in potential injurious acts such as "perpetrating, failing to prevent, or bearing witness to acts that transgress deeply held moral beliefs or they may experience conflict about unethical behaviors of others. Warriors may also bear witness to intense human suffering and cruelty that shakes their core beliefs about humanity . . . "[51] The following are other examples of the various ways inner conflict can manifest itself in a military setting:

- A Marine orders his team to open fire on a vehicle refusing to stop at a checkpoint, even though he can see it is filled with women and children.

- A Navy corpsman fails to save the life of his platoon commander after he was shot in the upper chest but believes he could have succeeded if he had had better training.

- A young sailor on deployment fails to be at his wife's side as she dies from breast cancer because his commanding officer said he was needed aboard ship during its underway period.

- A Marine squad in Afghanistan makes a pact to get brutal revenge on the Taliban fighters responsible for the cruel death of one of their own unit members.

- A Navy or Marine Corps leader promises to bring back the husband of a young bride and the father of her infant child but because of unexpected events, fails to do so.[52]

---

50 William P. Nash, et al, "Moral Injury: What Every Leader Should Know," Presentation, Navy and Marine Corps Combat Operational Stress Control Conference, San Diego, CA, 18 May 2010.
51 Brett T. Litz, et al., "Moral injury and Moral Repair in War Veterans: A Preliminary Model and Intervention Strategy," *Clinical Psychology Review* 29 (2009): 695–706.
52 Nash, "Moral Injury."

In these examples, a person's self-perception makes the difference between "moral courage vs. moral failure; ethical decision making vs. ethical violations; honor vs. dishonor; faithfulness vs. betrayal; purity vs. defilement (miasma, pollution); killer vs. murderer; pride vs. shame; forgivable vs. unforgivable; meaning vs. meaninglessness."[53]

What is distinctive about the spiritual domain that sets it apart from the moral, ethical, and psychological variables is the fractured transcendent relationship between the self and the outside world (to include the Divine). Robert Kane, in his book *Through the Moral Maze,* describes this fracture as the loss of one's "spiritual center": the place where one finds access to the divine and nearness to it. To lose one's spiritual center is to lose access to one's God.[54] This can result in spiritual alienation (from God and others) and the loss of meaning. Both of these concepts have been identified by clinicians as distressing issues to veterans seeking treatment for PTSD.[55] Spiritual injury is reflected in statements such as

"I was totally alone."

"I was not myself."

"I saw myself dead."

"I lost my innocence, sanity, and faith."

"Time stopped."

"Did I die there?"

"I became mean and cold."

"I was afraid."

---

53 Ibid.
54 Robert Kane, *Through the Moral Maze: Searching for Absolute Values in a Pluralistic World* (New York: Paragon House, 1994), 3–5.
55 Margaret Nelson-Pechota, "Spirituality and PTSD in Vietnam Combat Veterans," National Conference of Viet Nam Veteran Ministers website, http://www.warveteranministers.org/spirituality_intro.htm.

"I never talked about it."

"I reject religion."

"Nothing prepared me."[56]

Spiritual injury is clinically understood in relation to the moral and ethical dimension of a person's being. Clinical psychologist Dr. Brett Litz, a leading authority on PTSD, asserts that "potentially morally injurious events, such as perpetrating, failing to prevent, or bearing witness to acts that transgress deeply held moral beliefs and expectations may be deleterious in the long term emotionally, psychologically, behaviorally, spiritually, and socially (what we label as moral injury)."[57] Spiritual injury occurs as part of an inner conflict, where a person's worldview (spiritual, moral, or ethical) is in direct contrast to what he/she has experienced in life. Spiritual red flags (similar to medical symptoms) include loss of faith, negative religious coping, lack of forgiveness, and overwhelming guilt.

SPIRITUAL RED FLAGS

According to a study conducted with combat veterans in a residential treatment program for PTSD, clinician Kent Drescher surveyed 100 Vietnam War veterans and over 50 OIF/OEF veterans regarding spirituality and combat trauma.[58] His research indicated three areas in which spiritual injury was manifested: loss of faith, negative religious coping, and forgiveness.

Loss of Faith

Of those surveyed, 60 percent indicated they experienced a "significant loss of faith." The men were, on average, 24 years old

---

56 Kent D. Drescher, Peter E. Bauer, and Bill Carr, "Spiritual Injuries of War," PowerPoint PDF, http://www.careforthetroops.com/presentations/Peter%20 Bauer%20Source-Spiritual_Injuries_of_War.pdf.
57 Brett Litz, "Moral Injury and Moral Repair," (presentation, Navy and Marine Corps Combat Operational Stress Control Conference, San Diego, CA, 18 May 2010).
58 Drescher, "Spiritual Injuries of War."

when their loss of faith occurred. The percentage of those who reported a loss of faith within the last five years was 62 percent.

Loss of faith is associated with depression, social isolation, a sense of a foreshortened future, and a hightened awareness of life's fragility. Vietnam War chaplain William Mahedy, in his book, *Out of the Night: The Spiritual Journey of Vietnam Vets,* indicates, "It is not only about the loss of faith but also about the desperate struggle to find God in the midst of hell. In spite of lengthening shadows of despair and doubt, soldiers in the field often sought God, wanting nothing more than to experience touches of His mercy, forgiveness, and above all deliverance from evil."[59]

Negative Religious Coping

The results of the survey indicated that the majority of OIF/OEF combat veterans harbored negative religious coping thoughts. These thoughts were characterized by questions about God's presence and power, expression of chronic strong anger toward God, expression of frequent dissatisfaction with congregations and clergy members, and punitive interpretations of negative circumstances. Fifty-three percent of combat veterans responded that they felt God was punishing them for their sins or lack of spirituality. Likewise, 53 percent of combat veterans wondered whether God had abandoned them.[60]

Lack of Forgiveness

The survey results indicated that for OIF/OEF combat veterans, 80 percent had seldom or never forgiven themselves for things that they had done wrong. Seventy-four percent had seldom or never forgiven those that had hurt them and 50 percent had seldom or never known whether God had forgiven them.

Forgiveness is a broad term that can generally be defined as a "process of letting go of negative thoughts, feelings, and reactions

---

59 Mahedy, *Out of the Night,* 141.
60 Drescher, "Suggestions for Including Spirituality in Coping with Stress and Trauma."

toward the offender (and often toward oneself), as well as seeking to gain a more compassionate understanding of the offender."[61] It is not the same thing as pardoning, excusing, or condoning acts of the offender. The issue of forgiveness is one that can create tension. It suggests forgiveness should be bestowed upon an enemy (even though he/she has killed a soldier's friends and fellow comrades) or upon a military organization that deployed a soldier despite his/her wishes, a leader who made an unwise or unprofessional decision resulting in tragic consequences, or even God for allowing the things of combat to happen. It also suggests a person can forgive him/herself for the acts he/she committed or "perceived errors, mistakes, or lack of action."[62] Lack of forgiveness has been associated with more severe PTSD and more severe depression.

Guilt

An issue not specifically mentioned in the survey of combat veterans, but no less important to spirituality, is the concept of overwhelming guilt. Guilt can take the form of commission and omission—where an individual expresses guilt over actions that he/she had done or failed to do. The concept of survivor's guilt is also a factor in spiritual injury. It is when an individual survives where others did not.

The *Los Angeles Times* reported the tragic story of Marine Sergeant Jeff Lehner, whose deployment in Afghanistan left him with PTSD. During his time there, he witnessed the "unspeakable" and was unable to stop what he thought was morally wrong. The reporter, Ann Louise Bardach wrote,

> His case was compounded, his friends said, by strong feelings of "survivor's guilt" involving the crash of a KC-130 transport plane into a mountain in January 2002—killing eight men in his unit. He'd been scheduled to be on the flight and had been reassigned at the last minute. As part of the ground crew that

---
61 Ibid.
62 Figley and Nash, *Combat Stress Injury*, 306.

attended to the plane's maintenance, he blamed himself. Afterward, he went to the debris site to recover remains. He found his fellow soldiers' bodies unrecognizable. He also told me he was deeply shaken by the collateral damage he saw to civilians from U.S. air attacks—especially the shrapnel wounding of so many Afghan children.[63]

The article goes on to depict a man troubled and haunted by his memories in Afghanistan and his inability to find adequate help for his PTSD through an overwhelmed Veterans Administration. Sergeant Lehner's psychologist was so concerned about his behavior that she called his home one night and spoke to his father briefly before he handed the phone to Sergeant Lehner. As the phone line went static, the psychologist feared the worst and called the police. They arrived too late. Sergeant Lehner shot his father and himself. This murder-suicide is a tragic story and is a somber reminder of the reality of combat trauma and the power of overwhelming guilt. The question then becomes, "How can a damaged spirit be restored?"

SPIRITUAL REPAIR

The importance of incorporating spiritual practices into treatment of psychological disorders is paramount to restoration and healing. Edward Tick argues for this integration by stating that "the common therapeutic model . . . misses the point that PTSD is primarily a moral, spiritual, and aesthetic disorder—in effect, not a psychological but a soul disorder . . . such aspects can be healed only by strategies aimed at them directly in this context. For this reason, it is crucial that we expand our psychological focus to a more holistic view."[64]

---

63 Ann Louise Bardach, "For One Marine, Torture Came Home," *Los Angeles Times*, 12 February 2006, http://articles.latimes.com/print/2006/feb/12/opinion/op-bardach12.
64 Tick, *War and the Soul*, 108.

Purification and Cleansing

Purification and cleansing rituals are part of various religious traditions, both Western and Eastern. These traditions focus on the removal of uncleanliness and defilement as a part of worship through such means as repentance, cleansing, and transformation. For example, Christianity practices confession and baptism, Judaism offers the Ten Days of Repentance and Atonement, Islam incorporates ritual purification for prayer, Buddhism offers the wheel of karma to transform one's legacy after hurtful actions, and the Native American Indian tradition uses the sweat lodge for purification.[65] Purification and cleansing rituals can help combat veterans atone for the immorality and impurity they may feel as a result of their actions on the battlefield.

Spiritual Rituals and Practices

Spiritual rituals and practices keep one connected to his or her spiritual center. To have a spiritual center is to maintain connectedness to one's roots, maintaining a "historically defined sense of belonging to the cosmos. The idea of roots is built into the very meaning of the term 'religion' in English, which comes from the Latin re-ligio and literally means a 'linking backward' to one's origins. There is a connection here to the fact that the religious quest is concerned with the 'meaning of life.'"[66]

The search for the "meaning of life" is a pursuit to answer the basic questions of life itself: Who am I? Why am I here? What is fair/right? What is truth? What is love? Why is there suffering/evil? Where am I going? The sudden and overwhelming nature of a violent trauma can shatter the foundations that support the "meaning of life" for an individual. It is through spiritual rituals and practices that one can reconnect to his or her roots and find a sense of belonging and connectedness.

---

65 Ibid., 207.
66 Kane, *Through the Moral Maze*, 143–44.

Rituals then become symbolic actions by individuals that express meaning and purpose. Religious traditions have a variety of spiritual rituals and practices, such as worship, prayer, meditation, and reciting litanies. The benefits of incorporating rituals and spiritual practices can include improved self-esteem, enhanced motivation and commitment to change, service toward others, and a healthy social network that gives a sense of meaning, purpose, and value.[67]

Rituals that can be specific for warfighters include memorial services for fallen comrades and the simple act of reading poetry by war survivors in a group or including an empty chair as a reminder of those who were killed in combat. In a larger communal sense, it can be participating in Armed Forces Day, Memorial Day, Veterans Day commemorations, and POW/MIA remembrances, as well as visiting war memorials.[68]

Storytelling

A story is a "map for the soul," a living thing, a divine gift. It is thought that when a person tells his or her story and listens to others', that person comes "in touch with all three: life, divinity, and soul."[69] Storytelling is a way of preserving personal histories and reveals patterns and meanings that may have been missed. It has a way of bringing a community together and "transforms both the actor and listener alike into communal witnesses."[70] Thus stories become "powerful constructs that can resonate in both those that tell them and those who hear them." They have the ability to connect individuals as a means of making sense of a traumatic event.[71]

While it may be a natural tendency for warfighters to remain silent regarding their traumatic experiences on the battlefield, re-

---

67 *Understanding and Addressing Combat Stress for Caregivers*, 288.
68 Figley and Nash, *Combat Stress Injury*, 295–308.
69 Tick, *War and the Soul*, 217.
70 Ibid.
71 Roy Baumeister and Leonard Newman, "How Stories Make Sense of Personal Experiences: Motives That Shape Autobiographical Narratives," *Personality and Social Psychology Bulletin* 20 (1994): 676.

search has shown that storytelling is one of the most important components of effective therapy. Having the opportunity to tell one's story to someone who is caring, compassionate, and empathetic helps that individual move forward from the traumatic experiences. Storytelling is also "one of the most important rituals after returning from war."[72] The tradition of postwar storytelling links all the way back to the Greek tragedies of Sophocles' "Ajax" and "Philoctetes." Such stories reveal that the tragic results of war on an individual's body, mind, and soul are not simply a phenomenon of modern warfare, but have affected military personnel for at least 2500 years. In the telling of the story, healing begins to surface through the affirmation of the individual and the collective experience of all involved. In so doing, all learn to know themselves and others, thus shaping their identities, emotions, hopes, dreams, and desires.

Theodicy and Letting Go

The problem of suffering is one that has plagued humankind from perhaps its beginning. Theodicy asks, "How can God who is all loving and all powerful allow traumatic events to happen?" A soldier on the battlefield might ask, "Why did God allow innocent children to be killed in the line of fire?" or "Why did I survive and my buddy did not?" or "If God is so loving, why was I allowed to experience the things I did?" From a Judeo-Christian perspective, theodicy challenges the ideal of God's justice and care. Theologically, there are various ways the problem of suffering has been answered—whether God chooses not to intervene or provides a means for moral growth. The question of theodicy leaves room for the mystery of the Divine—there is much unknown and even unanswerable. While an individual may go on seeking answers, the act of voicing questions provides a framework to find meaning.

The concept of theodicy does not exist in the Buddhist or Hindu religious traditions. In these Eastern religions, all suffering is de-

---

72 Hoge, *Once a Warrior,* 116.

served.[73] However, the meaning in suffering (whether deserved or not) seems, from a human standpoint, to be unanswerable. The inward practice of meditation, particularly Buddhist *koan* tradition that teaches meditation on the unanswerable questions of life, has value in relation to spiritual recovery.

Noted Army psychiatrist, Charles Hoge, in his recent book, *Once a Warrior, Always a Warrior: Navigating the Transition from Combat to Home Including Combat Stress, PTSD, and mTBI*, urges returning warriors to let go of the unanswerable questions that ignite complex emotions that can start one down a path of "chronic hopelessness, guilt, shame, self-blame, (and) depression."[74] He praises the Zen Buddhist tradition of koan as a way of reprogramming the brain; asking (or meditating on) the unanswerable questions of war paradoxically "kicks the brain circuitry out of its propensity to get stuck in these types of questions." The end result is being unstuck from a cycle of unanswerable questions.

Meaning Making

Meaning making refers to the "process of working to restore global life meaning when it has been disrupted or violated."[75] A seminal work on the topic is Viktor Frankl's *Man's Search for Meaning*. Frankl, an Austrian psychiatrist and Holocaust survivor, developed logo therapy, which is a coping strategy that finds meaning even in the midst of tragic suffering. Finding meaning makes extreme circumstances bearable and can usher in a purpose for living. "We can discover this meaning in life in three different ways: (1) by doing a deed; (2) by experiencing a value; and (3) by suffering."[76]

---

73 Ian G. Barbour, *Religion and Science: Historical and Contemporary Issues* (San Francisco: Harper, 1997), 300–1.
74 Hoge, *Once a Warrior*, 233.
75 Crystal L. Park, "Religion and Meaning" in Raymond F. Paloutzian and Crystal L. Park, ed, *Handbook of the Psychology of Religion and Spirituality* (New York: Guilford Press, 2005).
76 Viktor Emil Frankl, *Man's Search for Meaning* (Boston: Beacon Press, 2006), 111.

Clinical research indicates that traumatic and damaging events are easier to endure and adjust to when "understood within a benevolent religious framework, and attributions of death, illness, and other major losses to the will of God or to a loving God are generally linked with better outcomes."[77] An example of this linking is seen in an interview with combat veteran Mark Clester, where he recalls his fourteen-hour ordeal in a firefight outside of Fallujah in 2003:

> We turned on this road, and we knew it was a bad situation. Immediately, we offered a prayer to God. It was a quick prayer, but we understood He was in ultimate control of the situation and we asked for his Divine intervention. In reality we should have been defeated. The IEDs should have gone off, but instead they failed to explode. We needed God and He was there. None of my Marines were killed that day and we know it was because of God. Throughout the whole ordeal I had peace knowing that God was providential.[78]

THE ROLE OF LEADERSHIP

The Marine leader has a role in the prevention of and recovery from stress injuries. As retired Lieutenant General Paul Van Riper, USMC, states, "An injured Marine is lost to the force, whether the injury is physical or psychological."[79] The same can be said for stress injuries that are ethical, moral, and spiritual in nature. When a Marine is injured, unit leadership is paramount to his or her recovery. This is accomplished by fostering a climate of community at the unit level and through training and supporting

77 Paloutzian and Park, *Handbook of the Psychology of Religion and Spirituality*, 309.
78 Lieutenant Colonel Mark Clester (Ret.) USMC, interview with CDR Judy Malana, 19 December 2010.
79 Jonathan Shay, *Odysseus in America: Combat Trauma and the Trials of Homecoming.* (New York: Scribner, 2002), 208.

competent, open-minded, and ethical military leaders who have the full support of their superiors.[80]

Dr. Jonathan Shay states in *Odysseus in America: Combat Trauma and the Trials of Homecoming* "that there are three keys to preventing psychological and moral injury in military service, and that these keys are in the hands of military line leaders and trainers. These are

- Positive qualities of community in the service member's ... unit, of which stability is the most important (cohesion);

- Prolonged, progressive training that works for what troops really have to do and face; and

- Competent, ethical, and properly supported leadership."[81]

With leadership being so critical to the prevention and recovery of these types of injuries, how does a Marine leader build unit cohesion and resilience, identify those at risk, take steps to mitigate potential injury, and use available resources? A successful strategy for dealing with stress injuries will be built around three key factors: (1) strong leaders who exhibit passionate, moral, ethical, and spiritual leadership; (2) implementation of the USMC's Combat Operational Stress Control (COSC) programs; and (3) utilization of the unit's chaplain.

Leadership Strategies

As with any other aspect of military leadership, having well-thought-out and effective strategies for accomplishing mission objectives is important and there are strategies that a leader must employ in addressing stress injury. First, as the 2007 Tri-MEF COSC Conference stated, "Combat Operational Stress Control is a leadership issue. Solid leadership, unit cohesion, realistic training, and high morale are vital to preventing and ameliorating

---

80 Ibid.
81 Ibid.

combat operational stress."[82] General James F. Amos, 35th Commandant of the USMC, states that "taking care of the Marines not only includes the proper treatment of the physical injuries associated with combat, but it also involves dealing with the mental injuries which can result from combat . . ."[83]

By its very nature, being a Marine is stressful. Marines are the nation's premier "shock troops," repeatedly thrust into tough and challenging operations around the world. Frequent deployments and family separations have always been a part of the USMC, and Marine leaders by necessity must be acutely aware of the need to manage stress in getting the job done while preserving warfighter verve. However, it has not always been possible to control combat and operational stress within the Marines' capacity to adapt. These types of injuries will always be a risk to Marines and impact Marine Corps operations.

It is imperative for Marine Corps leaders to be intensely aware of the conditions that expose Marines to the risk for stress injuries. Equally, every effort must be made to prevent stress injuries whenever possible, i.e., there must be a balance between repetitive and realistic training that prepares Marines for combat, while not creating stress injuries through that training. Similarly, it is imperative that stress injuries be quickly identified and appropriate care given to those in need.

Underpinning a Marine leader's ability to lead is the ethical, moral, and spiritual leadership displayed in his/her personal life on and off the battlefield. As Doug Crandall has so aptly shown, "A leader's integrity serves as a foundation for the moral and ethical execution of missions, which protects his or her subordi-

82 Conference Notes, Tri-MEF Combat Operation Stress Control (COSC) Conference, 5–7 September 2007, http://www.i-mef.usmc.mil/external/imef-01/ staff_sections/TriMEF_COSC_CONF_NOTES_Final.pdf.
83 U.S. Marine Corps Combat Development and Integration, *Combat Stress: A Concept for Dealing with the Human Dimension of Urban Conflict* (Quantico, VA: Marine Corps Combat Development and Integration, 2007), http://www.marines. mil/news/publications/Documents/Combat%20Stress.pdf.

nates' moral justification for fighting and sustains their will to win."[84] "Leaders are expected to reflect and uphold the morals, norms, and principles of conduct that are universal to the population they are leading. They must assess and reflect upon all conditions and possible outcomes prior to making a decision."[85]

Jonathan Shay argues in *Achilles in Vietnam* that a failure by leadership to display consistent ethical, moral, and spiritual leadership was a major contributor of PTSD injuries in Vietnam veterans. He argues that military units are social constructs built upon shared expectations and values. "The moral power of an army is so great that it can motivate men to get up out of a trench and step into enemy machine gun fire. When a leader destroys the legitimacy of the army's moral order by betraying 'what's right,' he inflicts manifold injuries on his men."[86]

In a similar manner, General James N. Mattis warns leaders not to neglect the spiritual dimension in their personal lives and in the lives of their subordinates, and warns of the impact such neglect can have. He says, " . . . your spiritual path is much more of your own choosing. Just make real sure that you don't dismiss this as something of idle interest or not that important, because with the physical and the mental, you can aspire and kick ass. You can sometimes put things on the spiritual level behind you; the problem is that we endanger our very country [when we do so]."[87]

The Marine Corps' success originates in its foundational values and the synergy created when individual, unit, and institution-

84 Doug Crandall, ed., *Leadership Lessons from West Point* (San Francisco, CA: Jossey-Bass, 2006), 259.

85 Kevin G. Bezy and Joseph Makolandra, "*Spiritual and Ethical Leadership,*" Connexions website, http://cnx.org/content/m26889/latest/. (Instructional Module was written and published by Kevin Bezy and Joseph Makolandra, doctoral students from Virginia Tech and is a chapter in a larger collection entitled, *21st Century Theories of Educational Administration.*)

86 Shay, *Achilles in Vietnam*, 196.

87 James N. Mattis, "Ethical Challenges in Contemporary Conflict: The Afghanistan and Iraq Cases" (Stutt Lectures, U.S. Naval Academy, 23 February 2006), 15, www.usna.edu/Ethics/Publications/ MattisPg1-28_Final.pdf.

al values, including the spiritual, are aligned. Consequently, the most powerful thing a Marine leader can do to help his or her Marines is "set the example." This is done primarily by aligning personal actions and beliefs with the shared values congruent to the Marine Corps. Leading by example makes values real, thus, one is able to encourage the heart and soul of those he or she is charged to lead. Because of the creditability garnered through leading by example, a leader can then implement a model of the human spirit that builds spiritual resilience, helping the spiritually injured recover.

The USMC Strategy to Address Stress Injuries

The second leadership strategy in addressing combat and operational stress is the USMC COSC program. Its intent is to maintain a resilient, ready force and promote the long-term health and well-being of Marines, sailors, and their families.[88] For unit leaders, it is essential that they understand and implement USMC COSC doctrine. The components of the doctrine are built around three tools: the COS Continuum, Combat & Operational Stress First Aid, and the "Five Core Leadership Functions."

The COS Stress Continuum is the foundation for all COSC and Operational Stress Control (OSC) doctrine, training, surveillance, and interventions in both the Marine Corps and Navy. It is evidence-based and consistent with the Marine Corps' culture and warrior ethos. The model highlights the shared responsibility leaders have with that of medical personnel, mental health professionals, and chaplains for force protection and force conservation.

The second aspect of the USMC COSC response is Combat and Operational Stress First Aid (COSFA). COSFA is a flexible, mul-

---

88 Greg Goldstein, U.S. Marine Corps Combat and Operational Stress Control Program Update (presentation, Navy and Marine Corps Combat & Operational Stress Control Conference, 18–20 May 2010), http://www.med.navy.mil/sites/nmcsd/nccosc/coscConference/Pages/2010/coscConference2010MediaDay1.aspx.

tistep process for the timely assessment and preclinical care of stress reactions or injuries in individuals or units. Its goals are to preserve life, prevent further harm, and promote recovery. There are two components: "Primary Aid," which focuses on safety and calming to save a life and thwart further injury; and "Secondary Aid," which guides individuals, peers, leaders, and caregivers to work together to promote recovery or facilitate appropriate referral for further evaluation or treatment.

Lastly, COSC is a leadership issue having "Five Core Leadership Functions." The USMC has identified these functions as strengthen, mitigate, identify, treat, and reintegrate.[89]

(1) Strengthen: Leaders design activities that enhance and build resilience, such as hard, realistic training that develops physical strength, mental strength, and endurance. It is leadership that strengthens unit cohesion by building mutual trust and support, which is an essential component of resilience. It is leadership that instills confidence and provides a model of ethical and moral behavior that strengthens the individual and unit as a whole.

(2) Mitigate: Mitigating stress is about balance. While there is the need to intentionally stress Marines in order to train and prepare them, leaders must ensure adequate sleep, rest, recreation, and spiritual renewal to allow for recovery. It is imperative to also identify stress reactions or injuries early, before they become entrenched. Leaders must know the individuals in their command, recognize when confidence has been shaken, and conduct after action reviews (AARs) in small groups at the completion of a mission to defuse stress, build confidence, acknowledge loss, and give permission to grieve. Leaders must live and enforce core values, and own up when they make mistakes.

---

89 FM 90-44, *Combat Stress,* 17–20.

(3) Identify: The leader must know the strengths and weaknesses of his or her Marines and sailors, and be aware of the challenges that they face within the command and in their personal lives. On a day-to-day basis, Marine leaders must know which stress zone each unit member is in. Leaders must know how to recognize stress reactions, injuries, and illnesses. They must be able to identify when the moral compass has been fractured and a member is left experiencing excessive blame or shame. The key to preventing these types of stress injuries is to balance competing priorities, while removing stressors not essential to training or mission accomplishment. Finally, one must set a good example by effectively managing one's own stress.

(4) Treat: Ensure that those with any type of stress injury or illness get help. Train Marines to take care of each other, by restoring meaning, making amends, and promoting forgiveness. The first line of defense to stress injuries is an observant buddy, who can help a fellow Marine navigate the stress. Marines should be encouraged to seek out a leader, chaplain, counselor, or corpsman when feeling excessive stress and get definitive medical or psychological treatment when they reach the ill or injured stress zones on the COSC Continuum.

(5) Reintegrate: Marines receiving treatment for stress-related injuries need to be welcomed back to the unit. This offers the injured an opportunity for redemption and acceptance. Doing so helps neutralize the stigma associated with getting help; dampens casual derogatory comments that, when left unchecked, may prevent other Marines from seeking help when needed; and exhibits actions that restore the confidence to the stress-injured. Communication with treating professionals such as medical and mental health professionals and chaplains is important. And above all, they must remember that healing is a slow process.

The Chaplain's Role in Preventing and Treating Spiritual Injuries

The third leadership strategy in addressing combat and operational stress is utilization of the unit chaplain. He or she is the principal professional resource and advisor to a Marine leader in addressing ethical, moral, and spiritual injuries. Unlike the embedded news reporters who are assigned to military units for short durations of time during war, chaplains are organic to the unit. This insider status enables chaplains to discern not only the religious needs of personnel, but also their moral, ethical, and spiritual well-being. Chaplains share in the same experiences and demands placed upon a unit, gaining credibility and useful insights into how to help individuals directly and advise the command in how best to target resources to an individual with stress-related injuries.

By virtue of the chaplain's position in the unit, he or she gains the ability to enter into the life of those in combat. Thus, the chaplain offers a humanizing point of reference to buffer against the moral disengagement of personnel who can be drawn into moral ambiguity by the emotional and ethical demands of asymmetrical warfare. Moral disengagement is the psychological, ethical, emotional, and spiritual process that creates the conditions in which moral, ethical, and spiritual injuries can occur.[90]

Navy Captain and Chaplain Robert Phillips points out, "Unlike other helping professionals such as psychologists or counselors, the chaplain's presence and human interaction are framed by a context of faith and recognition of the spiritual dimensions and struggle implicit in the warrior's response to the crisis. The point is not to place chaplains over against other forms of helping profes-

---

90 Robert J. Phillips, *The Military Chaplaincy of the 21st Century: Cui Bono?*, International Society for Military Ethics website, http://isme.tamu.edu/ISME07/Phillips07.html.

sional presence, but to nurture synergy to maximize assistance."[91] Thus, chaplains become one of the unit leader's most valuable resources for identifying, assessing, and managing the recovery of those suffering ethical, moral, and spiritual injuries.

Conclusion

The battlefield is a daunting place. The harsh environment and war zone stressors can overwhelm a soldier's coping mechanisms and lead him/her down a path of combat and operational stress injuries and illnesses, and potentially chronic PTSD. The moral domain of inner conflict is the hearth of spiritual injury, that is, a clash between what is believed about the world and the brutal reality of war, and the fact that the horrors of violence and suffering can cause hidden injuries to the soul.

Understanding the dynamics of spiritual injuries sustained on the battlefield is essential to the restoration and healing of warfighters. Although it is a concept that has been chiefly unexplored in the clinical realm, it has gained importance and is garnering more attention as large numbers of American service members continue to be affected by this war. The USMC sees Combat Operational Stress Control as a leadership issue. Strong, effective leadership is paramount in the treatment and recovery of Marines. While the military continues to march into battle, the warfighter is not without hope and restoration of the soul.

---

91 Ibid.

SOLDIER SPIRITUALITY IN A COMBAT ZONE
PRELIMINARY FINDINGS

FRANKLIN ERIC WESTER

It is in the national interest that personnel serving in the Armed Forces be protected in the realization and development of moral, spiritual, and religious values consistent with the religious beliefs of the individuals concerned. To this end, it is the duty of commanding officers in every echelon to develop to the highest degree the conditions and influences calculated to promote health, morals, and spiritual values of the personnel under their command.

-General George C. Marshall[1]

Growing conceptual agreement recognizes and seeks to engage spirituality as an element of character for members of the U.S. military. For example, spirituality, or the domain of the human spirit, is one of the three elements of the character development model for cadets at the U.S. Military Academy—along with the ethical and social domains.[2] Across the Army and the Department

An earlier version of this chapter was published in the *Journal of Healthcare, Science and Humanities* (vol 1, no 2, July 2011), a joint venture of the Bureau of Navy Medicine and Smithsonian Scholarly Press, and is in the public domain. The article can be found at http://nmvaa.org/mhrl/subPage.php?sp=13.
1 Stewart W. Husted, *George C. Marshall: The Rubrics of Leadership* (Carlisle, PA: U.S. Army War College Foundation Press, 2006), 179.
2 Don M. Snider, *Forging the Warrior's Character: Moral Precepts from the Cadet Prayer,* ed. Lloyd J. Matthew (New York: McGraw Learning Solutions, 2008).

of Defense (DOD), holistic fitness programs include spiritual fitness.[3] And in the areas of training, education, and development, leaders aspire to inculcate character development, including spirituality, to complement military training soldiers receive.[4]

In military leadership, spirituality and cognate constructs such as morals and values, as noted by General Marshall above, have long been viewed as integral aspects of command responsibility. Character development—including character development that addresses spirituality—is understood as a facet of military leadership. A strong spirit in military members may be viewed as instrumental in fostering ethical conduct and personal resilience. This paper examines spirituality and identifies correlations between spirituality and moral attitudes and behavior along with emotional and physical resilience.

From a wider perspective, military service has traditionally been viewed as offering potential to build character.[5] The services established Junior Reserve Officers Training Corps (JROTC) programs that have character building as a central tenant. In decades past, particularly when conscripted service was used, one popular idea advanced by some was the "option" of military service in lieu of short jail time for petty crimes. This was understood as a way to give offenders a chance to "grow up" and benefit from the structure of military life. In today's era of an all-volunteer (recruited) force, military life is still viewed as fostering personal discipline and maturity.

---

3 The Army-established Comprehensive Soldier Fitness website, available at http://csf.army.mil/; and Chairman of the Joint Chiefs of Staff Instruction 3405.01, *Chairman's Total Force Fitness Framework*, Appendix E to Enclosure A, September 2011, A-2, www.dtic.mil/cjcs_directives/cdata/unlimit/3405_01.pdf.
4 Joe Doty and Walter Sowden, "Competency vs. Character? It Must Be Both!" *Military Review* (2009): 72; and Training and Doctrine Command, *The U.S. Army Concept for the Human Dimension in Full Spectrum Operations, 2015–2024*, TRADOC Pamphlet 525-3-7 (Fort Monroe, VA: Department of the Army, 2008), 15.
5 Bill Shaw, "Military Service Builds Character," *Brazosport Facts*, 10 November 2008.

This paper examines results from the Army's Excellence in Character, Ethics, and Leadership (EXCEL) survey about spirituality and how it affects ethics and the resilience of soldiers.[6] These findings are based on a sample of more than 1,250 soldiers in a combat zone, and this paper offers a preliminary discussion of findings about spirituality and a three-factor construct of spirituality. The three-factor model emerged from the survey data by calculating fit indices of scores on 15 items (fit indices will be described below [see "Spirituality Defined for This Study"]). Higher mean scores of spirituality are examined taking demographic variables into account. Correlations between spirituality, ethics, and resilience are reported, showing how spirituality interacts with measurements of ethics and resilience. The findings also point to areas for further research.

The EXCEL survey presents an honest and thought-provoking perspective from soldiers in a combat zone. The interdisciplinary survey addresses more than 20 constructs including ethical attitudes, values and behavior, leadership, physical and emotional health, and spirituality. Items about spirituality were included within the larger, interdisciplinary research instrument. Spirituality, ethics, and resilience converge to give some contours of the interactions of these factors as elements of character in soldiers.

BACKGROUND OF THE ARMY EXCEL STUDY

In 2008, the U.S. Army initiated designs and plans for the Multi-national Forces, Iraq (MNFI) Survey-2009. The study was requested by General David H. Petraeus as he relinquished command of the Multinational Forces in Iraq in September of 2008. The study

---

6 Rightly and by design, individual religious beliefs and practices have been protected in the military with attention to the twin principles of avoiding the "establishment" of religion for soldiers and urging "free exercise" through a pluralistic military chaplaincy.

had the backing of the chief of staff of the Army and was implemented by the Center for the Army Profession and Ethic (CAPE) in collaboration with the Institute for National Security Ethics and Leadership at National Defense University, the U.S. Army Chaplains Corps, and a wide range of military and civilian academic partners. The study tests a wide range of constructs about the ethical attitudes and behavior of U.S. land forces. The intent of this study was to aid leaders in self-assessment, reflection, and continuous learning.

The survey was developed to examine findings from earlier reports conducted by the Mental Health Assessment Team (MHAT IV and V). In these reports, significant percentages of soldiers and Marines stated they would not report a fellow member of the military for "killing or wounding an innocent noncombatant."[7] The Army set a high priority on ethics and ethical decision making in the face of sustained operational demands. In a combat zone, ethical dilemmas abound, and soldiers are constantly faced with demanding challenges. Lapses like Abu Ghraib and other severe ethical failures make it evident that ethics training is an ongoing necessity.[8] Survey results reveal correlations between an

---

7 Office of the Surgeon, Multinational Forces, Iraq, and Office of the Surgeon General, U.S. Army Medical Command, *Mental Health Advisory Team IV Operation Iraqi Freedom 05-07 Final Report*, 17 November 2006, http://www.armymedicine.army.mil/reports/mhat/mhat_iv/MHAT_IV_Report_17NOV06.pdf. See p. 37 indicating 45 percent of soldiers would not report a unit member for killing or wounding an innocent noncombatant.

8 Charles J. Dunlap, Jr., BG, USAF, "The Joint Force Commander and Force Discipline," *U.S. Naval Institute Proceedings*, September 2005, 34–38. Charles Dunlap wrote of the effects that the Abu Ghraib prison abuse had on the military: "The highly publicized reports of the Abu Ghraib prison abuse scandal energized the Iraqi insurgency and eroded vital domestic and Coalition support. Most damaging was the negative reaction of ordinary Iraqis, a constituency whose backing is essential to strategic success. A 2004 poll found that 54 percent of them believed all Americans behave like those alleged to have taken part in the abuse. So adverse were the strategic consequences that it is no overstatement to say that Americans died—and will continue to die—as an indirect result of this disciplinary catastrophe."

individual's level of spirituality and two other constructs: ethics and resilience. The EXCEL research findings indicate spirituality measurably correlates with five factors of ethics, such as moral courage and moral confidence, as well as increased psychological and physical resilience.

SPIRITUALITY DEFINED FOR THIS STUDY

Definitions of spirituality have evolved over the past decades and vary across diverse faith groups and cultures. The word suggests a journey or process tied to spirit defined in this research project as a multidimensional, cohesive core of the individual expressed in beliefs, ideas, practices, and connections.[9] Going into this survey of soldiers, the working hypothesis was that spirituality could be assessed using three subscales: "Spirituality incorporates the three elements of a spiritual worldview, personal piety, and connection to a faith community." These are relevant, though not sufficient, factors of spirituality.

The three subscales in the design did not achieve acceptable levels calculating from fit indices using five items per subscale. What emerged from calculating fit indices of spirituality items confirmed that spirituality is indeed multidimensional, but along different subscales. These spirituality items clustered around three factors, but in a different combination: connection to others, religious identification, and hopeful outlook. These three factors do not account for all elements of spirituality. By analyzing data from the survey questions, a unifying construct of spirituality emerged along three subscales. With the exception of four ques-

---

9 Kenneth Pargament and Patrick Sweeney, "Building Spiritual Fitness in the Army: An Innovative Approach to a Vital Aspect of Human Development," *American Psychologist* 66 (2011): 58–64. Also, Patrick Sweeney, S.T. Hannah, and Don Snider, "The Domain of the Human Spirit" in *Forging the Warrior's Character: Moral Precepts from the Cadet Prayer*, ed. L.J. Matthews (Sisters, OR: Jericho, 2007), 23–50.

tions, all of the spirituality questions on the survey fell under one of these three subscales. Thus, the EXCEL study does not cover all dimensions of spirituality, but it does reveal a workable model of spirituality for the soldiers surveyed. Using this alternative measurement model, a three-factor substructure provides extremely strong fit indices.

Spirituality, like other constructs such as stress, social support, or self-worth, cannot be observed directly. Calculating a fit index is one statistical method used to assess how well specific items measure a particular factor. By statistically analyzing patterns of responses on several items, it is possible to calculate the relative "fit" of items around conceptual models. Two calculations of fit indices are shown. A fit index above .90 is considered extremely strong. Fit indices at .75 are acceptable. As shown, the fit of the three-factor model is much better than a one-factor model. Table 1 presents the fit indices of the items structured into three factors.

Factor Structure	χ^2	Normed Fit Index	Comparative Fit Index
3 Factors	335.12	.952	.958
1 Factor	1662.12	.759	.764

TABLE 1*. **Fit Index.**

*X^2: Chi-square, also called the discrepancy function, expresses the likelihood or goodness of fit. The lower the number, the closer the fit.

NFI: Normed Fit Indices. Expresses the covariance among items. Zero indicates no covariance and 1.0 is exact covariance. This calculates an adjustment to the non-normed index accounting for sample size and degrees of freedom.

CFI: Comparative Fit Indices. CFI expresses the fit of items to form a factor and is used to avoid underestimation of fit noted in small samples. This is a rather large sample so the fit index here is strong.

U.S. ARMY CHAPLAINCY AND DOD TERMS OF REFERENCE ON SPIRITUALITY

The three-factor construct of spirituality parallels and complements the definition of spirituality that the Army chief of chaplains employs, which is "a process transcending self and society that empowers the human spirit with purpose, identity, and meaning."[10] The three factors of the EXCEL model of spirituality connect to the three functions in the chaplaincy definition: empowering people with purpose, identity, and meaning. The chaplaincy definition also incorporates awareness of that which transcends self and society. Comparing the EXCEL model to the Army chaplaincy definition, connection to others relates to identity; religious identification relates to both identity and meaning; and hopeful outlook relates to purpose.

Another definition comes from the Chairman of the Joint Chiefs of Staff Instruction on the Total Force Fitness (TFF) framework. TFF addresses spirituality, defining it as "the expression of the human spirit in thoughts, practices, and relationships of connection to self, and connections outside the self, such as other people, groups, nature, and concepts of a higher order."[11]

Although these definitions all take a different view of spirituality, they overlap and incorporate common ideas. The three factors that fit the data from the EXCEL survey cluster along three identifiable constructs: connection to others, religious identification, and hopeful outlook. These factors, when present, correlate in the lives of soldiers to positive attributes and may act as a buffer against some psychological and physical risk factors. Each of the three factors is considered further and then examined in light of correlations between spirituality and subscales addressing ethics and resilience.

METHODS IN THE EXCEL STUDY

Survey Design

The EXCEL survey is a paper-and-pencil instrument survey that

---

10 Provided by e-mail from staff at the Center for Spiritual Leadership (CSL) at the U.S. Army Chaplain Center and School, Fort Jackson, South Carolina on 14 May 2010.
11 Chairman of the Joint Chiefs of Staff Instruction, *Total Force Fitness Framework-Spiritual Fitness Domain*, Enclosure B5 (working draft from the Office of the Joint Staff Chaplain, 23 March 2010). See also, *Chairman's Total Force Fitness Framework*, Appendix E to Enclosure A, A-2.

collects demographic and survey data primarily using Likert scales.[12] EXCEL addresses topics ranging from ethical attitudes, actions, and observed behaviors in others to leadership, attitudes about the Army, general physical concerns, attitudes, and well-being. The survey was designed in four versions: version A (which featured just core questions), version B (which featured core questions and spirituality items), version A-leader (which was administered only to leaders and featured core questions), and version B-leader (which was administered only to leaders and featured both core questions and spirituality items). Surveys were collected from 2,572 soldiers deployed in Iraq between 20 June 2009 and 24 July 2009.[13] To protect the anonymity of participants, data was collected from randomly selected units.

Survey Items

Fifteen items relating to spirituality were included in the EXCEL survey at the request of the Institute for National Security, Ethics, and Leadership (INSEL) at National Defense University and the U.S. Army Chaplain Corps. Items were selected from established surveys. All items were formatted using a five-point Likert scale in line with the layout of the larger survey.

Thirteen of the 15 items included in EXCEL were based on "Dimensions of Religion/Spirituality and Relevance to Health Research" by Haber et al. from the VA Palo Alto Health Care System. The purpose of the Haber study was to "identify unique religion/ spirituality (R/S) factors that account for variation in R/S mea-

---

12 Likert scales measure the degree of agreement or disagreement with a specific item (a single trait). Social science surveys use multiple items to assess an individual trait, and researchers use statistical analysis to develop factors for correlation studies.

13 S. Hannah, J. Schaubroeck, B. Avolio, S. Kozlowski, R. Lord, and L. Trevino, *ACPME Technical Report 2010-01, MNF-1, Excellence in Character and Ethical Leadership (EXCEL) Study.* (West Point, NY: Army Center of Excellence for the Professional Military Ethic [ACPME], U.S. Army Combined Arms Center, TRADOC, 2010.)

sures of interest to health research."[14] That research focused on identifying religious and spiritual items relevant in health care research through meta-analysis of personality and medical instruments. Haber et al. took many of their questions from other well-established studies, including the *Brief Multidimensional Measure of Religion and Spirituality* by Fetzer Institute / National Institution of Aging and R. L. Piedmont's *Development and Validation of the Spiritual Transcendence Scale: A Measure of Spiritual Experience*. In addition, Haber et al. used what they called two "classic measures with exceptional histories of use."[15] The first is the Spiritual Well-Being Scale by C. W. Ellison, which measures well-being associated with God and existentialism. The second is "The Age-Universal" version of Allport and Ross's Religious Orientation Scale.

These sources, combined with one of Haber's "Religion / Spirituality Motivation, Devotion, and Coping" questions and two MN-FI-specific questions, make up the 15 items. Appendix A provides a complete list of the 15 items and their sources.

In the design, the 15 spirituality items were to measure three dimensions of spirituality in individuals: spiritual worldview, prayer / personal piety, and connection to a faith community. These address private and personal spirituality, as well as the public aspects of spirituality, paralleling the approach in a study by Greenfield, Vaillant, and Marks.[16] Also, by matching leader

---

14 Jon Randolph Haber, Theodore Jacob, and David J. C. Spangler, "Dimensions of Religion/Spirituality and Relevance to Health Research," *The International Journal for the Psychology of Religion* 17 (2007): 271. Please refer to this article for information about the other studies from which Haber el al. derived survey questions.
15 Ibid.
16 Emily Greenfield, George Vaillant, and Nadine Marks, "Do Formal Religious Participation and Spiritual Perceptions Have Independent Linkages with Diverse Dimensions of Psychological Well-Being?" *Journal of Health and Social Behavior* 50 (2009): 196, examined a two-factor model of spirituality distinguishing "spiritual perceptions" as the inner perspectives of individuals and "religious participation" as public practice.

scores with scores of followers in their units, future analysis can examine spirituality within units and interactions between leaders and followers in multifactorial analysis.

Survey Participants

This paper focuses on data from version B and version B-leader. Of the 2,572 soldiers surveyed, 1,366 completed version B and version B-leader, which included the spirituality items shown in appendix A. Of 1,366 version B surveys, there were 1,263 valid responses, meaning surveys were sufficiently complete to be tabulated and analyzed. Table 2 presents a summary of demographics of version respondents. Note that 61 percent of respondents were under age 27, and 76 percent were grade E5 (sergeant) and below.

Based on a literature review, this is the largest sample to assess soldiers' spirituality in a combat zone. The Army does collect annual data on religious preference for soldiers but not qualitative survey data. The closest comparable sample probing aspects of spirituality numbered 800 in an unpublished thesis from World War II probing the effect of combat on religious belief and personal morality.[17]

From a review of relevant literature, surveys addressing spirituality and well-being most often sample populations in hospitals or other treatment facilities, college students, or congregational members. No comparable data was previously available about military personnel in a combat zone.

---

17 Mahlon W. Pomeroy, "The Effect of Military Service and Combat Experience on Religious Beliefs and Personal Morality" (master's thesis, Syracuse University, 1946). Pomeroy and a colleague collected data about the meaning and importance of faith in God and attitudes about prayer from 800 soldiers in hospital wards at Camp Kilmer, New Jersey, during January through March of 1946. He reports that 65,000 soldiers passed through Camp Kilmer some weeks. His major findings were that "men felt their religion meant more to them now than before the war," that "God evidently seemed more personal to the men now," and "34 percent indicate that they pray more now than before the war, and only 9 percent pray less."

Gender*	Male	Female	Unknown
	1123	130	13

Age	Number	Percentage	
18–22	378	27.7	
23–27	457	33.5	
28–32	219	16.0	
33–37	130	9.5	
38–42	77	5.6	
43–47	43	3.1	
48+	13	1.0	
Unknown	49	3.6	

Marital Status	Unmarried	Married	Unknown
	736	611	19

Army Component	Active Component	Reserve Component	
	909	428	29

TABLE 2. **Summary of Demographics.**

* Respondents did not answer the question on gender in 100 surveys out of the total that included the spirituality questions. We calculated our percentages and distributions on correlations that involved gender (only a few) using 1266 (instead of 1366) as the denominator.

Procedures

To obtain a representative sample, the Multinational Forces, Iraq-Inspector General (MNFI-IG) randomly selected two brigade-sized units from each of the four Army divisions then serving in Iraq. Two battalions were randomly selected within those brigades; from each of those battalions, three companies were randomly selected, and from each of the companies, three platoons were randomly selected. In addition to these troops, key leadership at the platoon, company, and battalion levels also participated in the survey, thus allowing the survey to assess the culture/climate individual leaders developed in their areas of responsibility. Battalion chaplains and chaplain assistants carried surveys to forward operating bases. They implemented survey administration protocols, distributing and collecting surveys in platoon-sized elements (20–40 individuals). Chaplains and chaplain assistants were chosen to administer the surveys because of their formal obligations for confidentiality and their role as trusted agents.

To protect privacy and ensure anonymity, respondents filled out the survey and returned it to the unit chaplains who served as survey administrators. Data collectors placed surveys in sealed folders immediately upon collecting them from participants. Chaplains used a coding scheme with each unit. This scheme randomly assigned a code to each unit and the code was written on the outside of each sealed envelope. Using these precautions, it was not possible to associate an individual's recorded data with that individual or his/her military unit, unless the individual failed to follow instructions and put his or her name on the survey. Chaplains were the only people to have access to both the unit designation and their data, and each chaplain had access to approximately 1/20th of the full sample's data. The paper surveys were transported from chaplains to the MNFI staff and shipped to CAPE at West Point, New York. From the time the chaplain turned in the sealed envelopes for shipping, neither the staff nor the CAPE knew the unit designations and thus were unable to determine the unit from which any survey set came. Data was entered, analyzed, and reported by code only.

All leaders surveyed were also asked to rate certain effects of leadership at the platoon, company, and battalion levels. Further, leaders were asked to evaluate the leadership and unit performance of subordinate leaders at the next level down from them. All soldiers completing the survey reported on their individual ethical behavior and beliefs, rated the ethical behavior of their immediate leaders and their peers, and evaluated the culture and climate in their respective units and their psychological and somatic conditions. All respondents receiving version B (leaders and followers) rated themselves on three factors of spirituality.

When the survey respondents completed their surveys, the chaplains and chaplain assistants collected the surveys and sent them to the MNFI-IG. The surveys were shipped to CAPE. The data was provided to the following individuals for further analysis: Colonel Sean T. Hannah, PhD, CAPE director, in conjunction with several leading university researchers, including (alphabetically) Dr. Bruce Avolio, University of Washington; Dr. Steve Kozlowski, Michigan State University; Dr. Robert Lord, University of Akron; Dr. John Schaubroeck, Michigan State University; and Dr. Linda Trevino, Pennsylvania State University. The draft technical report of the data was prepared by Dr. John Schaubroeck and Colonel Hannah with assistance from the following doctoral students at Michigan State University: Nikolaos Dimotakis, Katherine Guica, Megan Huth, and Chunyan Peng.[18]

THREE FACTORS OF SPIRITUALITY

Connection to Others

McMillan and Chavis, writing about inclusion with others, defined sense of community as "a feeling that members have of belonging, a feeling that members matter to one another and to the group, and a shared faith that members' needs will be met through their commitment to be together."[19] Spirituality is often expressed through community activities, such as worship and service to others. Spirituality acknowledges realities beyond the self, and soldiers report a connection to others as a dimension of spirituality. This factor correlates with intentions for ethical actions, moral attitudes, and a general increased ability to withstand the rigors of combat. Members of the military are familiar

---

18 Hannah, *ACPME Technical Report* (DRAFT), 1.
19 David McMillan and David Chavis, "Sense of Community: A Definition and Theory," *Journal of Community Psychology* 14 (1986): 9.

with feeling a common bond with each other, just as Shakespeare coined the famous phrase, "we happy few, we band of brothers."[20] But this sense of connection to others goes far beyond camaraderie or esprit de corps.

While esprit de corps is important, it is vital for a soldier to feel not only like she or he belongs to the unit but also belongs to the rest of the human race. Soldiers who integrate this perception at a deep level of their humanity recognize even their enemies are still part of humanity and deserve certain rights and protections. A connection to others may mitigate enemy abuses, POW mistreatment, and civilian casualties.

The following items comprise the subscale for the factor "connection to others":

Q.151 I feel that on a higher level all of us share a common bond.

Q.152 Although there is good and bad in people, I believe that humanity as a whole is basically good.

Q.154 Although individual people may be difficult, I feel a bond with all of humanity.

Religious Identification

Spirituality is not experienced in a vacuum. Soldiers who recorded a higher level of spirituality tended to connect that spirituality to some level of participation in recognized religious activity—prayer, prayer by others, and worship. Though definitions of spirituality are sometimes vague, the spiritual practices of soldiers can be quite clear and specific. For soldiers, practice is important, and practice is a prominent factor in their expression of spirituality. In correlating scores for "total spirituality," the two items most closely related to this score are those that express beliefs about prayer:

---

20 William Shakespeare, *Henry V*, Act 4, Scene 3, http://shakespeare.mit.edu/henryv/henryv.4.3.html.

Q.160 I believe my personal prayers help me during this deployment. (.794)

Q.161 I believe the prayers of my family and friends back home help me. (.786)

The EXCEL study data indicates when soldiers were surveyed concerning spirituality, their spirituality was most typically described with recognizable religious identifiers such as prayer, chapel attendance, and corporate worship, which are common to organized religion. In addition to the two items about prayer, three other items were used to measure this factor:

Q.155 My spiritual life is an important part of who I am as a person.

Q.159 I go to my place of worship (chapel, church, synagogue, temple) because it helps me connect with friends.

Q.162 I believe the presence and ministry of my unit chaplain brings value to the unit.

Religion and spirituality are sometimes complicated to discuss. As the instruction issued by the chairman of the Joint Chiefs of Staff points out, "Defining 'spirituality' in the Armed Forces is difficult because of the diversity of service members and their preferred spiritual practices and the confusion, ambiguity, and blurred lines that exist between understanding and defining 'spirituality' and religion."[21] The EXCEL study shows spirituality is experienced through religious identification. This underscores the need to ensure that individual soldiers have the opportunity to practice their respective beliefs with freedom and respect. Soldiers who make use of these opportunities have a higher level of spirituality and, as considered below, this translates into increased resiliency and a strengthened personal ethic.

---

21 *CJCSI Total Force Fitness Framework*, Enclosure B5.

Hopeful Outlook

A third factor of spirituality, called hopeful outlook, emerged from the survey. Hope, optimism, and positive outlook are notable given the conditions under which these surveys were collected—living in a combat zone.

This hopeful outlook was revealed through soldiers' responses to the following items:

Q.157 I feel a sense of well-being about the direction in which my life is heading.

Q.163 I feel good about my future.

Q.164 I have forgiven myself for things that I have done wrong.

This last item acknowledges the issue of guilt, which combat veterans face. Guilt can often become a debilitating symptom if not properly processed and dealt with. This will be discussed as an aspect of resilience.

FREQUENCY DISTRIBUTIONS ON SPIRITUALITY ITEMS

Responses of soldiers in the survey items about spirituality were distributed across a wide range of scores. Roughly one-third of respondents indicated they were not in agreement with these items about spirituality, one-third of respondents were neutral, and one-third of respondents were in agreement. Two frequency distribution graphs are included that illustrate the lowest and highest response patterns for spirituality scores. In both graphs, responses are grouped into three categories: strongly disagree/disagree; neutral; agree/strongly agree. Also, each graph depicts the distribution from version B and version B Leader surveys. Leaders tended to agree or strongly agree more with items measuring spirituality compared to the larger sample of respondents.

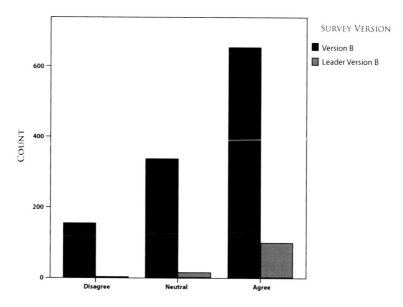

SURVEY VERSION

FIGURE 16. **I feel good about my future.**

Figure 16 shows the distribution of the highest scores on one of the spirituality items: Q164, I feel good about my future. In this distribution, 156 total respondents marked strongly disagree/disagree; 352, neutral; and 755, agree/strongly agree. This item reflects a prevailing positive outlook among the soldiers surveyed about their futures.

Figure 17 shows the distribution of the lowest scores on one of the spirituality items: Q159, If I have a problem or difficult situation, the people in my chapel community will comfort me and get me through it. In this distribution, 383 total respondents marked strongly disagree/disagree; 484, neutral; and 386, agree/strongly agree. This item indicates that even though it has the lowest scores for any of the items assessing spirituality, the response pattern is equally divided between those feeling connected and not connected to a church or chapel community with the largest group in the neutral range.

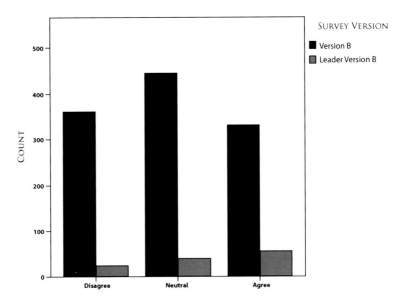

SURVEY VERSION

■ Version B
■ Leader Version B

FIGURE 17. **If I have a problem or difficult situation, the people in my chapel community will comfort me and get me through it.**

Regarding spirituality, a literature review identified no longitudinal studies that span the adult life cycle (from early adulthood to senior adulthood) that could provide conceptual descriptions of spiritual development. Most evidence of spiritual development comes from the study of individual lives[22] or is generalized from other fields such as analytic psychology,[23] moral development,[24] or faith development tied to a quest for meaning without regard to transcendence.[25]

In table 3, the three factors using subscales for spirituality and the spirituality total scores are listed with means from the Likert scale. The strongest correlations[26] (at the 0.01 level, 2-tailed) indicate

---

22 Eugene Bianchi, *Aging as a Spiritual Journey,* 2nd ed. (New York: Crossroad, 1987).
23 Carl G. Jung, *Man and His Symbols* (New York: Laurel, 1964).
24 Lawrence Kohlberg, *Essays on Moral Development, Vol. I: The Philosophy of Moral Development* (San Francisco, CA: Harper and Row, 1981).
25 James Fowler, *Stages of Faith* (New York: Harper and Row, 1981).
26 The significance of the correlations is as follows:
Strong > .350; Moderate .300 to .349; Modest .200 to .299; Slight .100 to .199.

- Higher spirituality scores correlated modestly with older respondents (.268)

- Higher spirituality scores correlated modestly with increased rank (.213)

- Higher spirituality scores correlated slightly with women (.121)

- Higher spirituality scores correlated slightly with higher education (.168)

- Higher spirituality scores correlated slightly with marriage (.073)

- Higher spirituality scores correlated slightly with having children (.145)

The cross-sectional data in this study indicate variables of age and rank produce the strongest statistically significant differences in all measures of spirituality but leave open the reasons for these differences.

	Connection to Others	Religious Identification	Hopeful Outlook	Total Spirituality Score
Mean (R=1-5)	3.0347	3.0343	3.4717	3.1517
Gender	.114**	.100**	.088**	.121**
Age	.242**	.232**	.181**	.268**
Education	.155**	.127**	.128**	.162**
Component	-.079**	-0.054	-0.023	-.064*
Married	.026	.063*	.093**	.073**
Children	.090**	.137**	.118**	.145**
Rank	.205**	.161**	.179**	.213**

TABLE 3. **Factor/Demographics.**

** Correlation is significant at the 0.01 level (2-tailed).
* Correlation is significant at the 0.05 level (2-tailed).
Notes: Range of Likert scale = 1-5 and N=1,223 to 1,263

In the EXCEL data, there are two additional items of note in the correlations. First, there was no statistically significant correlation between the number of deployments and any reported higher or lower total spirituality scores or scores on any of the three subscales. Second, an interesting and very strong correlation emerged in using single items about spirituality and the total spirituality score. The item that best correlates (.794) with the total spirituality score is belief in the benefits of personal prayer. This is nearly identical and closely followed (.786) by the item regarding belief in the benefits of prayers by family members and friends. The convergence of belief *about* prayer and the practice of prayer may be of particular interest. These responses on the belief in the effectiveness of prayer provide justification for chaplains and leaders to encourage soldiers' spiritual practice and growth.

FIVE FACTORS OF ETHICS CORRELATING WITH SPIRITUALITY

In addition to describing spirituality, this paper examines correlations between spirituality and two constructs: ethics and resiliency. Correlations between spirituality and five factors of ethics will be reported. Further below, resiliency will be analyzed describing correlations between spirituality and two factors: emotional and physical resiliency. In ethics, measuring individual responses indicated a positive correlation between spirituality and the following factors:

- Moral courage/ownership (.408, strong)
- Moral efficacy (.391, strong)
- Embracing Army values (.387, strong)
- Intent to report unethical conduct (.335, moderate)
- Soldier identification (.295, modest)

These five factors taken together could frame a useful approach to the ethical dimension of character. Using these to further specify the ethical dimension of character with soldiers may fit alongside the three-factor model for examining the domain of the human spirit or spirituality.

Factor/Spirituality Scale	Connection to Others	Religious Identification	Hopeful Outlook	Total Spirituality Score
Moral courage/ ownership	.335**	.277**	.380**	.408**
Moral efficacy	.331**	.257**	.380**	.391**
Embracing Army values	.318**	.286**	.345**	.387**
Report intentions	.309**	.232**	.283**	.335**
Soldier identification	.274**	.219**	.234**	.295**

TABLE 4. **Correlations Between Spirituality Scales and Ethics Variables.**[27]

Notes: N = 1107-1220. * p <.05. ** p <.01.

The correlations in table 4 above indicate probabilities < 0.01, and there are notably strong correlations between total spirituality scores and moral courage/ownership, moral efficacy, and embracing Army values. These correlations are all between .387 and .408, indicating a notable interaction in the character of individuals who identify with the Army values—and believe and intend to act on those moral ideas—with those who report beliefs and practices of spirituality.

Moral Courage/Ownership (.408)

The EXCEL study used seven items to assess personal moral courage and beliefs about ownership of moral responsibility. These items asked whether or not a soldier would address unethical acts. Each item was anchored on a five-point Likert scale ranging from *strongly disagree* to *strongly agree*.[28]

---

27 Dr. John Schaubroeck, "Correlational Analyses of Spirituality Scales Report" (unpublished, Center for the Army Profession and Ethic [CAPE], April 2010,) 1.
28 Hannah, *ACPME Technical Report* (DRAFT), 37.

"A majority (56 percent to 72 percent, depending on the ethical issue) of soldiers reported that they would confront others for unethical acts and would stand in the way of ethical misconduct as shown in table 26 [table 5 below]. Soldiers were most likely to agree that they would confront a peer, rather than a leader, if they observed that   person committing an ethical act. Soldiers were least likely to agree that they would not accept anyone in the unit behaving unethically, but even in this case the majority of soldiers agreed."[29]

	Percent (disagree or strongly disagree)	Percent (agree or fully agree)
I will confront my peers if they commit an unethical act	9.6	71.8
I will confront a leader if he/she commits an unethical act	10.8	69.1
I will always state my views about an ethical issue to my leaders	11.5	63.4
I will go against the group's decision whenever it violates my ethical standards	12.5	58.1
I will assume responsibility to take action when I see an unethical act	10.4	62.9
I will not accept anyone in my unit behaving unethically	12.9	55.7
I feel it is my job to address ethical issues when I know someone has done something wrong	13.2	56.0

TABLE 5. **Soldier Self-Reports on Personal Moral Courage/Ownership.**[30]

Notes: N=2572 individual soldiers. Effective sample size ranges from 2434 to 2468

(includes versions A & B).

In a forthcoming paper, Hannah and Avolio propose a psychological concept of moral potency comprised of moral courage/ownership and moral efficacy.[31] Moral potency is framed as the link between moral cognition (built out of awareness and understanding) with moral action.[32] Moral potency is proposed as the key valence in understanding an answer to the question "why do

---

29 Ibid.
30 Ibid.
31 Sean Hannah and Bruce Avolio, "Moral Potency: Building the Capacity for Character-Based Leadership," *Consulting Psychology Journal (forthcoming).*
32 James Rest, *Development in Judging Moral Issues* (Minneapolis: University of Minnesota Press, 1979).

leaders who recognize the right ethical decision or action to take still fail to act when action is clearly warranted?" Moral action is preceded by moral awareness and understanding, and perhaps it is in the area of moral potency where spirituality activates one's sense of identity, courage, and responsibility.

Moral Efficacy (.391)

Moral efficacy is, essentially, "one's confidence in his or her capabilities to organize and mobilize the motivation and cognitive resources needed to attain desired moral ends while persisting in the face of moral adversity."[33] Moral efficacy is important for individual soldiers who are facing complex moral dilemmas in the contemporary operating environment on a regular basis. Moral efficacy is developed over time in an individual's life and indeed is never completely developed. An integrated approach involving cognitive, affective, and social domains would likely enhance moral confidence.

Embracing Army Values (.387)

The American military is a values-based organization. These values are uniquely expressed by the "Army Values," "The Soldier's Creed," and "The Warrior Ethos" as outlined by the Department of Defense; its ideals are established within the Constitution of the United States of America. The Army values are presented as those attributes by which a soldier must live. The expectation is mandated across forces and applies regardless of the soldier's MOS or rank. There are seven values stipulated as vital to the success of the warrior, thereby facilitating success of the Armed Forces. These values are loyalty, duty, respect, selfless service, honor, integrity, and personal courage. Soldiers who reported that they had internalized the seven Army values to a great extent also reported lower levels of misconduct. They also reported higher levels of moral courage, that is, higher levels of intention to confront others for misconduct.

---

33 Snider, "The Domain of the Human Spirit," 82.

Intentions to Report Unethical Conduct (.335)

Six items assessed whether the respondent would report unit members if he/she observed unethical behavior directed toward a noncombatant. Each item was anchored on a five-point Likert scale with responses ranging from *strongly disagree* to *strongly agree*. Soldiers reported an intention to report a fellow unit member if that member was observed mistreating noncombatants as shown below in table 5. In particular, 70 percent would report a unit member for injuring or killing a noncombatant, while 57 percent would report "a buddy" for "abusing" a noncombatant. A minority of 15 percent stated they would not report a fellow unit member for these unethical behaviors.[34] Note that higher spirituality scores correlated with higher likelihood that soldiers would respond with their intention to report such misconduct.

Soldier Identification (.295)

Soldier identification means, in a word, internalization. The soldier internalizes the Army's values and identifies with the roles and responsibilities of being a soldier. These are the aims of character development as the Army furthers initiatives in the tiered learning model of "training-educating-development." The pamphlet, *US Army Concept of the Human Dimension in Full Spectrum Operations*,[35] discusses how the Army aims to have soldiers internalize Army values as part of their identity by linking physical, moral, and cognitive components.

THREE FACTORS OF RESILIENCE CORRELATING WITH SPIRITUALITY

Army medical research psychologists who investigate resilience

---

34 Hannah, *ACPME Technical Report* (DRAFT), 36.
35 Department of the Army, *U.S. Army Concept of the Human Dimension in Full Spectrum Operations, 2015-2024*, TRADOC Pamphlet 525-3-7 *(Fort Monroe, VA: Training and Doctrine Command, 2008)*, www.tradoc.army.mil/tpubs/pams/p525-3-7.doc.

(or "hardiness") define resilience as "the ability of adults in otherwise normal circumstances who are exposed to an isolated and potentially highly disruptive event such as the death of a close relation or a violent or life-threatening situation to maintain relatively stable, healthy levels of psychological and social functioning."[36] The term resilience has gained in usage in recent years, while military personnel have long been known for their endurance. For service members, resiliency includes not only sustaining themselves physically and emotionally while in combat but also coming home fit. "The final step in the long road home for the veteran is completing this initiation as a warrior. A veteran does not become a warrior merely for having gone to war. A veteran becomes a warrior when he learns to carry his war skills and his vision in mature ways. He becomes a warrior when he has been set right with life again."[37]

The effect of combat and the need to adapt upon returning home is reiterated by a university professor of philosophy who observes the effects of combat on veterans as students. She writes how war involves a " . . . shifting of habit and attitude. The point is that in putting on a uniform and going to war, a soldier grows skin that does not shed lightly. And even when it is time to slough that skin, after years of service, it does not come off easily."[38] Because combat affects soldiers on many levels, the need for resiliency is amplified—before, during, and after deployment.[39]

---

36 Paul T. Bartone, Mark A. Vaitkus, and Robert C. Williams, *Psychosocial Stress & Mental Health in the Forward-Deployed Military Community* (Heidelberg, Germany: U.S. Army Medical Research Unit-Europe, 1994); George A. Bonanno, "Loss, Trauma, and Human Resilience: Have We Underestimated the Human Capacity to Thrive After Extremely Aversive Events?" *American Psychologist* (2004): 20.
37 Edward Tick, *War and the Soul: Healing our Nation's Veterans from Post-Traumatic Stress Disorder*, 1st Quest ed. (Wheaton, IL: Quest Books, 2005), 251.
38 Nancy Sherman, "Soldiers' Moral Wounds," *The Chronicle of Higher Education*, 11 April 2010, 1–8.
39 The U.S. Army Medical Department first called their resilience program "Battle-Mind Training"; now it is calling the program simply "Resilience Training." For more information on the program, see https://www.resilience.army.mil/.

	Connection to Others	Religious Identification	Hopeful Outlook	Total Spirituality Score
Positive Affectivity	.339**	.321**	.424**	.442**
Negative Affectivity	-.157**	-.084**	-.215**	-.185**

TABLE 6. **Variable/Spirituality Scale.**

Notes: N = 1107-1220. * p <.05. ** p <.01.

Emotional Resilience

Regarding emotional resiliency, soldiers displayed the following correlations between their level of spirituality and emotional resilience:

- Higher spirituality scores correlated strongly with positive affectivity (.442, strong)

- Higher spirituality scores inversely correlated with negative affectivity (-.185, slight)

Positive affectivity reflects the extent to which a person feels enthusiastic, active, and alert. In table 6, positive affectivity correlated with spirituality and is similar to results from previous studies (see Greenfield et al.; Vaillant and Marks; Ellison and Fan; and Maselko and Kubzansky).[40] These indicate a potentially notable and strong linkage between spiritual perceptions and psychological well-being. Positive affectivity is generally viewed as a buffer against risks for depression, a serious variable in suicide risk. Also, the inverse correlation between spirituality and negative affectivity indicates some interaction between these constructs. Given that the soldiers surveyed were in a combat zone, the EXCEL survey found a surprisingly high level of hopeful outlook as well as other items reflecting positive views of the future regarding the soldier's situation in Iraq. Among the items describing hopeful

---

40 See References section of original publication in *Journal of Healthcare, Science and the Humanities* 1 (2011): 87–91.

outlook is one item assessing the perspective of soldiers who forgave themselves for actions that occurred during combat. This capacity to forgive oneself is related to emotional health in the period following combat deployment.

Resilience and Dealing with Guilt

Absolution from guilt is a core dynamic for combat veterans reentering life after war.[41] Encountering veterans as college students, one professor writes of how many combat veterans struggle with guilt. While researching for a recent book, Sherman found ". . . in virtually all of my interviews, guilt was the elephant in the room." She categorized the guilt soldiers experience into three forms: accident guilt, luck guilt, and collateral damage guilt. The first of these, accident guilt, is rather straightforward; it refers to the type of guilt veterans experience for mishaps that occurred in combat resulting in the loss of fellow soldiers or the lives of innocents. Although no one person can be found responsible in these types of situations, veterans still may blame themselves and experience accident guilt. Luck guilt is a form of guilt Sherman describes as a generalized form of survivor guilt. Sherman interviewed Marines who recently returned from Iraq and who were touring Annapolis. They felt genuine guilt about relaxing on a sailboat while their brothers were still in combat. The most troubling kind of guilt Sherman studied is what she calls collateral damage guilt, associated with the unintended killing of innocents by the actions of someone in combat.[42]

Physical Resiliency

A soldier's physical health is a large part of resiliency. During deployment, soldiers may endure a wide array of physical hard-

---

41 Larry Dewey, *War and Redemption: Treatment and Recovery in Combat-Related Post Traumatic Stress Disorder* (Farnham, UK: Ashgate, 2004), 201.
42 Sherman, "Soldiers' Moral Wounds," 1–8. See also Nancy Sherman, *The Untold War: Inside the Hearts, Minds, and Souls of Our Soldiers* (New York: Norton, 2010).

Variable/ Spirituality Scale	Connection to Others	Religious Identification	Hopeful Outlook	Total Spirituality Score
Somatic Complaints	-.140**	-.064*	-.154**	-.146**
Fatigue	-.162**	-.124**	-.160**	-.183**

TABLE 7. **Variable/Spirituality Scale.**

Notes: N = 1107-1220. * p <.05. ** p <.01.

ships. When they return home, it is essential for them to receive treatment for injuries and ailments incurred during deployment in order to prepare for future deployments. Since the ongoing process of deployment, redeployment, training, and subsequent additional deployments is a reality, resiliency is important. The correlation between a soldier's level of spirituality and his or her physical health is a vital link. The EXCEL study revealed an inverse relationship between a soldier's spirituality and somatic complaints and fatigue.

- Spirituality inversely correlated with physical and psychological fatigue (-.183)

- Spirituality inversely correlated with somatic complaints (-.146)

This study is consistent with other investigations that link spirituality with physical health. Among military populations, Frederick M. Dini, LCDR, SC, USN, wrote an unpublished masters-level thesis on a strategy for a military spiritual self-development tool and physical well-being.[43] Dini reports these studies show posi-

---

43 Peter C. Hill and Kenneth I. Pargament, "Advances in the Conceptualization and Measurement of Religion and Spirituality: Implications for Physical and Mental Health Research," *American Psychologist* 58 (2003): 64–74; Doug Oman and Carl E. Thoresen, "Do Religion and Spirituality Influence Health?"; Lis Miller and Brien S. Kelley, "Relationships or Religiosity and Spirituality with Mental Health and Psychopathology," in *Handbook of the Psychology of Religion and Spirituality*, ed. Raymond F. Paloutzian and Crystal L. Park (New York: The Guilford Press, 2005), 435–78.

tive correlations between spiritual development and health in the following areas: lower blood pressure, improved physical health, healthier lifestyles and less risky behavior, improved coping ability, less depression, faster healing, lower levels of bereavement after the death of a loved one, a decrease in fear of death, and higher school achievement.[44] These studies describe civilian populations. For military populations, physical health is potentially a life-and-death issue. A soldier's health and personal resiliency can very well mean the difference between coming home or not.

CONCLUSIONS AND IMPLICATIONS

This paper describes three factors that express aspects of spirituality and reports measurable correlations between spirituality, ethical attitudes and actions, and personal resilience. While spirituality is not identical to religious practice, the survey findings about soldiers in combat indicate beliefs about the benefits of prayer and participation in worship most strongly correlate with overall spirituality scores. The convergence of belief *about* prayer and the practice of prayer may offer a primary means for engaging soldiers regarding spirituality from a variety of religious perspectives. Also regarding spirituality survey scores correlate moderately with age and rank, and spirituality scores correlate slightly with gender (higher in women), education, having children, and marriage.

Spirituality positively correlates with both ethical attitudes and intentions. Spirituality strongly correlates with moral courage/ownership, moral efficacy, and embracing Army values. Spirituality moderately correlates with intention to report ethical violations observed in others and with soldier identification. These

---

44 William G. Huitt and Jennifer L. Robbins, "An Introduction to Spiritual Development" (paper presented at the 11th Annual Applied Psychology in Education, Mental Health, and Business conference, Valdosta, GA, October 2003), 6.

attitudes and intentions may be understood as an expression of character with spirituality as one dimension of character. From a leadership perspective, fostering moral potency may be a direct benefit for deepening spirituality as a dimension of character.

Spirituality correlates with indications of emotional and physical well-being. As reported in other research, there are apparent connections between spirituality and measures of emotional and physical health. These findings about soldiers provide data that support the Army's efforts to strengthen physical and emotional well-being.

Regarding character, mid-grade and senior noncommissioned officers (NCOs) offered perspectives as they presented personal reports of exemplary conduct observed or performed in close combat during an ethics and leadership program at Joint Forces Command.[45] The theme of the symposium was ethical decision making and high-performing teams. It involved approximately 100 combat-seasoned members of the armed forces, U.S. Special Operations Command, U.S. Joint Forces Command, civilian academics, and law enforcement leaders—all focused on ethical conduct in ambiguous and hostile situations. The NCOs observed that "members of the military operate both with highly trained skills and a human and moral core. This core of *character* is formed before and beyond the military. While in uniform, experiences can both test and potentially help develop moral strength."[46] This captures the essential context of how personal spirituality and significant family and community influences affect men and women in military service, both in terms of their moral awareness and understanding as well as their resilience under stress.

---

45 *Final Report* (limited distribution) from the Symposium on Ethical Decision-Making and Behavior in High Performing Teams, cohosted by Joint Forces Command, the Center for the Army Professional Ethic, and the Institute for National Security Ethics and Leadership, Suffolk, VA, 2–3 June 2010, 11.
46 Ibid.

The EXCEL study helps bring spirituality and its effects into the realm of legitimate study, worth scientific inquiry and further analysis. Though often categorized as the domain of anthropologists, psychologists, sociologists, and religious leaders, the topic of spirituality deserves to be brought into a wider, interdisciplinary line of effort. In efforts to develop ethics, resilience, and character, the benefits of including spiritual growth could be an area for continued research. Data from the EXCEL study could be analyzed to measure whether the amount of combat exposure, length of deployment, or frequency of deployments affect spirituality. A longitudinal study measuring spirituality during intervals of military service may indicate increases or decreases in factors of spirituality. Additionally, research could examine whether or not leadership styles, as measured in the EXCEL survey, correlate with spirituality scores among leaders or if the spirituality of leaders has effects on scores of spirituality among followers.

RECOMMENDATIONS FOR LEADERS AND CHAPLAINS

Leaders in military service can apply findings from the EXCEL study by acknowledging the value and positive impact of religious and spiritual activities on ethical behavior and resilience. Leaders can ensure troops have opportunities to practice their faith. Leaders interested in fostering character development can promote service members' participation in spiritual activities as a means of moral development within the limitations of regulations. They can provide adequate resources (funding, time on the training schedule) to unit chaplains to offer spiritual fitness training and activities. In speaking about ethical attitudes and behavior, leaders of all ranks can include spiritual values in reinforcing moral courage, responsibility, and reporting unethical conduct. Although this research was not structured to demonstrate a clear causal relationship between spirituality and ethics or resilience, there are correlations that indicate measurable interactions.

Chaplains can contribute to the moral, emotional, and physical strength of the force by assisting military members in strengthening their spirituality according to the faith tenets of those individual military members. Chaplains can foster relationships in faith communities as a means for reinforcing ethical behavior. In their religious training, chaplains can incorporate moral dilemmas and address what scriptures say about moral decision making. Since beliefs about prayer were prominent in the measures of spirituality in this study, chaplains can pray. Chaplains can provide instruction on prayer and conduct prayer services. In interacting with military members, chaplains can emphasize prayer as a means of resilience, as an item of personal protective gear. Effective chaplains will encourage connections "back home" with those who will offer prayers on behalf of military members. In their daily work, chaplains can provide scripture studies and instruction on the meaning and purpose of life and God working in spite of evil situations. Finally, chaplains can emphasize the practical application of love. Love is about selfless service and treating others with respect and dignity—others within the ranks, and even adversaries.

Appendix A

Excel Spirituality Questions with References[47]

1. I feel that on a higher level all of us share a common bond.

- *Question source:* Piedmont Spiritual Transcendance Scale

- *Original question:* I feel that on a higher level all of us share a common bond.

2. Although there is good and bad in people, I believe that humanity as a whole is basically good.

- *Question source:* Piedmont Spiritual Scale

- *Original question:* Although there is good and bad in people, I believe that humanity as a whole is basically good.

3. There is an order to the universe that transcends human thinking.

- *Question source:* Piedmont Spiritual Scale

- *Original question:* There is an order to the universe that transcends human thinking.

4. Although individual people may be difficult, I feel a bond with all of humanity.

- *Question source:* Piedmont Spiritual Scale

- *Original question:* Although individual people may be difficult, I feel an emotional bond with all of humanity.

---

47 Jon Randolph Haber, Theodore Jacob, and David J. C. Spangler, "Dimensions of Religion/Spirituality and Relevance to Health Research," *The International Journal for the Psychology of Religion* 77 (2007): 265–88.

5. My spiritual life is an important part of who I am as a person.

- *Question source:* Allport's Extrinsic Religion[48]

- *Original question:* Although I am religious, I don't let it affect my daily life.

- *Original question:* Although I believe in my religion, many other things are more important in life.

6. I feel deep inner peace or harmony.

- *Question source:* Existential Well-Being[49]

- *Original question:* I feel deep inner peace or harmony.

7. I feel a sense of well-being about the direction in which my life is heading.

- *Question source:* Existential Well-Being[50]

- *Original question:* I feel a sense of well-being about the direction in which my life is heading.

8. I have the sense of a larger of purpose in my life.

- *Question source:* Existential Well-Being[51]

- *Original question:* I have been able to step outside of my ambitions and failures, pain and joy, to experience a larger sense of fulfillment.

9. I go to my place of worship (chapel, church, synagogue, temple) because it helps me to connect with friends.

- *Question source:* Fetzer/NIA Religious Support[52]

- *Original question:* I go to my place of worship (church, synagogue, temple) because it helps me to make friends.

- *Original question:* I go to my (church, synagogue, temple) mostly to spend time with my friends.

---

48 "The Age-Universal" version of Allport and Ross's Religious Orientation Scale, as reported by Haber, 278.
49 "Spiritual Well-Being Scale" by C. W. Ellison, as reported by Haber, 277.
50 Ibid.
51 Ibid.
52 "Brief Multidimensional Measure of Religion and Spirituality" by Fetzer Institute/National Institution of Aging, as reported by Haber, 278.

10. I believe my personal prayers help me during this deployment.

- *Question source:* R/S Motivation, Devotion, & Coping

- *Original question:* How important is it to you to be able to turn to prayer when you are facing a personal problem?

11. I believe the prayers of my family and friends back home help me.

- *Question source:* This question was created by the Chaplain Corps to determine the recognized level of spiritual support from home.

12. I believe the presence and ministry of my unit chaplain brings value to the mission.

- *Question source:* This question is a military-centric question created to meet the specific needs of the Chaplain Corps.

13. I feel good about my future.

- *Question source:* Existential Well-Being

- *Original question:* I feel good about my future.

14. I have forgiven myself for things that I have done wrong.

- *Question source:* Existential Well-Being

- *Original question:* I have forgiven myself for things that I have done wrong.

15. If I have a problem or difficult situation, the people in my chapel community will comfort me and get me through it.

- *Question source:* Fetzer/NIA Religious Support

- *Original question:* If you were ill, how much would the people in your congregation help you out?

- *Original question:* If you had a problem or difficult situation, how much comfort would the people in your congregation be willing to give you?

The Spiritual as the Nexus for the Ethical and the Legal

Jeffrey S. Wilson

In a recent discussion with senior noncommissioned officers at the U.S. Army Sergeants Major Academy Directorate of Training and Development, Army chaplain Major Mark Johnston suggested that there are "[t]hree questions that challenge the whole purpose of spiritual fitness [in the Army]: what it is, can we train it, and if so, how."[1] The fact that even the chaplains are not sure of the answers to these questions illustrates the cognitive dissonance in the U.S. Army with regard to the concept of spiritual fitness and the core concepts that underlie it. Indeed, without a proper understanding of the terms "spirit" and "spirituality," a coherent conception of ethics, law, and their relationship cannot be constructed. This chapter attempts to answer Major Johnston's three questions. In order to do so, it first examines the term "spiritual fitness" and reveals a philosophical incoherence in the official institutional documents governing the use of the term. Second, it examines the relationship between the philosophical incoherence of the term "spiritual fitness" as used in official organizational documents and the cognitive dissonance with regard to the term "spirit" and the concepts of "spiritual fitness" and "spirituality"

---

1 Eric B. Pilgrim, "Spiritual Fitness: What Is It, Can We Train It and If So, How," U.S. Army Hooah4Health.com, accessed 31 December 2010, http://www.hooah-4health.com/spirit/FHPspirit.htm.

in professional discourse (journals, books, and blogs). How this cognitive dissonance weakens the conceptual basis of professional identity in the Army and inhibits the evolution of a professional ethic that explicitly displays the links between spirituality, morality, and law will be emphasized. Third, it will argue for an official working definition of the term "spirit" and the concepts of "spiritual fitness" and "spirituality." A definition of such terms and concepts will clarify the Army's conceptual basis of professional identity and enhance the evolution of a professional ethic that explicitly displays the links between spirituality, morality, and law. Finally, two intertwined historically centered objections will be considered that may seriously challenge this view and a reply to those objections will be offered in the conclusion.

SIGNIFICANCE OF THE ISSUE UNDER DISCUSSION

Since spirituality plays a key role in the philosophical underpinning of any logically coherent construction of ethics and law, both legal and ethical education must begin with an education in the spiritual. Therefore, legal and ethical education that ignores the spiritual is fragmented in a way that renders it incoherent in both theoretical and practical realms. This fragmentation is particularly destructive in the military organizations of liberal democratic nation-states, whose essential *telos* (ultimate end or purpose) demands that every person in the organization be able to articulate justificatory arguments for the use of potentially lethal force in defense of values, beliefs, institutions, and fellow citizens. I center the discussion here on the term "spiritual fitness" because at the time of writing there is no institutionally driven dialogue within the U.S. Army concerning the root terms "spirit" and "spirituality." I will begin with an exploration of spiritual fitness.

Spiritual Fitness: The Concept and the Problem

The notion of spiritual fitness is receiving much attention at the time of this writing. This is due in part to the ongoing effort within the Army to refine its ethic, solidify its conceptual framework for professional identity, and generate an Army-wide discussion of the relationship between these abstract concepts and organizational effectiveness. The first explicit appearance of spiritual fitness in Army doctrine only occurred in September 1987, when Department of the Army Pamphlet (DA PAM) 600-63-12, *Spiritual Fitness*, was released. Over a decade earlier, the Army undertook the comprehensive revision of its normative ethical framework that resulted in the Army values and the warrior ethos. The institution officially manifested awareness of the fact that "life as a human being and as a soldier [the term "soldier" was not yet officially capitalized in all official Army documents] depends upon both physical and emotional states of being."[2] The definition of spiritual fitness offered in 1987 is still in use today, although the Army Health Promotion Program, in which it resides, is now articulated in Army Regulation (AR) 600-63. In the May 2007 version of AR 600-63, spiritual fitness is defined as the "development of the personal qualities needed to sustain a person in times of stress, hardship, and tragedy. These qualities come from religious, philosophical, or human values and form the basis for character, disposition, decision making, and integrity."[3]

The definition of spiritual fitness has presented Army leaders with a conceptual quandary in the logical structure of the path to it. By reframing the definition of spiritual fitness into a syllogism, one can see how it connects a number of complex philosophical concepts in a puzzling way:

---

2 Department of the Army, *Spiritual Fitness*, DA PAM 600-63-12 (Washington, DC: Department of the Army, 1987), 1.

3 Department of the Army, *Army Health Promotion*, AR 600-63 (Washington, DC: Department of the Army, 2007), 37.

(1) If a person is to be spiritually fit, then a person must develop the qualities needed to sustain him or her in times of stress, hardship, and tragedy.

(2) If a person is to develop the qualities needed to sustain him or her in times of stress, hardship, and tragedy, then a person must develop some sense of character, disposition, decision making, and integrity.

(3) If a person is to develop some sense of character, disposition, decision making, and integrity, then a person must access some set of religious, philosophical, or human values.

(4) Therefore, if a person is to be spiritually fit, then a person must access some set of religious, philosophical, or human values.

This syllogism, and the definition of spiritual fitness that inspires it, manifest two fallacies in its substance: the fallacy of vagueness and the fallacy of the false dichotomy. It is possible to commit the fallacy of vagueness when an expression allows "for a continuous range of interpretations [because] the meaning [of key terms in the expression] is hazy, obscure, and imprecise."[4] When the definition claims that certain qualities "form the basis for character, disposition, decision making, and integrity," one is conceptually lost: is not *integrity* itself a *disposition* and an aspect of *character*? Is not *decision making* a skill that manifests a *disposition*?

It is possible to commit the fallacy of the false dichotomy when one "presents two nonjointly exhaustive alternatives as if they were jointly exhaustive."[5] When the definition posits that certain qualities "come from religious, philosophical, or human values," one is again conceptually lost: if *religious* and *philosophical* values are not *human* values, then what are they? The structure of the definition implies that the former two categories of value are

---

4 Patrick J. Hurley, *A Concise Introduction to Logic*, 10th ed. (Belmont, CA: Wadsworth, 2008), 76–77.
5 Ibid., 676.

somehow separate from each other and from the latter; however, it is unclear how or why.

Given the philosophical incoherence with the definition of spiritual fitness as currently stated in official U.S. Army documents, it is little wonder that the U.S. Army Chaplain's Corps, who is the "proponent for Army moral leadership training,"[6] seems to be confused about how to proceed with such a mission. Interestingly, the formal charter for the moral leadership training mission does not mention spiritual fitness per se, yet specifies that moral leadership training must focus on the "education and application of current Army values"[7] and on "the virtues and values that were formative in the shaping of America and are still present in the contemporary military setting."[8] The moral leadership training rubric "recognizes the inherent dignity of all people, the value of the state, the virtues of leadership, selfless citizenship, and duty [and also] examines the religious and spiritual connections associated with ethical decision making, personal values, and personal relationships."[9] The rubric for moral leadership training manifests the same kinds of philosophical incoherence. The definition of spiritual fitness fails to explain the dichotomy between "religious" and "spiritual," and to conceptualize a difference between "virtues" and "values." Finally, it fails to recognize a "personal" sphere of value that is somehow discrete from a professional sphere without an articulation of how soldiers should understand that separation.

Another separation that is implied but not articulated in depth is that between church and state. The religious connotations inherent in the discussion of spiritual fitness and morality are troubling for a government institution in a pluralist society such as the

---

6 Department of the Army, *Army Chaplain Corps Activities,* AR 165-1 (Washington, DC: Department of the Army, 2009), 2.
7 Ibid., 31. The seven Army values as of the time of this writing are loyalty, duty, respect, selfless service, honor, integrity, and personal courage.
8 Ibid.
9 Ibid.

United States. The fact that the U.S. Army chief of chaplains is the proponent for moral leadership training and is heavily involved in spiritual fitness education only adds to popular misconceptions of the essential goals and ambitions of both institutional processes. For example, in December of 2010, the Wisconsin-based Freedom from Religion Foundation (FFRF) took the Army to task in an open letter to Secretary of the Army John McHugh. FFRF posed specific questions about the spiritual fitness component of an assessment instrument administered to soldiers periodically to assess what the Army calls "comprehensive soldier fitness." FFRF alleged that the Army was, in effect, endorsing a link between being a person of faith and being a good soldier.[10] Research into the links between what the Army calls spiritual fitness and overall health often feature rubrics that manifest the same kinds of philosophical questions, if not always the same kinds of philosophical problems. For example, the HOPE rubric[11] endorsed by the American Academy of Family Physicians defines the term "religion" as one of many paths to "spirituality." Yet the HOPE rubric recognizes how difficult it is to separate "religion" and "spirituality" when looking for specific correlations between patient attitudes and patient condition.[12]

In the Sergeants Major Academy discussion, Chaplain Johnston contends that "spiritual fitness is a basic upon which the Army values flourish," and "the fruit of spiritual fitness ought to be selfless service, ought to be loyalty, and ought to be personal courage."[13] One can reasonably suppose that the chaplain

10 Adelle M. Banks, "Army Faces Questions over 'Spiritual Fitness' Test," *The Christian Century*, 6 January 2011, http://www.christiancentury.org/article/2011-01/army-faces-questions-over-spiritual-fitness.htm.
11 H=sources of hope, strength, comfort, meaning, peace, love, and connection; O=the role of organized religion for the patient; P=personal spirituality and practices; E=effects on medical care and end-of-life-decisions.
12 Gowri Anandarajah and Ellen Hight, "Spirituality and Medical Practice: Using the HOPE Questions as a Practical Tool for Spiritual Assessment," *American Family Physician* 63 (2001), http://www.aafp/2001/0101/p81.html.
13 Pilgrim, "Spiritual Fitness."

sees all seven Army values as fruits of spiritual fitness, and here he suggests one plausible approach to reframing spiritual fitness in a way that is philosophically coherent. However, he is constrained by the lack of a definition for "spirit" and "spirituality" that would enable a new syllogism. Such a syllogism could link the goals of Army moral leadership training with the concept of spiritual fitness and the existing Army values in a coherent and easily communicable way. Chaplain Johnston correctly noted that "we don't know how to define spiritual fitness easily, we do not know how to train it well, and we do not know how to facilitate that training so that it becomes a real part of life."[14] This essay offers a path forward with three components: a suggested definition of "spirit," a suggested definition of "spirituality," and a suggested model for illustrating a nexus between spirit, spirituality, morality, and law. This chapter's objective is to enable a more philosophically coherent rubric for spiritual fitness and for moral leadership training in the U.S. Army that makes the Army values the bedrock for spiritual fitness. Though articulated in Army-centric language, this framework is applicable to other services struggling with similar philosophical problems within the framework of their own service cultures.

TERMS

Spirit and Spirituality

The *Oxford English Dictionary* contains 24 distinct ways to define "spirit," but the one that is most appropriate to this discussion is the following: "a particular character, disposition, or temper existing in, pervading, or animating, a person or set of persons; a special attitude or bent of mind characterizing men individually or collectively."[15] This definition of "spirit" clearly decouples the concept from religion and other supernatural metaphysics and

---

14 Ibid.
15 *Oxford English Dictionary, 2nd edition*, s.v. spirit, http://dictionary.oed.com.

focuses attention on a quality that can describe the intellectual ("bent of mind"), emotional ("temper"), and psychological ("attitude") orientation of an individual or a group. These intellectual, emotional, and psychological orientations can properly be called "things of the spirit;"[16] further, "attachment to or regard for" spiritual things "as opposed to material or worldly interests" can properly be called "spirituality."[17]

Nexus

This essay uses the biological definition of "nexus," understood as "an area of fusion or close contact between two adjacent cell membranes, which is characterized by low electrochemical resistance,"[18] as the conceptual linkage ("area of fusion") between the spiritual, the moral, and the legal in an institutional ethic.

A NEW CONCEPTUAL FRAMEWORK FOR SPIRITUAL FITNESS AND MORAL LEADERSHIP

The soldier's spirit can be understood as a manifestation of a disposition to live a life in accordance with the Army values of loyalty, duty, respect, selfless service, honor, integrity, and personal courage. Attention to and disposition toward authentic expression of the Army values, therefore, displays an emotional, psychological, and intellectual spirituality, in that adherence to the Army values trumps "material or worldly interests" in the way one lives. This holistic (emotional, psychological, and intellectual) spirituality manifests no connection to a supernatural metaphysics. Such a values-based spirituality grounds the soldier's orientation to morality and law—two separate spheres in which a soldier must make informed judgments that will hopefully express the worldview encompassed by the Army values in a consistent manner. It is easy to see, then, the relationship between a coherent spirituality and a coherent orientation to morality and law.

---

16 Ibid., s.v. spirituality.
17 Ibid., s.v. spirituality.
18 Ibid., s.v. nexus.

Chaplain Johnston notes how challenging it is for the U. S. Army to articulate and give new soldiers reasons for internalizing an organizational sense of spirituality.[19] He rightly emphasizes that the U.S. Army must begin the spiritual fitness discussion with new soldiers at "the very foundational level."[20] In addition, the Army must adopt a method for defining "spirit" and "spirituality" from day one of basic training in a secular framework that rests upon the Army values. It is important for the Army to move into a discussion of spiritual fitness that serves as a nexus for an orientation toward morality and law. Such a move would avoid the philosophical incoherence that has led to the problems the U.S. Army faces in defining, training, and facilitating a discussion of spiritual fitness and its relation to moral leadership.[21]

SPIRITUAL AS NEXUS OF MORAL AND LEGAL: REFINING AND EXTENDING THE SYLLOGISM

By incorporating all that has been discussed thus far, a new syllogism can be articulated that links concepts of spirit, spirituality, spiritual fitness, morality, and law together. Such a new syllogism establishes a conceptual relationship between manifestation of spirit as understood in a particular way and manifestation of the qualities that define a good soldier. This syllogism is presented in a way that seeks to reinforce Chaplain Johnston's assertion that "good soldiers are soldiers who maintain hope, who maintain serenity, who seek to respect life and are responsible":[22]

(1) If a soldier manifests the dispositions and attitudes of the Army values, then a soldier manifests spirit.

(2) If a soldier manifests spirit, then a soldier manifests a spirituality that trumps material and worldly concerns.

19 Pilgrim, "Spiritual Fitness."
20 Ibid.
21 Ibid.
22 Ibid.

(3) If a soldier manifests a spirituality that trumps material and worldly concerns, then a soldier manifests "development of the personal qualities needed to sustain a person in times of stress, hardship, and tragedy."[23]

(4) If a soldier manifests development of the personal qualities needed to sustain a person in times of stress, hardship, and tragedy, then a soldier is spiritually fit.

(5) If a soldier is spiritually fit, then a soldier "recognizes the inherent dignity of all people, the value of the state, the virtues of leadership, selfless citizenship, and duty."[24]

(6) If a soldier recognizes the inherent dignity of all people, the value of the state, the virtues of leadership, selfless citizenship, and duty, then a soldier manifests the qualities of a moral leader.[25]

(7) If a soldier manifests the qualities of a moral leader, then the soldier is a good soldier.

(8) Therefore, if a soldier manifests the dispositions and attitudes of the Army values, then the soldier is a good soldier.

The new syllogism will enhance what social psychologist and philosopher Mihaly Csikszentmihalyi calls "flow" in soldiering— the "optimal experience"[26] of being a soldier in the U.S. Army— to a degree not possible for a person who is not spiritually fit. I believe that the spiritually fit soldier is not merely a better soldier in an extrinsic sense of being more active, reflective, and participatory in all aspects of professional practice, but a better soldier in an intrinsic sense, one who is able to thoroughly appreciate the experience of soldiering. If, as Csikszentmihalyi argues, the

---

23 Department of the Army, *Army Health Promotion*, 37.
24 Department of the Army, *Army Chaplain Corps Activities*, 31.
25 Ibid. Note that the term "moral leader" is used in this essay to denote a person with the qualities listed who can lead by example, whether or not the person holds a leadership position in a technical sense within an organizational hierarchy.
26 Mihaly Csikszentmihalyi, *Flow: The Psychology of Optimal Experience* (New York: Harper Perennial, 1990), 3.

"optimal state of inner experience is one in which there is order in consciousness,"[27] then a soldier who is spiritually fit sees that there is no dichotomy between the person he is and the soldier he is. In a true flow experience, there is no meaningful sense of having to put on the Army values when one puts on the uniform and take them off while living life outside of the nominal duty day. I believe that the syllogism offered here recognizes Csikszentmihalyi's observation that "[e]verything we experience—joy or pain, interest or boredom—is represented in the mind as information [and that control of] this information [will enable us to] decide what our lives will be like,"[28] even under conditions of "stress, hardship, and tragedy."[29] The connection between being spiritually fit and being a good soldier enables even new soldiers to see how they can use the framework of the Army values to "join all experience into a meaningful pattern [in order to feel] in control of life [in a way that] makes sense."[30]

GOOD SOLDIERS AND THE OODA LOOP

In addition to offering a syllogism that is intended to connect the Army values to good soldiers, this essay also suggests that the graphic representation of human cognition in John Boyd's OODA Loop offers a complementary vehicle for enabling focused discussions in Army schooling and unit training. John Boyd was a U.S. Air Force fighter pilot who, as a colonel in the 1970's, revolutionized the way defense analysts think about the relationship between thought and action, particularly in the realm of air combat.[31] Standing for "observe-orient-decide-act," the OODA Loop has been applied to numerous professional applications

---

27 Ibid., 6.
28 Ibid.
29 Department of the Army, *Army Health Promotion*, 37.
30 Csikszentmihalyi, *Flow*, 7.
31 Robert Coram, *Boyd: The Fighter Pilot Who Changed the Art of War* (New York: Little Brown, 2002), 4–10.

outside the military, particularly in organizational management, knowledge management, and leadership theory.[32] Although Boyd himself never published his theory concerning how the OODA Loop could enable one to understand the past and enable competitive advantage in the present, Boyd's ideas have been promulgated by the scores of people—military and civilian alike—who heard him personally present graphics-laden briefings over a roughly 20-year period.[33] The graphic below (figure 18) is one of hundreds available that represent Boyd's OODA Loop concept, chosen here because of its clarity and for its explanatory notations provided by the knowledge management consulting firm material it is drawn from.[34]

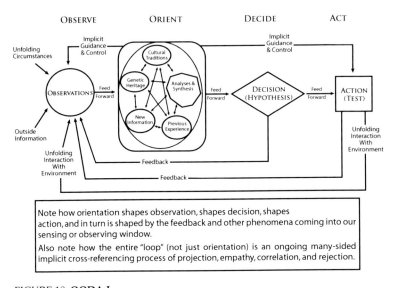

FIGURE 18. **OODA Loop.**

32 Ibid., 334.
33 Ibid., 327–44. Boyd died in 1997.
34 Joe Firestone, "OODA LOOP," Joe Firestone's Blog on Knowledge and Knowledge Management, http://kmci.org/alllifeisproblemsolving/archives/the-ooda-loop-and-double-loop-learning/.

This essay argues that the U.S. Army should insert the conceptual link between the Army values and the concept of a good soldier into the soldier's OODA Loop, mainly in the "orient" phase. Thus the good soldier would acquire deeper understanding of certain "cultural traditions," particularly those relating to how the basic documents of American government recognize the inherent dignity of all people and the value of the state. As the soldier matures and develops, the tools provided him in the training and education programs through which the Army values are communicated will enable him to use the experience of others (his teachers and soldiers he sees in case studies derived from actual situations) and his own experience as data points in an ever-deepening analysis and synthesis process through which decisions are made.

The OODA graphic notes that "orientation shapes observation,"[35] and in a way this phrase epitomizes the heart of the argument in this essay. Through this process, the U.S. Army can inform soldiers' orientation and thus enable them to reframe their past experience and learning in light of new knowledge of the Army as an institution. Soldiers will acquire new knowledge and a better understanding of human dignity and equality. These are important notions as they are at the foundation of the basic documents that soldiers swear to uphold and defend. With this, the U.S. Army will enable the formation of good soldiers, while contributing to the development of good citizens, neighbors, and friends. Then, once they have internalized the Army values and are spiritually fit, they will be able to orient a lens on particular decisions, even those made under conditions of great stress. Soldiers will acquire an appreciation for and will be able to apply moral and legal constructs that are in line with those values. An environment based on dignity and respect will affirm decisions made in their light. Further observation of the world will take place with that input in the literal and figurative memory bank of orientation in the next

---

35 Ibid.

decision situation the soldier faces, thereby reinforcing the habit of orienting one's behavior in all areas of life in light of an ever-deepening sense of intrinsic human dignity and equality. The various documents discussed in this essay demonstrate—some more directly than others—that the bedrock of American national, as well as political, identity is the commitment to intrinsic human dignity and equality. The argument offered here suggests and explicates ways to teach soldiers about that commitment by reframing the concepts of spirit, spirituality, and spiritual fitness in a more naturalistic way than they have been framed within the U.S. Army in recent years.

OBJECTIONS

An objection to the view offered here is that it does not take into adequate account the explicitly religious spirituality offered in the foundational documents of the U.S. government or of the explicitly religious (though pluralistically so) spirituality of the American society from which soldiers come, and for which they defend. In a society in which political independence is but one manifestation of the "separate and equal station to which the Laws of Nature and of Nature's God" entitles a people, further acknowledges "unalienable Rights" with which people are "endowed by their Creator" and which posits a "firm reliance on the protection of divine Providence," one might argue that it would be appropriate to infuse any organizational sense of spirituality in a government institution with the same supernatural metaphysics, albeit in a general way.[36] AR 165-1, *Army Chaplain Corps Activities*, frames the role of the Army chaplaincy as representative of "the unique commitment of the American social and religious culture that values freedom of conscience and spiritual choice as proclaimed in the founding documents."[37] Yet, in authorizing the Army

---

36 *U.S. Declaration of Independence*, http://www.archives.gov/exhibits/charters/declaration_transcript.html.
37 Department of the Army, *Army Chaplain Corps Activities*, 1.

chaplaincy, the U.S. Congress "recognize[d] the necessity of the Chaplain Corps in striking a balance between the establishment and free exercise clauses" of the U.S. Constitution.[38]

One might object also that the benefit to the U.S. Army in adopting a more naturalistic concept of spiritual fitness, though doing so might reduce cognitive dissonance with regard to spiritual fitness, comes at the cost of an essentially American identity. To frame the important historical and philosophical roots in these related objections, I turn to two foreign commentators on the American identity, writing a century apart: Alexis de Tocqueville and Denis W. Brogan.

Alexis de Tocqueville, a nineteenth-century French political writer and historian, thought that any society/community committed to the intrinsic moral equality of every person would experience a complicated relationship with spirituality: the greater the sense of equality and pluralism, the more complicated the complexity. He believed that "complete religious independence"—understood as an absence of the idea of God in the practical life of a community—and "entire political freedom" were incompatible, saying that "if a man is without faith, he must serve someone and if he is free, he must believe."[39] In *Democracy in America,* de Tocqueville highlights the way religion can impose a "healthy restraint upon the intelligence" which is "very useful for [fostering a sense of human] happiness and importance in this [world]."[40] Because "times of enlightenment and equality"[41] generate an intellectual environment where "everything in the domain of [human] intelligence is shifting,"[42] religion responds to human desire for "a

---

38 Ibid.
39 Alexis De Tocqueville, *Democracy in America and Two Essays on America* (London: Penguin Classics, 2003), 512.
40 Ibid., 511.
41 Ibid., 513.
42 Ibid., 512.

firm and stable state in [the] material world;"[43] one which, in de Tocqueville's view, can prevent a political community of free and equal persons from sliding into the paralyzing suffocation of self-gratification.[44] This can arise in the absence of the kind of religious and political authority that characterized the Europe from which the Americans fled.[45]

However, de Tocqueville cautions against religions that attempt to "stretch their powers beyond matters religious"[46] and embark on a project of "restraining the free flight of the human mind in every respect"[47] in an effort to define and codify the very conception of spirituality within an explicitly political context.[48] Of all the ways that Alexis de Tocqueville displays his admiration for American society, he shows appreciation for the American way of keeping religion (and the general sense of the spiritual writ large) within "proper bounds . . . within a circle of religious matters [alone]."[49]

Roughly a century after de Tocqueville wrote *Democracy in America*, Denis W. Brogan, a British intellectual, wrote in *The American Character* that America is "built like a church on a rock of dogmatic affirmations."[50] This is evidence of not only a firm, but a "lively conviction [in the American character] of divine interest and direction."[51] Even before the well-known allusions to a supernatural metaphysical commitment in the Declaration of Independence, the immediate predecessors of the founding fathers uttered equally "famous assertions of faith in things unseen."[52] Brogan poses rhetorical questions as he wonders at the source of

---

43 Ibid.
44 Ibid., 511.
45 Ibid., 511–12.
46 Ibid., 513.
47 Ibid., 511.
48 Ibid., 513.
49 Ibid.
50 Denis W. Brogan, *The American Character* (New York: Knopf, 1944), 128.
51 Ibid., 129.
52 Ibid., 128–29.

the deep and abiding faith in men such as Lord Baltimore, who, "in 1649, noted the evils arising from 'the inforcing [sic] of the conscience in matters of Religion' and so came out for the toleration of all Christians—this in an age when the Inquisition was still going strong."[53] Further, Brogan asks "[w]ith what Hebraic confidence in their mission did the people of Massachusetts in 1780 acknowledge 'with grateful hearts the goodness of the great Legislator of the universe, in affording us . . . an opportunity . . . of forming a new constitution of civil government, for ourselves and posterity and devoutly imploring His direction in so interesting a design[?]'"[54] In fact, Brogan thinks that the whole idea of a "government of laws and not of men"[55] is but one in a long series of articulated "aspirations [and] hopes [which, although they may seem] extravagant [and] meaningless . . . to the critical, have been fighting words, hopes, and beliefs leading to action [for a polity that, uniquely of all polities past and present, cherishes the idea of] absolutes in ethics."[56]

When Brogan asserts that an essence of American identity is the idea that "good is good, even if they quarrel over what, in the circumstances, *is* good [Brogan's italics],"[57] he might as well be referring to the conversation over the notion of spiritual fitness in the U.S. Army today. What is not a matter of contention is the idea that spiritual health is a nexus of physical health and psychological health, and as such has a nonmaterial but comprehensible reality. What are matters of contention are the essential nature of spiritual health and the characteristics and dimensions of administrative mutability with regard to that essence. The view that has been offered here posits that the essential nature of spiritual fitness is philosophical coherence with regard to a set of values

---

53 Ibid., 129.
54 Ibid.
55 Ibid.
56 Ibid., 130–31.
57 Ibid., 132.

in the choices a person makes in all areas of his or her life. Such philosophical coherence can enable a person to manifest in his or her own being exactly what AR 600-63 states spiritual fitness is: the "development of the personal qualities needed to sustain a person in times of stress, hardship, and tragedy."[58] An alternate view, though, might suggest that the essential nature of spiritual fitness in an American context is betrayed by this essay's view because it decouples faith in a supernatural metaphysics—something that Alexis de Tocqueville and D.W. Brogan have argued is essential to American identity in a psychological as well as a social and political sense—bedrock for the "development of the personal qualities needed to sustain a person in times of stress, hardship, and tragedy."[59] This alternate view might suggest that relying on a secularized conception of spiritual fitness would be meager and inadequate sustenance indeed "in times of stress, hardship, and tragedy."[60]

REPLY AND CONCLUSION

In considering a response to the objections recognized here, I revisited two especially powerful narratives of American character that have influenced my own life and the lives of many military professionals: those of U.S. Air Force Lieutenant Colonel Barry B. Bridger and U.S. Navy Vice Admiral James B. Stockdale. Both men were shot down, imprisoned in North Vietnam for several years, and withstood torture that included prolonged periods of solitary confinement. Both men cite American values as the spiritual and psychological strength that enabled them to survive physically and morally. They maintained moral integrity and helped the other prisoners do the same. In an October 2010 address at Grand Forks Air Force Base, North Dakota, Bridger

---

58 Department of the Army, *Army Health Promotion*, 37.
59 Ibid.
60 Ibid.

said that the story of his survival is a "story about the power of American values," which centered on the unwillingness "to let each other down, because when it felt like we lost everything else, we only had one duty left [and that was] to care for each other."[61] In his well-known 1978 essay for *The Atlantic Monthly* entitled "The World of Epictetus," Stockdale, too, noted that the "object of our highest value was the well-being of our fellow prisoners."[62] Stockdale wrote that to "love our fellow prisoners was within our power," and noted how the best way to manifest that love was to "keep [one's] conscience clean [and refrain from making] that first compromise."[63]

At first glance, these two testimonies seem to reinforce the syllogism I offer here, linking spiritual fitness with a defined set of values that in turn define character without an appeal to supernatural metaphysics. However, a closer read of Bridger's and Stockdale's testimonies reveals that, in both cases, it was the manifestation of a specifically religious sense of spirituality—faith—that most clearly signaled to the community of prisoners (and to the enemy, for that matter) how well they were holding up under pressure. As the senior American, Stockdale rallied his men under an ethic of "compassion, rehabilitation, and forgiveness," [telling] fellow prisoners that it was "neither American nor Christian to nag a repentant sinner to his grave" [explicitly referring here to his efforts to head off an attitude of self-pity in their situation and to give everyone the best possible chance to return with a degree of honor intact].[64] Similarly, Bridger notes that his most intense period of torture resulted from a "religious rebellion" in which

61 Rachel Waller, "Former Vietnam POW Shares His Story with Grand Forks Airmen," Air Force Print News Today, 27 October 2010, http://www.amc.af.mil/news/story_print.asp?id=123228286.
62 James B. Stockdale, "The World of Epictetus," in *Vice and Virtue in Everyday Life: Introductory Readings in Ethics,* ed. Christina Sommers and Fred Sommers (New York: Harcourt, Brace, Jovanovich College Publishers,1993), 664.
63 Ibid., 670.
64 Ibid., 663.

he and 35 fellow prisoners attempted to openly display their faith in God—the punishment was several months in solitary confinement.[65] Thus, it is reasonable to conclude that, although it is clear that the values upon which the prisoners relied in order to maintain their sanity and their moral integrity were not explicitly religious in essence, it might well have been that their expression in a religious context was the most effective evidence that the men were in fact spiritually fit. This explanation would fall under the AR 600-63 definition: that it was their religious affiliation and expression more than anything else that testified to their development of "the personal qualities needed to sustain a person in times of stress, hardship, and tragedy."[66]

Stockdale's phrasing of a core tenet of conduct explicitly recognizes an implicit dualism in the general character of American spiritual identity—a dualism that both Alexis de Tocqueville and Denis W. Brogan recognized and commented upon. When Stockdale admonished his men that it was "neither American nor Christian" to fail to forgive prisoners who had somehow collaborated with their captors but had since repented, he reflected and advanced a connection between the two. He had good reasons to believe that they would be recognized and given moral weight by the men whose behavior he was supposed to inspire. Is it any wonder, then, that particular ways of thinking about spirituality in American institutions such as the U.S. Army continue to reflect and advance (intentionally or not) the same dualism today?

The discussion in the U.S. Army about spiritual fitness remains generally useful and continually evolving. Such a discussion is important for the integration of physical, psychological, and emotional health for soldiers and their families. In *Politics and Passion*, Michael Walzer observes that "[c]onsidered as individual men

---

65 Waller, "Former Vietnam POW Shares His Story."
66 Department of the Army, *Army Health Promotion*, 37.

and women, none of us are fully autonomous, and none of us are fully integrated into and bound by any of our groups. We are each unique, one and only one; and we are at the same time tied closely to specific others in ways we sometimes resist, sometimes embrace."[67] When the ties fall into one of the first three of Walzer's four categories of involuntary association (familial and social, cultural, political), the embracing of such association is usually easy to see and understand.[68] Involuntary moral associations, though, are often harder to recognize and embrace—especially when the moral import of such associations seems, once recognized, to define who we are in ways that transcend any or all of the other three.

One could suggest that Americans might share an involuntary moral association with religious spirituality, especially in its Judeo-Christian form. The Army and the U.S. government could as well consider more explicitly recognizing in official policy such association. Yet doing so might exacerbate tensions between advocates of more individual liberal autonomy on the one hand and advocates of more communitarian concerns on the other. It can be argued, however, that this tension may be at the heart of what it means to be an American in the first place. I cannot decisively reply to this objection, that the reduction in institutional cognitive dissonance that might be achieved by adopting the framework I offer here might come at the cost of alienating the institution's members from involuntary moral associations that they recognize and value. However, I recognize the coherence of that objection as I continue to reflect on the issues discussed in this essay.

I would like to recall Major Johnston and his suggestion that there are "[t]hree questions that challenge the whole purpose of spiritual fitness [in the Army]: what it is, can we train it, and if so,

---

67 Michael Walzer, *Politics and Passion* (New Haven: Yale University Press, 2004), 140.
68 Ibid., 3–8.

how."[69] I noted that the fact that even the chaplains are not sure of the answers to these questions illustrates the cognitive dissonance in the U.S. Army with regard to the concept of spiritual fitness and the core concepts that support it—"spirit" and "spirituality"—concepts without which I believe coherent conceptions of ethics, law, and their relationships cannot be constructed. In offering one way of answering Major Johnston's three questions, I have neither eliminated the cognitive dissonance nor decisively refuted what I see to be the most serious objections to the view I offer. However, I think I have provided a deductive argument that links the Army values, spiritual fitness (defined in a naturalistic, institutionally focused way), professionalism (the idea of a good soldier), and character (the idea of a good person). The ongoing discussion within the Army and between the Army and the other services is neither simple nor anywhere close to being over. Yet such a discussion can be inspirational to service personnel as well as to the polity from which they come and for which they serve.

In a 1970 lecture at the U. S. Air Force Academy entitled "The Military in the Service of the State," British General Sir John Winthrop Hackett observed that "the major service of the military institution to the community . . . it serves may well lie neither within the political sphere or the functional. It could easily lie within the moral [as a] well from which to draw refreshment for a body politic in need of it."[70] In that spirit, the U.S. Army can, through continued open and honest dialogue, be a catalyst for a more reasoned discussion in the polity writ large of spirituality and its importance in the United States.

---

69 Pilgrim, "Spiritual Fitness."
70 General Sir John Winthrop Hackett, *The Military in the Service of the State* (Colorado Springs, CO: USAF Academy Press, 1970), 19.

THE RELIGIOUS FACTOR IN MILITARY LEADERSHIP

PAULETTA OTIS

Marines are often portrayed as individuals who serve the United States in uniform and exemplify the highest moral principles. These principles are often assumed to be rooted in religious tradition and the personal application of a faith tradition. In fact, many leaders express the hope that the positive values of religion in individual service personnel will automatically result in ethical behavior in military operations. Military commanders, leaders, chaplains, and ethicists consistently support training regimes to teach moral and ethical principles and decision making for individuals based on confidence in that assumption.

The traditional, albeit often subconscious, assumption is that personal religiosity and ethical behavior are positively related. There are few that publicly challenge this basic idea. The result is that the relationship between personal faith, religious institutions, and the ethical behavior of uniformed personnel is seldom studied in any serious, systematic way. There is little research to answer questions such as can personal morality be measurably different in incoming personnel depending on their religious identity? Does a soldier's faith system or belief in personal responsibility to a "higher authority" make him a more ethical or moral professional? Are spiritual or religious commanders more ethical in their decision making? Are religious service members more aware of the short- and long-term consequences of ethical standards? Are military members more subject to indecision, guilt, PTSD, or leaving the USMC as a result of seeing or experiencing a "lack of ethical behavior?" Can the "raw material" be changed (for good or bad) by military teaching/training?

The short answer to each of the above questions is we don't know; there is no specific research on whether a self-defined "spiritual" or "religious individual" is more moral or behaves more ethically than one who is not.[1] Academics, commanders, and chaplains pontificate on the questions, but clear answers are simply not there. There are three very good reasons for this: (1) in some cases, we do not want to know; (2) there is no basic framework for addressing some of the research questions; and (3) although the United States' population (as reflected in the military population) is self-defined as "religious," the separation between institutionalized religion and institutionalized warfare is an important factor in just how these questions are handled within military circles.

This essay focuses on the relationship between personal spirituality and ethical behavior, religious identity and ethical behavior, and how either individual faith or religious institutions relate to the mission and requirements of the U.S. Department of Defense. Suggestions are also made for further research. The author posits four fundamental assumptions: (1) religion and war concern life and death; (2) military personnel, including commanders, reflect the values and behaviors of a society; (3) military forces, as a specialized professional part of a society, are given specific responsibility for warfare by the government; and (4) the United States' military and its leadership reflect the values, principles, and prac-

---

1 For that matter, rigorous studies that address differences between ethnic, race, gender, age, service, or military specialty in relationship to ethical behavior are notably absent for the same reason.

tices of American society, including religious or "faith-based" values; they are constrained by law and principle with regard to specifics of that relationship.[2]

In anthropological or sociological terms, the United States government (USG)/ Department of Defense (DOD) replaced the social control mechanisms of the institutions of religion (the church, mosque, temple, synagogue) with the institutions of the military.[3] This implies that, although values and beliefs are held by individual soldiers, sailors, and Marines, the enforcement and reinforcement of those values are sustained by institutional commitment. The DOD's institutional commitment to moral and ethical behaviors is a direct reflection of American society, traditions of the military, and the laws and policies of the state. The balance between military effectiveness and moral commitment to ethical behavior is, and should be, a subject of conversation and contention. New issues of "just war" and "justice in war" are a result of changing social, environmental, economic, and political environments and challenge the application of existing moral and ethical principles in "new" contexts. The ethical principles taught may or may not be based on religious principle or secular consequentialism, but the effect is that they provide a basis for unity of command, principled leadership, common values, and standards of behavior for all service personnel.

It is important to clarify concepts used in this essay. Generally speaking, when DOD members speak of individual "religiosity," they use the terms, "faith-based" or "spirituality." The term "religion" is used to refer to an institutional affiliation, such as "Christian, Muslim, Jewish" or in more detailed assessment to Christian

---

2 Pauletta Otis, "Religion and War in the Twenty-First Century," in *Religion and Security: The New Nexus in International Relations*, ed. Robert A. Seiple and Dennis R. Hoover (Lanham, MD: Rowman and Littlefield, 2004).

3 James A. Beckford, *Social Theory and Religion* (Cambridge: Cambridge University Press, 2003).

denominations, such as "Baptist, Methodist, Lutheran." In point of fact, the DOD has traditionally recognized religious affiliation as a basis for death notification, ceremonies, and burial services and, derivatively, for assigning chaplains to provide such services.

RELIGION AND WARFARE

That there is a relationship between religion and war is unquestionable; both concern the basic issues of life and death. Every known religion deals with questions about war as a part of the theological/ideological inquiry: What is the value of life? How should life be lived? Military members are forced to think about "just war" and "justice in war" simply as part of their day jobs, for example, when is taking life in "war" required? How should that be done? Who lives and who dies? If there is a dearth of institutionalized handling of the topics of spirituality, religion, and warfare as evidenced in moral and ethical behaviors, it is not because individual soldiers are unaware of the relationship. If the leadership does not verbalize this relationship, it is often because commanders recognize the importance of the things that unite as more significant than the things that divide.

It may be useful to understand religion in two ways: as individual belief and/or as a group adherence to a theology with attendant beliefs and behaviors that provide coherence and meaning for the group. Religion, for individuals, is very often assumed to be a matter of personal beliefs and behaviors rooted in a faith tradition. Religion is assumed to link the individual to the transcendent, i.e., individuals are responsible to a "higher power" for earthly behaviors. Individuals take what they know of a faith tradition and apply it to their lives—in personal adaptation to circumstances. Morality and ethics become a matter of personal conscience. The problem with this, of course, is that it is not "predictive"—individuals can understand the ideal, consider themselves responsible to a higher power, and still not behave in ways that reflect the social norms of morality and ethics held by the larger society. The society's responsibility becomes one of teaching and enforcing codes of behavior with or without a socially normed "body of received truth."

Religion for groups includes the ideas/theology, codes, beliefs, and behaviors that reflect the ideals of the society and provide social guidance for individuals' behavior. Theology provides "ideal" behavior with reference to the transcendent power and provides guidance for individual behaviors that correspond to those ideals, including means and methods of social control that provide incentive and disincentive for behaviors in order to support group survival. For anthropologists and sociologists, religion for a society provides coherence through theology and contributes to long-term adaptability. This means that the codes, beliefs, and behaviors are "theological" and adapted to ethnic, national, and group requirements for survivability. An example might be the "theology of Christianity" as adapted to Russia, Greece, England, Nigeria, and the United States: the theology is basically the same, but the group adaptations are quite different when studied in relationship to specific cultures.

Figure 19 illustrates the relationship between "levels of analysis" of religious factors: theology (ideology with a "God factor") provides ideals in relationship to the transcendent; codes provide rules of behavior for societies with attendant rewards and punishments; beliefs are cultural adaptations of religious theology/ codes; and actual behaviors are what we "see" and can measure concerning religious behavior, including individual and group morality and ethics.

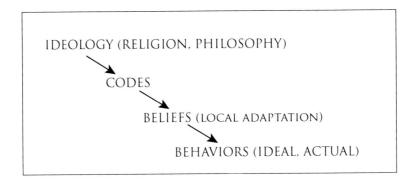

FIGURE 19. **Relationship Between "Levels of Analysis" of Religious Factors.**

The U.S. military has a theoretical dilemma: individuals who say they have a faith tradition do not necessarily see themselves responsible to the larger "religious" traditions that "norm" the ethics of a society; religious institutions provide norms, but different religious institutions provide competing norms in the public square (for example, DOD).

PERSPECTIVES ON THE RELATIONSHIP BETWEEN INDIVIDUAL FAITH, INSTITUTIONALIZED RELIGION, AND DOD POLICY

There are three basic perspectives on religion in the U.S. Armed Services:

(1) The U.S. military is a direct reflection of the United States as a Christian country, and therefore all DOD policies and the activities of individual servicemen and women may be interpreted in the light of adherence to Christian values, beliefs, and behaviors. Those values are reflected in moral and ethical behavior of individual servicemen and supported by commanders. Churches and other religious organizations are encouraged to support the U.S. defense "establishment" and foreign policy, and to encourage religious activities within DOD;

(2) The U.S. military is entirely secular wherein religion has no place in the secular state so that morals and ethics can and should be taught entirely separate from religion. Churches, mosques, synagogues, and related organizations have no role whatsoever in U.S. military activities; and

(3) The U.S. military reflects the values and behaviors of a religious constituency that is plural, diverse, and contentious, so that religious teaching should not be excluded from the public square but the military's requirement for unit/service cohesiveness should remain the priority. Under this third model, an individual's rights to religious beliefs and behaviors are encouraged insofar as they do not damage the "whole," and religious institutions are not allowed to interfere in the conduct of military operations.

Each of the above perspectives has differing implications regarding the relevance of religion as related to leadership. If perspective 1 is correct, then military leaders can and should be adherents in religious organizations and lead in actions that support the overlap of religion and military activities; if perspective 2 is correct, then military leaders/commanders should not refer to or support any religiously factored belief or behavior in military activities, regardless of their own personal beliefs or faith tradition; and if perspective 3 is correct, the appropriate separation of church and state as premised in the U.S. Constitution should be followed, which implies that personal faith is essential for military leaders but that institutions of church and state should not overlap any more than is appropriate and legal.

Perspective 1: The United States is a Christian country and DOD is a reflection of that fact; ethics should be taught and enforced as related to Christian principles.

Many people in the United States believe that the United States is a Christian country and that the armed services reflect that fact.[4]

---

4 Amy Patterson Neubert, "Religious Diversity Increases in America," Purdue University website, 19 October 2010, www.purdue.edu/newsroom/research/2010/101019StraughnReligion.html.

The logic and evidence in support of this contention is largely based on an interpretation of history and tradition, the fact that many individuals in the U.S. armed forces are Christian, that "just war" and "justice in war" principles are based on Christian principles, and that the United States is a Christian country by numerical percentage. Note that in table 8, approximately 68 percent of respondents considered themselves "Christian." The evidence that the United States is a Christian country by history and tradition seems incontestable; the evidence that the United States military has a Christian agenda is more problematic. It is even more difficult to assess whether calling oneself a "Christian" is predictive of higher moral and ethical standards than adherents of other religions or ideologies.

Religious Preference	Military	Civilians
All preferences	100	100
Protestant	35	45
Catholic/Orthodox	22	26
Other Christian	11	3
Atheist/no religion	21	19
Jewish	<1	1
Muslim/ Islam	<1	1
Buddhist/ Hindu	<1	2
Other religions	11	3

TABLE 8. **Percentage of Religious Preferences of the U.S. Population and Military Personnel, 2001.**

Insofar as history and tradition are concerned, it is true that most of the founding fathers were Christian by identity or definition, if not by personal adherence. The populations of the first 13 states also had a Christian identity, and many of the states prohibited other religions. The principles of "just war" and "justice in war" are based on the writings of St. Augustine and St. Aquinas; they have at their core a system of values that relate specifically to Judeo-Christian traditions. Religious beliefs, generally Christian, have had a role in shaping the leadership of the country and its military. The symbols and rituals of war are identifiably Christian. Further, Christianity, in its various forms, helped shape the battlefields of the Revolutionary War, Civil War, Spanish-American War, the Philippine War, and the First and Second World Wars.[5]

In terms of the U.S. population, a recent report issued by Purdue University stated that many Americans believe that "being a Christian" is a key aspect of "being American."[6] A similar poll conducted by the Public Religion Research Institute found that 42 percent of Americans believe that "America has always been and is currently a Christian nation."[7] The research also indicated that those who were Christian were more likely to answer this question in the affirmative than those who were not. Non-Christians consistently reported that it was not necessary to be a Christian to be a "good American." Pew Research Center's Forum on Religion and Public Life asked different questions from another perspective and used a different metric. It found (1) that the number of mainline Protestants and Catholics make up approximately 60 percent of the population; (2) that there is no direct relationship between income, education, and religious affiliation; (3) that Americans ages 18 to 29 are considerably less religious than older Americans yet remain fairly traditional in their religious beliefs and practices; (4) that contrary to common misperception, only 6 percent of voters in the last election were contacted by religious groups about issues or their vote; (5) that nearly 72 percent say they attend religious services at least a few times a year—roughly

5 Edwin S. Gaustad, *A Documentary History of Religion in America* (Grand Rapids, MI: William B. Eerdsman Press, 1982).
6 Neubert, "Religious Diversity Increases in America."
7 Public Religion Research Institute, American Values Survey,1–14 September 2010; findings based on a random sample of 3,013 adults, http://publicreligion. org/research/2010/10/religion-tea-party-2010/.

one-quarter say they seldom or never attend religious services (27 percent); and (6) that there were few changes in these statistics in the past decade.[8]

No research was found that asked whether being a good "Christian" or being "religious" was a necessary qualification for service in the armed forces or whether being self-identified as a good "Christian" produced Christian ethical behavior. Military commanders invariably avoid the question in favor of a character assessment: is the soldier, Marine, or airman a person of good character in terms of ethics and morals (noted as honor, loyalty, and commitment)? Although these may be attributes of a good "Christian," they may also be the attributes of a good Muslim, Jew, Hindu, or member of any other faith. It is also noted that there are Christians (Muslims, Jews, Hindus, etc.) who do *not* have good character. The major conclusion is that military leadership focuses on the attributes of character, not on religious affiliation, when making a judgment about leadership, ability to command, or what "makes a good Marine."[9]

The complementarity of Christian identity, U.S. citizenship, and military service poses several inherent problems.[10] The first is the principle of inclusive pluralism—the belief that all, regardless of religion, are entitled to equal treatment. If American identity and that of its military forces are exclusively Christian, then the

8 Pew Research Center's Forum on Religion and Public Life has a well-respected research capability and is the source of much of what we know about religion in the United States and how it affects the public domain. See http://www.pewforum.org/.

9 The debate over gays in the military was framed by some commentators as being a "religious" or moral/ethical issue. The conversation in military circles tended to focus on military effectiveness (author's perspective only).

10 In an informal survey of 25 Marine Corps officers at Command and Staff College, it was unanimously reported that (1) promotion and advancement were not related in any way to religious preference; (2) they had not experienced proselitization; and that (3) it was not necessary to be a good Christian to be a good officer (January 2011). Please contact author for details.

participation of non-Christians in the military should be precluded. If the United States is a Christian country, then the Defense Department should have rules and responsibilities that directly mirror biblical principles, and leaders should have a test of religion before taking command. Foreign and defense policy should be held to the rigorous religious standards of Christianity. All "others," i.e. other countries, cultures, and religions should likewise be assessed primarily with regard to their religious identity and how "close" they are to Christian principles.[11]

There are selected individuals and groups who not only believe that the United States is a Christian country, but that the U.S. military should self-consciously and vigorously promote Christian identity. They believe that the U.S. military cannot be a moral or ethical organization without complete correspondence of Christianity and U.S. military identities. This advocacy is reflected, but not officially supported, by a number of authors who believe that they speak for the DOD. Stephen Mansfield, the author of *Faith of the American Soldier*, exhorts soldiers to follow the example of the Crusaders in their religious devotion; *Shariah: The Threat to America: An Exercise in Competitive Analysis*, edited by Lieutenant General William (Jerry) Boykin, defines the enemy in religious terms.[12] Publishing companies recognize the "salability" of religion and warfare and publish a large number of books wherein the authors portray heroes as *both* Christian and military.[13]

---

11 David Reiff, "The Crusaders: Moral Principles, Strategic Interests, and Military Force," *World Policy Journal* 17 (2000): 39–47.

12 These arguments are reminiscent of the Cold War arguments discussed by Lori Lyn Bogle in *The Pentagon's Battle for the American Mind* (College Station: Texas University Press, 2004). Stephen Mansfield, *Faith of the American Soldier* (New York: Jeremy P. Tarcher/Penguin, 2005); Lieutenant General William (Jerry) Boykin, ed., *Shariah: The Threat to America: An Exercise in Competitive Analysis* (Washington, DC: Center for Security Policy Press, 2010).

13 A cursory glance at lists of recent publications, especially during the period of 2001 to 2011, indicates an amazing increase in the number of books published on the subject of religion relating to warfare.

Perspective 2: The United States is a strictly secular state and the military should reflect secular arrangements; ethics must be taught separately from any religious tradition.

The traditional separation of state and church has worked very well for the United States.[14] The writers of the U.S. Constitution, acknowledging the religious wars of Europe, bowing to practical politics, and respecting the rights of individuals to practice their own religions, were extremely careful not to establish an official church that would link the power and resources of the government with the power and resources of a religion. Whether the framers meant for the United States to ignore religion entirely and rely on the scientific premises of the liberal state is unknown, but certainly the Constitution is not built around religion or religious institutions.[15] The debate between the secular Constitution and the sacred Constitution will not be settled here, and suffice it to note that the military has had an uneven history in this regard but that the principles of a secular military have been the most workable for military forces.[16]

The world's experiences in the twentieth century weighed heavily on the side of the benefits of a secular approach. In World War I and II, there was an unsavory connection between religion, national fascism, and militarism. Religion, albeit universally found

14 The United States became a country subsequent to the vicious religious-political European wars of the seventeenth and eighteenth centuries. As a result, the framers of the U.S. Constitution, historically aware and uncommonly prescient, set out new guidelines for the relationship between church and state—commonly referenced as the Establishment Clause and Free Exercise Clause in the First Amendment. The country continues to struggle with the various ramifications of this arrangement and continues to use the courts as a virtual battleground.
15 Isaac Kramnick and R. Laurence Moore in *The Godless Constitution* (New York: W.W. Norton, 1997) take the view that the U.S. Constitution is a strictly secular document.
16 One of the most insightful books on this subject is Jonathan H Ebel, *Faith in the Fight: Religion and the American Soldier in the Great War* (Princeton: Princeton University Press, 2010).

as part of culture and society, came to be seen around the world as dangerous when linked to politics (or war) in any way.[17]

After the Cold War, with its endless arguments over the role of ideology, the debate about the causes of warfare focused on "idealism" or "realism." The Cold War warriors maintained that the role of ideas was dominant in predictive analysis, and the realists maintained that self-interest was more explanatory. Interestingly enough, the concept of "idealism" did not pertain to religious issues. Religion was simply left out of the discussion. The "realists" eventually took center stage and focused on the elements of material power that could be measured, such as the size of a military, capability assessment, economic resources, and weapons. Ideas, ideologies, and theology (when it was even mentioned) were seen as supporting, but not leading, international affairs. Religion as explanatory either in its own capacity or as a part of culture, was seen as an artifact of history.

Ironically, the scientific secular approach seemed insufficient to deal with the issues surrounding interrogation, torture, rendition, and the other matters that depend on individual and institutional values, morality, and ethics. From a strictly secular, self-interest power approach, the use of torture was a rational act. Those from a "secular humanist" perspective called for morality based on common humanity, humanitarian law, and even scientific behavioral analysis, but those arguments did not seem to win the day.[18]

Insofar as military services are concerned, religious issues are handled by commanders and settled as military issues rather than as intrinsically "religious." Although religious diversity has been a reality since 1775, religious differences have been downplayed in favor of military discipline throughout the course of U.S. mili-

17 See Leonard W. Levy, *The Establishment Clause: Religion and the First Amendment* (New York: Macmillan, 1986).
18 Intelligence Science Board, *Educing Information: Interrogation: Science and Art* (Washington, DC: National Defense Intelligence College Press, 2006), www.ndic.edu/press/3866.htm.

tary history. The codification of religious accommodation of differing religions was only officially institutionalized in 2009 with a Department of Defense Directive (DoDD) dated 5 February 2009, stating that DOD will

> Promote an environment free from personal, social, or institutional barriers that prevent Service members from rising to the highest level of responsibility possible. In this environment, Service members shall be evaluated only on individual merit, fitness, and capability. Unlawful discrimination against individuals or groups based on race, color, religion, sex, or national origin is contrary to good order and discipline and counterproductive to combat readiness and mission accomplishment and shall not be condoned.

In summary, the idea of the United States as a secular state is premised on the First Amendment's restrictions on the federal government with regard to establishment of an official church and protection of religious freedoms. The U.S. Department of Defense generally subscribes to those principles both on constitutional grounds and with the idea that a good military cannot afford religious divisions.

Perspective 3: Religion is supported for individuals, but religious institutions are constrained by law; ethics should be taught as reflecting religious principles, but not in relationship to specific religious institutions.

The third perspective on the relationship of religion and the U.S. military may have the most explanatory power when seen from American political and religious history and law. This position rests on the assumption that people/the polity can vary widely in their religious viewpoints, identities, and beliefs under conditions of free speech, assembly, petition, and religion but that the Department of Defense is constrained by law and policy with regard to religious factors both within the services and in relationship to foreign and defense policy.

Religion is protected along with speech, petition, and assembly—ideas and communication—as fundamental to freedom and liberty: "Congress shall make no law respecting an establishment of religion, or prohibiting the free exercise thereof; or abridging the freedom of speech, or of the press, or the right of the people peaceably to assemble, and to petition the Government for a redress of grievances."[19] Acknowledging the risk inherent in competing power structures, the writers maintained that the institutions of religion and government should be "separate," i.e., no "establishment" of religion. This organization of church-state and faith-polity in the United States can be illustrated in figure 20:

U.S. SECULAR AND SACRED ARRANGEMENTS

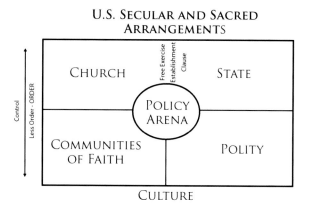

FIGURE 20. **U.S. Secular and Sacred Arrangements.**

Most of the founders believed that religious belief was conducive to good citizenship and public morality, and was the basis of the successful working of a nation; however, they balked at any official church establishment.[20] George Washington, in his role as general and then president, set a useful precedent in practicing a

---

19 Alfred Stepan, "Religion, Democracy and The Twin Tolerations," *Journal of Democracy* 11 (2000): 37–57.
20 Daniel L. Dreisbach, Mark David Hall, and Jeffry H. Morrison, eds, *The Founders on God and Government* (Lanham, MD: Rowman & Littlefield, 2004).

personal faith, supporting religious diversity and pluralism, and encouraging good morals among soldiers and citizens, yet not publically advocating for a specific church or faith tradition. [21]

The people of the United States, the government, and the Department of Defense have, in general, maintained this position since 1793. The influence of religion on the Pentagon is through the moral and ethical principles of the individuals in uniform; there is no need for religion to take a more formalized role if it means that the mission will be compromised through dissention in the ranks. Religious influence as found in military hymns, rituals, and symbols, is hidden in plain sight.

Department of Defense directives are clear about precluding religious preference in the workplace, providing for individual worship and religious practice *when* and *if* it is secondary to mission definition, and accommodating individuals, not religious institutions. An example is that the religious place of worship on a military base (regardless of the religion) is technically a "community center" and therefore is multifunctional.

Another example is that chaplains are recruited from denominations or various religions, but they must serve all service personnel regardless of the service member's denominational affiliation simply because they are uniformed personnel. The U.S. military chaplains ensure religious freedom for all service personnel and have a ministry of presence that is meant to ensure that commanders are always aware of the ethical and moral dilemmas of warfare. [22]

---

21 Michael Novak and Jana Novak, *Washington's God: Religion, Liberty and the Father of Our Country* (Cambridge, MA: Basic Books, 2006).

22 The government may fund religious activity if done in order to accommodate the religious needs of people who, because of government action, no longer have access to religious resources. This, in effect, requires religious pluralism—the accommodation of all persons with regard to religious needs. The caveat is that of "military necessity," i.e. the religious accommodation cannot impede other official requirements, provide an undue burden on the government, or promote a specific religion. See Joint Publication 10-5, *Religious Affairs in Joint Operations, chaplain. ng.mil/Docs/Documents/jp1_05.pdf.*

The consequences of this position of personal faith and accommodation of religious preferences, yet separation of church and state, in the Pentagon is clearly anathema to those who would like to see religion as the dominant force in both the U.S. military and in relationship to the deployment of forces in the international arena. They believe that by not having a more visible role, religion will lose its power, Christianity will be defeated by the enemy, and moral and social decay are certain. Those who believe that the military should be strictly secular are equally convinced that all religious services, accommodation, symbols, and ceremonies should never involve religion in any way, shape, or form. The reality is far more complex than either of these simplistic viewpoints.[23]

The delicate balance between too much and too little is always a homeostatic process—new problems, leaders, understandings, political pressures, and international events conspire to make it a "work in progress." Nevertheless, both military doctrine and military practice provide clear guidelines for legal behavior within the services with regard to religious accommodation rather than religious differentiation.

Figure 21 illustrates that the Department of Defense is singularly insulated from some of the more contentious debates in American society. Both law and policy, as defined in the Constitution, by Congress, and by the president, constrain practices with regard to religion.

SUMMARY: PERSONAL ETHICS, RELIGIOUS FACTORS, AND MILITARY LEADERSHIP

The U.S. military tends to mirror the rest of the U.S. polity with regard to religious belief. It differs in that military organiza-

---

23 These positions were clearly in evidence and were the basis for serious dissention in the recent reports from the U.S. Air Force Academy events of 2009–11. For more information, contact the author of this essay.

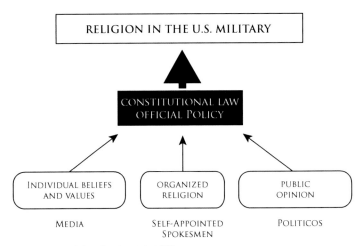

FIGURE 21. **Religion in the U.S. Military.**

tion demands setting priorities during times of war. Unity of command, unit morale, ethical behavior, and attention to the Law of Armed Combat are critical to mission success, force protection, and saving lives.

Some of the major findings in this study are

- Individuals in the U.S. military are religious in that most service personnel have faith traditions and believe in a "higher power."

- The profile of religious adherence in the U.S. military reflects to a substantial degree the profile of the U.S. population in general.

- The traditional belief in separation of church and state in the U.S. military, along with a need for unit cohesion to accomplish a military mission, has produced a system that discourages divisive beliefs and practices.

- An interpretation of the U.S. military as led or unduly influenced by institutionalized Christianity is unfounded.

- An interpretation of the U.S. military as strictly "secular" is naïve and unfounded.

- The U.S. military, in recognition of and with respect for individual religious rights and with respect for the constitutional separation of church and state, has instituted policies and culturally normed practices that emphasize unity and de-emphasize religious differentiation.

- Ethics is taught as individual responsibility and reinforced by the institutions of the U.S. military; ethical principles and standards are derivative from traditional Western Christian principles as interpreted in light of warfare requirements.

THE WAY FORWARD

The United States—as reflected in both military and civilian institutions—has developed a unique system to address the positive influences that religion has to offer without taking on the "divisiveness" of religious institutions. Any accusation that the U.S. Department of Defense is "conspiracy driven" either by being overly evangelical or overly secular is not verifiable in any systematic research. Neither "liberal" nor "conservative" parties wish to destroy the fundamental requirement for military cohesiveness. The Department of Defense has a system it can live with; the system is far from perfect and under consistent scrutiny but has found the delicate balance between religion in warfare and military requirements. Given the above statement, it is relatively safe to assume that the future will look much like the past: traditional religious values will continue to underlie military ethics, and the changing war environment(s) will require continual and constant attention to ensure that military ethics conform to basic principles. It is also important to remember that no matter how this is put into practice, the "devil is in the detail" will be, and should be, contentious. This is consistent with the idea of democratic governance in the public domain. It is also important for every service member as it helps define the enduring moral and ethical base of his/her profession and provides guidance for military conduct in a changing world.

The leadership in the Department of Defense has the obligation to ensure that ethics and guidance for practices that conform to ethical principles be part of the "situational awareness" at every level of command, in every operation, for every military specialty. Professional military educators, ethicists, and chaplains may be assigned a formal responsibility for keeping ethics in teaching and training environments, but this does not absolve each and every military member from an obligation to know and act according to ethical principles in the profession of arms.

Spiritual Leadership in the Battle Space
Who is in Charge? A Personal Essay
Arnold Fields

Let me begin with a few confessions: First, I do not profess to be an expert on spiritual leadership and find it an extraordinary concept to even try to define. It is a term for which I have searched high and low to find a common definition and have concluded that none exists. However, for the purpose of this chapter, I define the term as follows: *Spiritual Leadership is the use of an amalgamation of natural, learned, and spiritual qualities in order to influence ethical behaviors of self and others in the interest of morally acceptable objectives and outcomes.* I emphasize the term *amalgamation* to avoid the assumption that any one of the aforementioned qualities should be dominant for a person to be an effective spiritual leader. In other words, it is arguable that a warrior may be successful in the battle space by implementing only natural and learned qualities, without the potential advantage of spiritual influence. Moreover, worship in the U.S. Armed Forces demands religious neutrality and individual choice of chaplains. In this regard, the commander's legal role and the chaplain's religious role should be distinctive; however, it remains the commander's overall responsibility to ensure the environment exists wherein individual spiritual needs can be pursued.

My first combat experience took place during the Gulf War of 1991, also known as Operation Desert Storm. Prior to this engagement, I had often pondered how I would ultimately respond to the real, unequivocal effects of combat if the opportunity ever emerged. Similarly, I was concerned about my troops and their worries about combat and how they would ultimately respond under fire. About two days prior to Desert Storm's ground war, I visited the foxholes of as many troops as possible. Everything I had read or had been told suggested that as much as 30 percent of the battalion could be lost to the enemy as we penetrated the extensive minefield the Iraqi army had constructed. My Marines and sailors were obviously concerned about their welfare, but it was also clear to me that they were ready for the fight. I emphasized visiting the lead company because it would be the most at risk and subject to casualties while leading the fight through the minefield.

There had been much informal commentary among the troops about the war's potential effects and the injuries they could incur. The officers spoke freely of their concerns, often in small groups while sipping cups of coffee. With smiles and nervous laughter, they rhetorically discussed their potential injuries and considered them in a sort of hierarchy: "Would I prefer to lose an arm or a leg, and if a leg, which one? And if an arm, which one? What about an eye versus a leg?" My personal worries were not unlike those of my officers, although I did not express them in order to avoid demoralizing the battalion.

I also escorted press reporters who visited our base camp and introduced them to our battalion surgeons whose unexpurgated show-and-tell briefings left nothing to the imagination regarding anticipated injuries, triage, and treatment. The reporters were obviously shocked by the briefings and the impending threat.

If spiritual leadership were ever to be a factor in the battle space, this was certainly the moment.

In order to begin a discussion on the idea of spiritual leadership, one must consider a common point of departure, which is this: military personnel are in denial if they do not believe that commanders are spiritual leaders in the battle space. The modern battle space demands a holistic approach to leadership and ethics, and spiritual leadership is as much the commander's personal responsibility as is good order and discipline in the ranks. I had been naïve at the beginning of Desert Storm and would ultimately find that the battalion expected more than just my tactical leadership as we maneuvered in the face of the formidable Iraqi army.

Modern conflict is occurring in an asymmetrical environment where military operations almost always involve combined U.S. and allied forces and supporting contributors. Iraq and Afghanistan are two of the most recent examples where multiple nations have brought their resources to bear upon multilateral, as well as unilateral, interests and strategic objectives. Military commanders and diplomats are expected to provide leadership in this widely diverse environment made up of people representing different ethnicities, cultures, socioeconomic classes, and religions.

Human understanding has long been an imperative for the warfighter and for success in conflict. It is more often manifested in the intelligence preparation of the battle space (IPB), but there should be a more intrinsic consideration of human understanding that transcends the enemy's battlefield capabilities and peers into his soul. The ancient warrior Sun-Tzu wrote, "One who knows the enemy and knows himself will not be in danger in a hundred battles. One who does not know the enemy but knows himself will sometimes win, sometimes lose. One who does not know the enemy and does not know himself will be in danger in every battle."[1] I believe this philosophy is as much a lesson in basic life

---

1 Sun-Tzu, *The Art of War: Planning Attacks*, Chapter 3, http://www.sonshi.com/sun3htm.html.

and human understanding as it is the battle space imperative Sun-Tzu recommended.

The Marine Corps Manual, 1980 edition, defines leadership as "the sum of those qualities of intellect, human understanding, and moral character that enables a person to inspire and to control a group of people successfully."[2] This is the definition to which I was introduced when I enlisted in 1969, and I believe that this definition, together with its intent, remains as relevant today as it did more than 40 years ago. Each leadership challenge that I encountered on active duty and as a civilian has involved intellect, human understanding, moral character, or a combination of all three of these qualities. These are important qualities for success, but human understanding has emerged as the most useful to me. The asymmetrical imperatives of the modern-day battle space have necessitated greater human understanding not just as a fundamental human aspiration, but as a matter of survival and mission accomplishment. In retrospect, the cultural familiarization and training my Marines and I received in 1970 was helpful before we deployed to the Mediterranean, but it was largely one-dimensional and passive. It consisted of lectures, pictures, short stories, and movies. The entire training element was brief and was completed in less than one day. Today, there is almost total immersion where certain knowledge and cultural skills are mandatory and a prerequisite to deployment.

These factors are an integral component of the still-emerging strategic communications (SC) concept, which acknowledges that winning in conflict requires more than tactical or operational success. Strategic communications has been described as the orchestration and / or synchronization of actions, images, and words to achieve a desired effect and is considered an extremely impor-

---

2 *Marine Corps Manual, W/Ch 1–3* (Washington, DC: Headquarters, Department of the Navy, 1980), Glossary A-2, http://www.usmc.mil/news/publications/Documents/MARINE%20CORPS%20MANUAL%20W%20CH%201-3.pdf.

tant joint concept. There are nine SC principles, four of which I find relevant to spiritual leadership in the battle space: "leadership driven"; "credible" (involving perceptions of truthfulness and respect); "understanding" (defined as a deep understanding of others); and "pervasive" (where every action sends a message).[3]

The revolution in technology and communications over the past few years has made understanding the human, thus also the spiritual, dimension of leadership even more important. Leaders at all levels are challenged by the need to get the facts, provide updates, fix the problem, and convince everyone that rules were obeyed, ethical standards were upheld, and all related activities met the expected standards of good order and discipline. As a battalion commander during Operation Desert Storm, I did not feel that my division commander, fellow battalion commanders, or I had to be seriously concerned about asymmetry as a factor in accomplishing the mission. In general, we employed conventional tactics, techniques, procedures, and equipment, and were successful. We knew who the enemy was, where the enemy was located, and could recognize the uniform. The forward edge of the battle area (FEBA) was well defined and the distinctive rear remained mostly secure throughout the war.

We have recently entered a stage on the continuum of conflict and national security where military roles and other governance dimensions are much more diffused and therefore less well defined. For example, in a tragic incident, one of my civilian staff members, a Department of State employee, was assassinated while performing duties as senior advisor to the Iraqi government. I was the employee's direct supervisor, considered him my "troop," and elected to escort his remains home to his family in the United States. Ironically, during over 34 years of active military service, I never had cause to take on such a task. The Army's protocol for

---

3 U.S. Department of Defense, *Principles of Strategic Communications*, 2008, www. au.af.mil/info-ops/documents/principles_of_sc.pdf.

handling the remains was basically the same as those for several Marines who had been recently killed in action. They were all transported in the same vehicle from a military hospital to a makeshift morgue facility at the Baghdad International Airport where they were processed for further movement. Each set of remains was eventually placed in a flag-draped metal casket and preserved for transportation. There was no distinction between the deceased warriors and the civilian, all of whose remains were flown on the same U.S. Air Force C5 aircraft to Dover Air Force Base in New Jersey. A Marine honor detachment met the aircraft when it landed at Dover and the remains were ceremoniously removed using protocol, which, again, in no way distinguished the civilian remains from the deceased Marines. This episode was a very poignant example to me of warfare's changing face and gave cause to recall a question I had posed to Department of Defense senior leaders while I was on active duty: Who is the warrior? Is it still the traditional person dressed in a well-defined uniform, trained in a common set of skills, indoctrinated in a certain battlefield code of ethics, and sent off to war? Or has the evolution of war and conflict created a battle space condition in which the duties and expectations of civilians directly contributing to the mission cannot be easily distinguished from those of the traditional warrior? I believe it is the latter, which suggests to me that military leaders should be more adaptable, trained accordingly, and prepared to take on uncommon leadership dimensions, including spiritual leadership.

The Department of Defense defines battle space as "the environment, factors, and conditions that must be understood to successfully apply combat power, protect the force, or complete the mission. This includes the air, land, sea, space, and the included enemy and friendly forces; facilities; weather; terrain; the electromagnetic spectrum; and the information environment within the operational areas and areas of interest."[4] Battle space is conceptual and is not assigned by a higher commander; commanders determine their own battle space based on their mission, the enemy and their concept of operations, and force protection. They use experience and understanding to adapt their battle space as the situation or mission changes. According to MCDP 1-0, *Marine Corps Operations*, the battle space is normally composed of an area of operations (AO), an area of influence, and an area of interest.[5] Keeping this definition in mind, there is apparently no shape, form, or permanence to the battle space. Ownership is ambiguous and may therefore overlap with the interests of adjacent commanders. One thing is clear: commanders must have influence in their battle space if the mission is to be accomplished and must be able to effectively coordinate and de-conflict issues with adjacent commanders. Actual ownership is a matter of perspective because objectives are being pursued in a space where tactical, operational, and strategic initiatives are simultaneously underway. As this continuum plays out, anyone from the president, who is always in command, to a fire team leader, could be perceived as being "in charge." When viewed from this perspective, the battle space offers a fluid, unpredictable, amorphous, and dangerous environment where not even terrain is a constant. I asked my church pastor, Dr. Michael Bledsoe, to share a perspective on spiritual leadership in the battle space; he offered the following:

> The person in charge in the battle space is that person who masters not only the variables of terrain, numbers, etc., but the person in charge is that one who has charge over not only his

---

4 *DOD Dictionary of Military Terms*, http://www.dtic.mil/doctrine/dod_dictionary/.

5 U.S. Marine Corps, MCDP 1-0, *Marine Corps Operations* (Washington, DC: Department of the Navy, 2001), 4-3, www.dtic.mil/doctrine/jel/service_pubs/mcdp10.pdf.

body and his tools, but is possessed by those ineffable qualities of the spirit that include a resolute courage, an abiding devotion to comrades, a fearlessness in the face of confrontation, and an enduring belief that s/he is part of something greater so that even if s/he were to die, that greater purpose and higher cause will continue.[6]

There are examples of commanders whose strength in leadership had prepared the unit so well that when the leader was lost early in battle, the mission continued and intended objectives were accomplished. According to educational psychologist Dr. Bruce Tuckman, some organizations are able to reach this high level of maturity, which he refers to as the "performing phase" of group development. In his model published in 1965, Dr. Tuckman explains that teams progress through four distinct developmental stages:

(1) Forming, which requires a high degree of dependence on the leader for guidance and direction;

(2) Storming, where team members vie for position as they attempt to establish themselves in relation to each other and the leader and are inclined to challenge their fellow team members as well as the team leader;

(3) Norming, characterized by agreement and consensus that largely form among the team members whose roles and responsibilities appear to be clear, and the members respond well to facilitation by the leader; and

(4) Performing, where the team is more strategically aware, knows clearly what it is doing and why, and requires no participation from the leader.[7]

---

6 Michael Bledsoe, pastor, Riverside Baptist Church, and professor, Howard University School of Divinity, e-mail message to author, 15 February 2011.
7 Mind Tools, Essential Skills for an Excellent Career, "Forming, Storming, Norming and Performing: Helping New Teams Perform Effectively, Quickly," Mindtools.com, http://www.mindtools.com/pages/article/newLDR_86.htm.

Dr. Tuckman may not have had spiritual leadership in mind when he wrote about these phenomena, but there is a relationship between Dr. Tuckman's 1965 view and Dr. Bledsoe's 2011 perspective regarding the leader's enduring impact and effect. Some organizations eventually achieve an advanced stage of organizational development in which the leader's physical presence may not be necessary for the mission to be accomplished.

Dr. Louis W. Fry, professor of management at Tarleton State University in central Texas, offers the following:

> Spiritual leadership is a causal leadership theory for organizational transformation designed to create an intrinsically motivated, learning organization. The purpose of spiritual leadership is to create vision and value congruence across the strategic, empowered team, and individual levels, and ultimately, to foster higher levels of organizational commitment and productivity. Spiritual leadership comprises the values, attitudes, and behaviors that one must adopt in intrinsically motivating self and others so that both have a sense of spiritual survival through calling and membership—i.e., they experience meaning in their lives, have a sense of making a difference, and feel understood and appreciated.[8]

He further argues that humility and integrity are essential components the spiritual leader must possess and suggests that humility allows him or her to first consider the importance of others rather than him or her self. Thinking too highly of ourselves prevents us from fully caring about others and setting the example we would wish them to emulate. People repudiate those who are dishonest and lack integrity. Every military professional is expected to have integrity; it is one of the traditional 14 leadership traits that

---

8 Louis W. Fry, "Spiritual Leadership and Army Transformation: Theory, Measurement, and Establishing a Baseline," *The Leadership Quarterly* 16 (2005): 835–62, http://www.tarleton.edu/Faculty/fry/SLTSpIssueArmy.pdf.

have been an integral part of Marine Corps leadership doctrine. Warriors may not consider themselves to be humble, but they are expected to be unselfish and to lead by example. This often means ensuring that the welfare of the troops is in place before the leaders consider their own needs and comfort. Marines know that this is particularly true in the field where the officers are expected to always be the last to eat. Marines also expect their leaders to set the example at all times, and there is no tolerance for substandard physical performance. As a leader, during physical training and testing, I endeavored to be the first on the pull-up bar to perform 20 pull-ups, the first to complete the desired 80 sit-ups in two minutes, and the first to complete the three-mile run. But as might be expected, I was not consistently able to be the most proficient in performing these events; someone was always younger, faster, and more agile. However, the important point is that the troops expect the leader to be present and to participate even when being clearly outperformed by junior troops. In my view, the leader's mere presence and willingness to be tested with the troops—and outperformed by them—is a form of humility.

As we examine these ideas about what constitutes an effective leader, it becomes apparent that the characteristics and desired outcome of spiritual leadership are not unlike those of a good leader in general. Each leader in the battle space is in search of behaviors that do the right thing, work together for the benefit of the whole, are pleasing to the nation, and ensure mission accomplishment. As we take note of these facts, it is fair to say that spiritual leadership requires the understanding and application of distinctive attributes that must be practiced if spiritual leaders are to be successful. They should be accountable not only to their commanders and troops but also to a higher spiritual power.[9] This requires a faith not only in the warrior's training, equipment, and

---

9 Henry and Richard Blackaby, *Spiritual Leadership: Moving People on to God's Agenda* (Nashville, Tennessee: Broadman and Holman Publications, 2001), 16–20.

ethos, all of which are profound in modern warfare, but also in the "force" of spiritual or other origin, which I believe drives humanity toward good and away from evil. It is for this very pursuit that so many wars have been and likely will be fought. First Lieutenant David Stuckenberg, a pilot in the United States Air Force, offers the following:

> Faith gives both leaders and subordinates essential character qualities not imparted through organized instruction, which are vital elements to a disciplined and restrained military force. To avoid a tangential debate about definitions of what faith means and the differences between faith and religion, I will pause to provide the best definition I have heard, from Hebrews 11: "Faith is the substance of things hoped for and the evidence of things not seen." Both my personal view and what I have witnessed when men and women claim to have faith is that they ultimately believe there is a higher power in the universe which has established governance for man and that we will ultimately answer to him for our individual actions. In other words, people of faith believe in accountability and consequences whether good or bad will result from their actions.[10]

The spiritual leader's task is to move people toward goodness and to believe that every act of leadership is intended for this purpose. Even Machiavelli may have hinted at this by accepting that the state as he knew it could not stand outside the confines of religion because it was bound to use God for political ends.[11] The military operational mission imposed by legitimate government authorities should be perceived as a means to accomplish an objective that will bring about a positive impact on the nation or on humanity in general. Every war in which the United States has engaged

---

10 First Lieutenant David Stuckenberg, active duty U.S. Air Force pilot, e-mail message to author, 16 December 2010.
11 Lauro Martines, *Fire in the City: Savonarola and the Struggle for the Soul of Renaissance Florence* (New York: Oxford University Press, 2006), 6.

has been in the interest of what was believed to be a just cause. Freedom, democracy, and a sense of well-being have usually been the desired end state. The overall effect is that spiritual leaders inspire and influence not only the religious but the secular as well, whether or not they are under the leader's direct control or authority. I believe that leaders in the United States Armed Forces *are* spiritual leaders because they are asked to engage only in *just* wars and conflicts, and they have been trained to employ leadership doctrine and traits that are founded upon truths and universally expected and accepted behaviors. Personally, I would find it difficult to be considered a spiritual leader without having at least a measure of faith. And I would be challenged to confidently follow a leader for whom faith is not a relevant factor in the battle space.

I developed my personal faith early in life and it has strengthened over the years as a result of studying the Christian Bible, attending religious services, and resolving various leadership and personal challenges I have encountered. My spiritual journey has been very enlightening and satisfying; however, despite this positive life experience, I still have difficulty offering a spiritual rationale for killing in the name of freedom—and am inclined to believe most warriors have similar concerns even though they may dutifully follow through in carrying out their oath of service. Specifically, I find the Ten Commandments to be very clear and unequivocal. But for me, the spiritual leader in the battle space, the sixth commandment has been the most troubling. It has not been a problem in my private life because I have never had a personal desire to take the life of a human being. However, as a Marine leader sworn to protect and defend the Constitution, I was prepared to carry out my oath to the fullest extent necessary and expected. Nonetheless, fighting and killing in the name of freedom remain a moral dichotomy for me and, I believe, for our nation as well. How can genuine spiritual leadership take place in the battle space where every soldier is trained to be a quick and efficient killer? This poses a serious dilemma and, in my view, except through personal faith, understanding, and spiritual leadership, there is apparently no ready remedy at any level in our country by which to fully resolve this question so central to understanding the morality of war. Although the Bible may cite instances where God apparently sanctioned killing, such as in the account of David and Goliath,[12] stories like these do not always mitigate or relieve the emotional scars and posttraumatic stress disorder (PTSD) that trouble soldiers who have personally killed or supported a teammate in killing an enemy.[13] Despite these ongoing concerns, citizens continue to demand peace and a secure environment in which to live, work, raise families, and contribute to the body politic and, as such, have supported the U.S. Armed Forces and the conflicts in which they have been tasked to engage. Death and destruction are always expected outcomes. Government planners and policy makers at the strategic and operational levels may debate war's morality, efficacy, and lethality, but it is more often at the tactical level and the young warrior's trigger finger that the morality question is ultimately resolved. It is he who is left with the need for decision and immediate action as the enemy advances and the warrior's life is in question. I doubt that there is any ultimate remedy other than faith by which the warrior may find comfort when faced with life or death situations in the battle space. It takes time to develop faith resilient enough to withstand the many challenges all warriors undoubtedly face in the battle space.

---

12 1 Samuel 17: 1-58.
13 Mary E. Card-Mina, "Leadership and Post Traumatic Stress Syndrome," *Military Review* (January-February 2011).

Conclusion

I played a de facto role as the battalion's spiritual leader, a thought that did not enter my mind until well after the Desert Storm war. Yes, I had tried my best to physically, mentally, and spiritually prepare the Marines and sailors for the horrors of war, although the sum total of my actions at the time was not considered spiritual leadership. As the realization of an imminent ground attack into Kuwait became evident, so did the aphorism, "There is no atheist in foxholes." It suggests that in times of extreme stress or fear in warfare, all people will believe in or hope for a higher power. I observed that the attendance rate at religious services began to increase as the air war continued and the Iraqi government refused to compromise. I was comforted by an almost daily stream of visits to my tent during which handwritten notes of encouragement and Bible verses were conspicuously placed, usually while I was away.

The battalion chaplain certainly did his share, always engaging the troops and carrying out a robust religious activities program in the field. He was in fact the battalion's religious leader. However, I believe the Marines and sailors expected me—the battalion commander—to provide the *spiritual leadership*. I have labored over these reflections, wondering how I might have improved the battalion's spiritual readiness and my overall effectiveness had I better understood spiritual leadership and accepted it up front as the commander's role and responsibility. Moreover, I could have done more to ensure that there was a spiritual dimension to the whole of my Marine Corps leadership experiences. In other words, I was the de facto spiritual leader and should have more formally accepted this role. I, the battalion commander, should have been the "stream of traffic" delivering spiritual messages of support and encouragement to the troops' tents and foxholes rather than the troops having come to me. They had apparently recognized me as their spiritual leader; however, the leadership training I received during my career had not conditioned me to anticipate this specific aspect of battle space leadership, which I now believe is the commander's personal responsibility.

The modern battle space is a complicated environment in which to pursue the nation's objectives and to ensure that "good" always prevails. Strong leadership is important and successful leaders are best characterized by their ability to employ their intellect, human understanding, and moral character in a way that accomplishes the mission by positively influencing the diverse human dimensions of the battle space. I believe that spiritual leadership is a battle space imperative that helps to unify an otherwise diverse and asymmetrical environment. It is the commander's calling to accept this uncommon responsibility, as it is a fundamental element of the holistic leadership I now believe each warrior expects.

— Contributors —

Lieutenant Colonel Carroll J. Connelley is currently teaching law and ethics at the U.S. Naval Academy, where he is serving as the deputy director of the Leadership, Education, and Development Division. He received his commission from the U.S. Naval Academy in 1995 and served in various leadership positions and deployments. Since completing his juris doctorate, he has deployed to Iraq as the principle legal advisor to RCT-2, become the command judge advocate for The Basic School, deployed in support of CJTF-HOA as the operational law SJA, and served as the Law of War Program director for the Lejeune Leadership Institute at Marine Corps University.

Dr. Paula Holmes-Eber is professor of operational culture at Marine Corps University. She holds a BA in psychology from Dartmouth College and an MA and doctorate in anthropology from Northwestern University. She is the author of three books: *Operational Culture for the Warfighter: Principles and Applications* (2008 with Barak Salmoni); *Applications in Operational Culture: Lessons from the Field* (2010 with Patrice Scanlon and Andrea Hamlen); and *Daughters of Tunis: Women, Family, and Networks in a Muslim City (2002).*

Lieutenant Colonel Brian Christmas is a graduate of the Amphibious Warfare School and a distinguished graduate of the Marine Corps Command and Staff College. He is currently the director of warfighting at the Marine Corps' Command and Staff College, after having served in a wide variety of positions while deployed in support of both Operations Enduring Freedom and Iraqi Freedom, including command of 3rd Battalion, 6th Marines while deployed in support of OEF in Marjeh, Afghanistan.

Geoffroy Murat is a doctoral candidate at the Raymond Aron Center of the EHESS (L'École des hautes études en sciences sociales or School for Advanced Studies in the Social Sciences) in Paris. He is doing a comparative thesis on the teaching of military ethics in American and French military academies, thanks to a stipend from the French ministry of defense. Mr. Murat is also the founder of Nicomak, a company that provides ethical training for companies and schools. His primary focus of research involves the analysis of ethics pedagogy and the effect various approaches have on the individual and collective behavior of service members.

Major Clinton Culp (Ret.) enlisted in the U.S. Marine Corps in 1985 and upon receiving his commission in 1997, was assigned to 2d LAR as a platoon commander. He has experienced two deployments to support Operation Enduring Freedom (OEF): first, as an Afghan National Army advisor while officer in charge of the Mountain Leader Course; second, as the liaison officer to the UN Joint Election Management Body with 6th Marines. He deployed in support of Operation Iraqi Freedom as the weapons company commander for 3/6. After serving as University of Idaho's Marine officer instructor (MOI), he retired in 2009. He is currently working toward a doctorate in the pedagogy of ethics and character development.

Captain Emmanuel Goffi, French Air Force, is a specialist in military ethics. He is a lecturer in international relations at the French Air Force Academy and teaches international relations at the DSI (Diplomatic Studies Institute) in Marseille. He holds a master's degree in international affairs and a research master's in comparative military history and geostrategy. He is currently working on a doctorate in political science at the Institute of Political Studies in Paris (Sciences Po)—CERI. He is the author of *Les armées françaises face à la moral: Une réflexion au cœur des conflits modernes* (Editions l'Harmattan, 2011).

Dr. Peter Bradley teaches psychology and ethics at the Royal Military College of Canada. He is an active researcher in the areas of ethics and mental health in the military. A former lieutenant colonel, he retired from the Canadian Forces in 2004 with 33 years of service in the infantry and personnel branches.

Dr. Paolo Tripodi is professor of ethics and ethics branch head at the Lejeune Leadership Institute, Marine Corps University. He served as professor of strategic studies at the Marine Corps War College and was the MCU Donald Bren Chair of Ethics and Leadership. He has held academic positions at the U.S. Naval Academy (Annapolis, MD); the Catholic University of Chile (Santiago, Chile); and the Nottingham Trent University (Nottingham, UK). Dr. Tripodi trained as an infantry officer and received his commission with the Italian Carabinieri upon graduation from the School of Infantry and Cavalry in Cesano, and the Carabinieri Officer's School in Rome, Italy.

Dr. Clyde Croswell is president of Community-L, Inc., a human resource and organization development firm, and adjunct faculty with George Washington University and Shenandoah University. His research interests and consulting practice focus on mindfulness, leadership complexity, self-organizing systems, affective learning, and applying twenty-first-century affective neuroscience, phenomenology, and science of mind to organizational transformations. Lieutenant Colonel Croswell (Ret.) is a veteran with 29 years of service in the U.S. Marine Corps, and he has consulted to the Virginia Bankers, The Marine Corps University, American Woodmark Corporation, OakCrest Companies, The U.S. Army Corps of Engineers, and Educational Field Services.

Lieutenant Colonel Dan Yaroslaski, U.S. Marine Corps, is currently the future plans officer at Marine Forces Cyberspace Command. He has led Marines in combat operations in both Iraq and Afghanistan. LtCol Yaroslaski has participated in multilateral training exercises and planning events with military and civilian leaders from more than 15 foreign countries. His latest deployment was as the leader of a team of advisers assigned to the Afghan National Army. His areas of interest are primarily the role of ethical leadership and cultural understanding in the development of military organizations. He holds masters degrees in military studies, operational studies, and strategic studies.

Major Winston S. Williams is a professor in the international and operational law department at The Judge Advocate General's Legal Center and School (TJAGLCS). Maj Williams has served in the U.S. Army since 1998. His most recent assignments include senior legal observer/controller at the Joint Readiness Training Center (JRTC), Fort Polk, Louisiana; and trial counsel/operational law attorney for 3rd Brigade Combat Team (BCT), 82nd Airborne Division, Fort Bragg, North Carolina. While serving with 3rd BCT, he spent 15 months in Iraq advising commanders on rules of engagement, detention operations, and rule of law. Other assignments include chief of administrative law, 82nd Airborne Division; executive officer, D Company, 35th Engineer Battalion, Fort Leonard Wood, Missouri; and assault and obstacle platoon leader, B Company, 44th Engineer Battalion, 2nd Infantry Division, Republic of Korea. Maj Williams holds a BS in civil engineering from Florida A&M University, a JD from University of Tennessee College of Law, and an LL.M. in military law (international and operational law specialty) from TJAGLCS.

Laurie R. Blank, Esq., is the director of the International Humanitarian Law Clinic at Emory University School of Law. Professor Blank is the codirector of a multiyear project on military training programs in the law of war and coauthor of *Law of War Training: Resources for Military and Civilian Leaders* (2011). She is also the coauthor of an upcoming casebook on the law of war, *International Law and Armed Conflict: Fundamental Principles and Contemporary Challenges in the Law of War* (2013) and the project director of the *Rules of War and Tools of War* project. Professor Blank was a program officer in the rule of law program at the United States Institute of Peace and is the author of numerous articles and opinion pieces on topics in international humanitarian law.

Jamie Williamson, Esq., serves as the legal advisor for the Washington delegation of the International Committee of the Red Cross (ICRC). In this capacity, he is responsible for legal support to the ICRC activities in the United States and Canada, with particular focus on Guantanamo and military operations in Afghanistan and Iraq. He was the ICRC regional legal advisor based in Pretoria, South Africa and served with the UN ad hoc international criminal tribunals in Tanzania and the Netherlands, as well as at the special court for Sierra Leone, where he worked on the first international judgments on the crimes of genocide and war crimes in noninternational armed conflicts. From 2002, Mr. Williamson headed the Chambers legal support section of the ICTR Appeals Chamber based in The Hague. He has published numerous papers, including noteworthy contributions on repression of war crimes, international justice, and the challenges to international humanitarian law in modern-day conflicts.

Lieutenant Colonel Chris Jenks is an active duty judge advocate in the United States Army. He received an LLM with distinction from Georgetown University, an LLM in military law from the Judge Advocate General's Legal Center and School (TJAGLCS), and a JD from the University of Arizona College of Law. LtCol Jenks began his Army service in 1992 as an infantry officer following graduation from the U.S. Military Academy at West Point. He served in the Republic of Korea as the chief of international and operational law for U.S. Army units operating near the demilitarized zone and North Korea. He then deployed to Mosul, Iraq, where he provided law of armed conflict advice for the employment of artillery, close air support, and direct fire weapons during enemy engagements. LtCol Jenks has published articles on international criminal law, human rights law, and government contractors as well as spoken on those same topics at universities in the United States and Australia and with the armies of several different European, African, and Middle Eastern countries.

Lieutenant Colonel Kenneth L. Hobbs is a U.S. Air Force judge advocate and serves as the legal advisor for NATO School, Oberammergau, Germany. LtCol Hobbs advises an organization staffed by personnel from 25 nations that provides military instruction each year to more than 10,000 students from over 60 countries. He also provides academic support as a course director and lecturer on legal and operational issues at NATO School, as part of Mobile Education and Training Teams, and as a guest lecturer at other educational institutions in Europe and Asia. LtCol Hobbs holds a BA from Hendrix College, a JD from St. Mary's University School of Law (San Antonio, TX), and an LLM in military law from TJAGLCS. LtCol Hobbs began his military service in the U.S. Army Reserve in 1988 as a combat engineer.

CDR David Gibson, Chaplain Corps, USN, has 27 years of military service, having had numerous assignments with the USN, USMC, and USCG. He holds a doctorate from Regent University (Virginia Beach, VA) in organizational strategy and leadership, and is ordained by the Church of God (Cleveland, TN). CDR Gibson served as the technical representative for the team that designed, developed, and delivered the USN Chaplain Corps/Navy Bureau of Medicine and Surgery (BUMED) training in Combat and Operational Stress Control (COSC), a three-year program that used leading experts in the field to develop the Department of the Navy's response to PTSD, TBI, and combat/operational stress and trained over 3500 attendees. He was a member of the USMC TRI-MEF COSC Working Group, and has presented at DOD and VA Conferences on COSC. He is also an instructor for the "Combat and Operational Stress Control First Aid Course."

Lieutenant Commander Judy Malana is a chaplain who has served in ashore, afloat, and forward deployed units with the Navy, Marine Corps, and Coast Guard. She is an alumna of the University of California, Los Angeles (BA, design); Bethel Theological Seminary (masters of divinity); the National Graduate School (MS, quality systems management); and Marine Corps University (masters of military studies). She completed advanced graduate studies at Salve Regina University and is a doctoral candidate in humanities. LtCdr Malana is currently a student at the Naval War College, working toward an MA in national security and strategic studies.

Colonel Chaplain Franklin Eric Wester, USA, (Ret.) is the assistant to the presiding bishop of the Evangelical Lutheran Church in America for federal chaplaincy ministries overseeing all military, veterans affairs, and federal prison chaplains. He served as senior military fellow at the Institute for National Security Ethics and Leadership, National Defense University (Fort McNair,

Washington, DC). He holds masters degrees from the U.S. Army War College, New Brunswick Theological Seminary, and Trinity Lutheran Seminary. His published articles include "Preemption and Just War" (2004) and "Armed Force in Peace Operations" in *Parameters*. He served as executive assistant to the joint working group co-locating Army, Navy, and Air Force chaplain schools into a single campus at Fort Jackson, South Carolina; as command chaplain, U.S. Army Reserve Command; and on staff for the Army Chief of Chaplains.

Lieutenant Colonel Jeffrey S. Wilson, USA, has devoted much of his 25 years in service to ethics training and education. He holds a BA in philosophy from Western Illinois University, an MA in philosophy from the University of Illinois at Urbana-Champaign, and a doctorate in educational leadership, management, and policy from Seton Hall University (New Jersey). LtCol Wilson has taught ethics at West Point for 10 years and serves as an adjunct instructor in philosophy at Mount Saint Mary College and Saint Thomas Aquinas College. His scholarship focuses on character education and applied ethics.

Dr. Pauletta Otis is professor of security studies at the Command and Staff College, Marine Corps University. She served as professor of international strategic studies at the Department of Defense's Joint Military Intelligence College and was a senior research fellow for religion in international affairs at Pew Forum (Washington, DC). Formerly a tenured full professor of political science and international studies at Colorado State University-Pueblo (1989–2004), her doctorate was awarded from the Graduate School of International Studies at the University of Denver (1989). Selected publications include *Handbook of Religion and Security* (Routledge Press, Fall, 2012); "Ethics in the Times of War" in *Ethics: Beyond War's End* (Georgetown University Press, 2012); "Anthropology and Arms Control" in *Arms Control*, edited by

Paul Viotti (ABC Clio Press, 2012); "The U.S. Military Chaplaincy: Ministry of Presence and Practice" in the *Journal of Religion and Security* (December 2009); "Religion in Information Operations" in *Ideas as Weapons* (2009); "The U.S. Military Chaplaincy: Ministry of Presence and Practice" in *Religion and Security* (2009); "Armed with the Power of Ideas" in *Armed Groups: National Security, Counterterrorism and Counterinsurgency* (Naval War College, 2008).

Arnold Fields is a retired Marine Corps major general who served more than 34 years on active duty. He is a combat veteran and commanded a Marine infantry battalion during Operation Desert Storm. Gen Fields also served as commanding general, Marine Corps Bases Hawaii; deputy commanding general, Marine Forces Europe; and director of the Marine Corps Staff. He retired from the Marine Corps in 2004 and served with the Department of State in Iraq; the Department of Defense as deputy director of the Africa Center for Strategic Studies; and accepted a presidential appointment as special inspector general for Afghanistan reconstruction. He is a member of the Board of Visitors of Marine Corps University and a Fellow of the National Academy of Public Administration.